T0155924

OXFORD SERIES IN OPTICAL AND IMAGING SCIENCES

Methods in Theoretical Quantum Optics

OXFORD SERIES IN OPTICAL AND IMAGING SCIENCES

1. D. M. Lubman (ed.). *Lasers and Mass Spectrometry*
2. D. Sarid, *Scanning Force Microscopy With Applications to Electric, Magnetic, and Atomic Forces*
3. A. B. Schvartsburg. *Non-linear Pulses in Integrated and Waveguide Optics*
4. C. J. Chen. *Introduction to Scanning Tunneling Microscopy*
5. D. Sarid. *Scanning Force Microscopy With Applications to Electric, Magnetic, and Atomic Forces, revised edition*
6. S. Mukamel. *Principles of Nonlinear Optical Spectroscopy*
7. A. Hasegawa and Y. Kodama. *Solitons in Optical Communications*
8. T. Tani. *Photographic Sensitivity: Theory and Mechanisms*
9. D. Joy. *Monte Carlo Modeling for Electron Microscopy and Microanalysis*
10. A. B. Shvartsburg. *Time-Domain Optics of Ultrashort Waveforms*
11. L. Solymar, D. J. Webb, and A. Grunnet-Jepsen. *The Physics and Applications of Photorefractive Materials*
12. R. M. Haralick. *Mathematical Morphology: Theory and Hardware*
13. J. R. Schott. *Remote Sensing: The Image Chain Approach*
14. P. Mouroulis and J. Macdonald. *Geometrical Optics and Optical Design*
15. S. M. Barnett and P. M. Radmore. *Methods in Theoretical Quantum Optics*

Methods in Theoretical Quantum Optics

STEPHEN M. BARNETT
Department of Physics and Applied Physics
University of Strathclyde, Glasgow

and

PAUL M. RADMORE
Department of Electronic and Electrical Engineering
University College London

CLARENDON PRESS • OXFORD

This book has been printed digitally and produced in a standard specification in order to ensure its continuing availability

OXFORD
UNIVERSITY PRESS

Great Clarendon Street, Oxford OX2 6DP

Oxford University Press is a department of the University of Oxford.
It furthers the University's objective of excellence in research, scholarship,
and education by publishing worldwide in

Oxford New York

Auckland Cape Town Dar es Salaam Hong Kong Karachi
Kuala Lumpur Madrid Melbourne Mexico City Nairobi
New Delhi Shanghai Taipei Toronto
With offices in
Argentina Austria Brazil Chile Czech Republic France Greece
Guatemala Hungary Italy Japan South Korea Poland Portugal
Singapore Switzerland Thailand Turkey Ukraine Vietnam

Oxford is a registered trade mark of Oxford University Press
in the UK and in certain other countries

Published in the United States
by Oxford University Press Inc., New York

Oxford is a registered trade mark of Oxford University Press
in the UK and in certain other countries

Published in the United States
by Oxford University Press Inc., New York

Reprinted 2005

ISBN 0-19-856361-2

For Claire and Margaret

PREFACE

This book is aimed at those readers who already have some knowledge of mathematical methods and have also been introduced to the basic ideas of quantum optics. Our hope is that it will be attractive to students who have already explored one of the more introductory texts such as Loudon's *The quantum theory of light* and are looking to acquire the mathematical skills used in real problems. Our book is not primarily about the physics of quantum optics but rather presents the mathematical methods widely used by workers in this field. We necessarily discuss the physical assumptions which lead to the models and approximations employed but the main purpose of the book is to give a firm grounding in those techniques needed to derive analytic solutions to relevant model problems. We have tried, where space and continuity of thought have allowed, to insert all or most of the steps in calculations. The quicker and more experienced reader may pass over these rapidly, and the less experienced may, we hope, be saved a little effort. Our aim is to assist the latter reader without exasperating the former! We havc included a chapter on the quantum mechanical foundations of the subject and a number of appendices which either summarize the more important topics in mathematics required in the book or present the lengthier calculations not included in the chapters. A very selective bibliography is given at the end of the book and includes relevant books and a few substantial papers. This is obviously in no sense an exhaustive list and we have made no attempt to represent the vast literature of journal papers on quantum optics. The bibliography is intended only as a starting point in exploring this literature.

We are grateful to Graeme Harkness for producing the figures, to David Pegg and Franco Persico for their critical reading of and constructive comments on selected chapters, and to Chris Jones for an outstanding third-year project at UCL, material from which is included in Chapter 6. We are also grateful to former students, Gordon Yeoman, Norbert Lütkenhaus, Tony Chefles, and Sonja Franke, for exposing weaknesses in some of our early explanations. This book would not have been completed without the patience and continued encouragement of OUP and above all the encouragement and understanding of Claire Gilson and Margaret Leszczyńska. Thanks for putting up with us.

Glasgow S.M.B.
London P.M.R.

February 1997

CONTENTS

1
FOUNDATIONS

1.1 Introduction

Fundamental issues in physics and especially in quantum theory are often most clearly highlighted and studied with the aid of simple models. In quantum optics, there are many such models describing quantum effects in the interaction between matter and the electromagnetic field, a number of which have been realized experimentally. The description of either the matter or the field may be classical or quantum depending on the phenomena being studied. Much research in quantum optics has been centred on effects which distinguish between the fully quantum field and the classical field interacting with matter treated quantum mechanically.

This book is not primarily an introduction to the physics of quantum optics but rather aims to provide the mathematical methods widely used in the subject. We necessarily discuss some of the physical principles on which the models and approximations are based but concentrate on techniques needed to obtain analytic solutions. Throughout we make extensive use of some of the basic ideas from quantum mechanics and this first chapter presents these, specializing to the description of atoms and radiation. These ideas will be developed later in the book, especially in Chapter 3.

1.2 Basic quantum theory

The state of a quantum system is completely specified by its state vector, the ket $|\psi\rangle$. If $|\psi_1\rangle$ and $|\psi_2\rangle$ are possible states then the superposition

$$|\psi\rangle = a_1 |\psi_1\rangle + a_2 |\psi_2\rangle \tag{1.2.1}$$

is also a state of the system, where the amplitudes a_1 and a_2 are complex numbers. This superposition principle is one of the most fundamental concepts in quantum theory leading to the ideas of probability amplitudes and of incompatible observables. The bra $\langle\psi|$ is an equivalent representation of the state given by

$$\langle\psi| = a_1^* \langle\psi_1| + a_2^* \langle\psi_2|, \tag{1.2.2}$$

where a_1^* and a_2^* denote the complex conjugates of a_1 and a_2, respectively. The overlap between, or inner product of, two states $|\psi\rangle$ and $|\phi\rangle$ is the complex number $\langle\psi|\phi\rangle$ or its complex conjugate $\langle\phi|\psi\rangle$, analogous to the scalar or dot product of two vectors. If this overlap is zero, then the states are said to be orthogonal. The inner product of a state $|\psi\rangle$ with itself is real and strictly greater than zero so that

$$\langle\psi|\psi\rangle > 0. \tag{1.2.3}$$

If this inner product is unity, so that $\langle\psi|\psi\rangle = 1$, then the state is said to be normalized. If the states $|\psi_1\rangle$ and $|\psi_2\rangle$ in (1.2.1) are orthonormal, that is both orthogonal and normalized, then the amplitudes a_1 and a_2 are given by the overlaps

$$\langle\psi_1|\psi\rangle = a_1 = \langle\psi|\psi_1\rangle^*, \tag{1.2.4}$$

$$\langle\psi_2|\psi\rangle = a_2 = \langle\psi|\psi_2\rangle^*. \tag{1.2.5}$$

If $|\psi\rangle$ is itself normalized, then $|a_1|^2 + |a_2|^2 = 1$ and we interpret $|a_1|^2$ and $|a_2|^2$ as the probabilities that the system will be found in states $|\psi_1\rangle$ and $|\psi_2\rangle$, respectively. The generalization of (1.2.1) to a superposition of many possible states $|\psi_n\rangle$ is

$$|\psi\rangle = \sum_n a_n |\psi_n\rangle, \tag{1.2.6}$$

where if $|\psi\rangle$ is normalized and the states $|\psi_n\rangle$ are orthonormal then

$$\sum_n |a_n|^2 = 1, \tag{1.2.7}$$

consistent with the interpretation of $|a_n|^2$ as the probability that the system is found to be in the state $|\psi_n\rangle$.

The description of a system is completed by the introduction of operators. An operator $\hat{A}$ acting on any state of the system produces another state which, in general, will not be normalized. The Hermitian conjugate $\hat{B}^\dagger$ of an operator $\hat{B}$ is defined by the requirements that

$$(\hat{B}^\dagger)^\dagger = \hat{B}, \tag{1.2.8}$$

$$(\hat{B} + \hat{C})^\dagger = \hat{B}^\dagger + \hat{C}^\dagger, \tag{1.2.9}$$

$$(\hat{B}\hat{C})^\dagger = \hat{C}^\dagger\hat{B}^\dagger, \tag{1.2.10}$$

$$(\lambda\hat{B})^\dagger = \lambda^*\hat{B}^\dagger, \tag{1.2.11}$$

where $\hat{C}$ is any other operator and λ is a complex number. An operator $\hat{A}$ satisfying $\hat{A} = \hat{A}^\dagger$ is an Hermitian operator, and an observable A is associated with an Hermitian operator $\hat{A}$. The eigenvalues λ_n of $\hat{A}$ satisfy the eigenvalue equation

$$\hat{A}|\lambda_n\rangle = \lambda_n|\lambda_n\rangle, \tag{1.2.12}$$

where the $|\lambda_n\rangle$ are the eigenstates. The conjugate equation to (1.2.12) with λ_n replaced by λ_m is

$$\langle\lambda_m|\hat{A}^\dagger = \langle\lambda_m|\hat{A} = \lambda_m^*\langle\lambda_m|, \tag{1.2.13}$$

since $\hat{A}$ is Hermitian. Taking the overlap of (1.2.13) with $|\lambda_n\rangle$ gives

$$\langle\lambda_m|\hat{A}|\lambda_n\rangle = \lambda_m^*\langle\lambda_m|\lambda_n\rangle \tag{1.2.14}$$

and, similarly, taking the overlap of (1.2.12) with $\langle\lambda_m|$ gives

$$\langle\lambda_m|\hat{A}|\lambda_n\rangle = \lambda_n\langle\lambda_m|\lambda_n\rangle. \tag{1.2.15}$$

Subtracting (1.2.15) from (1.2.14) gives

$$(\lambda_m^* - \lambda_n)\langle\lambda_m|\lambda_n\rangle = 0, \tag{1.2.16}$$

so that if $m = n$, we see from (1.2.3) that the eigenvalue λ_n must be real, while if $\lambda_m \neq \lambda_n$ then the states $|\lambda_m\rangle$ and $|\lambda_n\rangle$ are orthogonal and may be chosen to be orthonormal with

$$\langle\lambda_m|\lambda_n\rangle = \delta_{mn} \tag{1.2.17}$$

(see Appendix 1). Hermitian operators therefore have real eigenvalues associated with orthonormal eigenstates. Measurement of A yields one of the real eigenvalues of $\hat{A}$ so that if the normalized state is

$$|\psi\rangle = \sum_n a_n|\lambda_n\rangle, \tag{1.2.18}$$

then the probability that a measurement of A gives the result λ_n is $|a_n|^2$. If all possible states can be expressed in the form (1.2.18) then the set $\{|\lambda_n\rangle\}$ is said to be complete. If two orthonormal eigenstates $|\lambda_m\rangle$ and $|\lambda_n\rangle$ have the same eigenvalue λ then the states are said to be degenerate and the probability of obtaining the result λ is $|a_m|^2 + |a_n|^2$. The mean value $\bar{A}$ of A found from measurements on an ensemble of identically prepared systems is the expectation value of $\hat{A}$ given by

$$\bar{A} = \langle\hat{A}\rangle = \sum_n \lambda_n|a_n|^2 = \langle\psi|\hat{A}|\psi\rangle. \tag{1.2.19}$$

The statistical spread of the results is most commonly expressed in terms of the variance

$$\begin{aligned}\Delta A^2 &= \langle\psi|(\hat{A} - \langle\hat{A}\rangle)^2|\psi\rangle \\ &= \langle\psi|\hat{A}^2|\psi\rangle - (\langle\psi|\hat{A}|\psi\rangle)^2,\end{aligned} \tag{1.2.20}$$

or the uncertainty $\Delta A = \surd(\Delta A^2)$. This uncertainty is zero if and only if $|\psi\rangle$ is an eigenstate of $\hat{A}$.

The commutator of two operators $\hat{A}$ and $\hat{B}$ is defined to be

$$[\hat{A}, \hat{B}] = \hat{A}\hat{B} - \hat{B}\hat{A}. \tag{1.2.21}$$

If $\hat{A}$ and $\hat{B}$ are Hermitian and this commutator is zero, then the observables A and B are said to be compatible and the operators $\hat{A}$ and $\hat{B}$ have a common complete set of eigenstates. Some or all of the eigenstates of incompatible observables will be different. The commutator (1.2.21) of two Hermitian operators is a skew-Hermitian operator in that $[\hat{A}, \hat{B}]^\dagger = -[\hat{A}, \hat{B}]$. The commutator of an operator $\hat{A}$ and the operator product $\hat{B}\hat{C}$ is easily seen to be

$$[\hat{A}, \hat{B}\hat{C}] = \hat{B}[\hat{A}, \hat{C}] + [\hat{A}, \hat{B}]\hat{C}, \tag{1.2.22}$$

or $[\hat{B}\hat{C}, \hat{A}] = \hat{B}[\hat{C}, \hat{A}] + [\hat{B}, \hat{A}]\hat{C}$. The anticommutator is defined to be

$$\{\hat{A}, \hat{B}\} = \hat{A}\hat{B} + \hat{B}\hat{A} \tag{1.2.23}$$

which is clearly Hermitian if $\hat{A}$ and $\hat{B}$ are Hermitian. The uncertainties associated with the observables A and B for any given state are bounded by the uncertainty principle

$$\Delta A \cdot \Delta B \geqslant \tfrac{1}{2}|\langle [\hat{A}, \hat{B}] \rangle|. \tag{1.2.24}$$

States for which the equality in (1.2.24) holds are termed minimum uncertainty states for A and B.

The outer product of two states $|\phi_1\rangle$ and $|\phi_2\rangle$ is the operator $|\phi_1\rangle\langle\phi_2|$ or its Hermitian conjugate $|\phi_2\rangle\langle\phi_1|$. This outer product is an Hermitian operator if and only if $|\phi_1\rangle = |\phi_2\rangle$. The operator $|\phi_1\rangle\langle\phi_2|$ acting on a state $|\psi\rangle$ produces the state $\langle\phi_2 | \psi\rangle |\phi_1\rangle$, that is the state $|\phi_1\rangle$ multiplied by the complex number $\langle\phi_2 | \psi\rangle$. If $\{|\lambda_n\rangle\}$ is a complete orthonormal set of eigenvectors of an Hermitian operator $\hat{A}$ then

$$\hat{A} = \sum_n \lambda_n |\lambda_n\rangle\langle\lambda_n|. \tag{1.2.25}$$

The identity operator **1** is the Hermitian operator which when acting on any state $|\psi\rangle$ gives the state $|\psi\rangle$. It follows from (1.2.25) that

$$\mathbf{1} = \sum_n |\lambda_n\rangle\langle\lambda_n| \tag{1.2.26}$$

since, using (1.2.17) and (1.2.18), we find

$$\begin{aligned} \mathbf{1}|\psi\rangle &= \sum_m |\lambda_m\rangle\langle\lambda_m| \sum_n a_n |\lambda_n\rangle \\ &= \sum_n a_n \sum_m \delta_{mn} |\lambda_m\rangle \\ &= |\psi\rangle. \end{aligned} \tag{1.2.27}$$

The resolution of the identity (1.2.26) is, in fact, an alternative statement of the completeness of the set $\{|\lambda_n\rangle\}$. The function $f(\hat{A})$ of the Hermitian operator $\hat{A}$ is defined to be the operator

$$f(\hat{A}) = \sum_n f(\lambda_n)|\lambda_n\rangle\langle\lambda_n|, \tag{1.2.28}$$

so that $|\lambda_n\rangle$ is an eigenstate of $f(\hat{A})$ with eigenvalue $f(\lambda_n)$.

The evolution of a state $|\psi(t)\rangle$ is governed by the Schrödinger equation

$$\mathrm{i}\hbar\frac{\partial}{\partial t}|\psi(t)\rangle = \hat{H}|\psi(t)\rangle, \tag{1.2.29}$$

where $\hat{H}$ is the Hamiltonian. This Hermitian operator may, depending on the system and the model being used to study it, be time dependent. The expectation value of a time-independent operator $\hat{A}$ changes with time owing to the evolution of $|\psi(t)\rangle$, so that

$$\frac{\mathrm{d}}{\mathrm{d}t}\langle\hat{A}\rangle = -\frac{\mathrm{i}}{\hbar}\langle\psi(t)|[\hat{A},\hat{H}]|\psi(t)\rangle. \tag{1.2.30}$$

This change in $\langle\hat{A}\rangle$ arises from transitions between the eigenstates $|\lambda_n\rangle$ of $\hat{A}$ induced by $\hat{H}$, and can only occur if $\hat{A}$ and $\hat{H}$ do not commute. If $\hat{A}$ commutes with $\hat{H}$, then $\langle\hat{A}\rangle$, and all higher moments of $\hat{A}$, are constant and we then say that $\hat{A}$ and its associated observable A are constants of the motion. Substituting the eigenstate expansion (1.2.18) into the Schrödinger equation (1.2.29) and taking the overlap with $|\lambda_m\rangle$ gives the amplitude equations

$$\dot{a}_m(t) = -\frac{\mathrm{i}}{\hbar}\sum_n\langle\lambda_m|\hat{H}|\lambda_n\rangle a_n(t). \tag{1.2.31}$$

This set of coupled differential equations is equivalent to the Schrödinger equation and from the solution we can construct $|\psi(t)\rangle$. The formal solution of the Schrödinger equation (1.2.29) is

$$|\psi(t)\rangle = \hat{U}(t)|\psi(0)\rangle, \tag{1.2.32}$$

where $\hat{U}(t)$ is a unitary operator for which

$$\hat{U}^\dagger = \hat{U}^{-1}, \tag{1.2.33}$$

so that $\langle\psi(t)|\psi(t)\rangle = \langle\psi(0)|\psi(0)\rangle$. It follows from (1.2.25) that the time evolution operator $\hat{U}(t)$ also satisfies the Schrödinger equation

$$\mathrm{i}\hbar\frac{\partial\hat{U}(t)}{\partial t} = \hat{H}\hat{U}(t), \tag{1.2.34}$$

and hence if $\hat{H}$ is time independent then

$$\hat{U}(t) = \exp(-\mathrm{i}\hat{H}t/\hbar). \tag{1.2.35}$$

The Hermitian conjugate operator $\hat{U}^\dagger(t)$ is given by $\hat{U}^\dagger(t) = \exp(\mathrm{i}\hat{H}t/\hbar)$ from which, using the definition (1.2.28) of the function of an Hermitian operator, it is straightforward to show that $\hat{U}\hat{U}^\dagger = \mathbf{1} = \hat{U}^\dagger\hat{U}$ and hence, as stated in (1.2.33), that $\hat{U}$ is unitary. An alternative to using the Schrödinger equation is to work in the Heisenberg picture in which $|\psi(t)\rangle = |\psi(0)\rangle$ remains unchanged while operators evolve so that $\hat{A}(t)$ satisfies the Heisenberg equation

$$\frac{\mathrm{d}}{\mathrm{d}t}\hat{A}(t) = \frac{\mathrm{i}}{\hbar}[\hat{H}, \hat{A}(t)] + \frac{\partial}{\partial t}\hat{A}(t), \tag{1.2.36}$$

where the partial derivative term accounts for any explicit time dependence. If $\hat{A}$ has no explicit time dependence and commutes with $\hat{H}$ at all times then $\hat{A}$ and the corresponding observable A are constants of the motion.

If we do not have enough information to specify the state vector of a system but know only the probabilities P_n that the system is in a normalized state $|\psi_n\rangle$, then the mean value of A is

$$\bar{A} = \sum_n P_n \langle \psi_n | \hat{A} | \psi_n \rangle. \tag{1.2.37}$$

In such cases, it is useful to introduce the density matrix ρ which is the Hermitian operator

$$\rho = \sum_n P_n |\psi_n\rangle\langle\psi_n|. \tag{1.2.38}$$

The states $|\psi_n\rangle$ need not be mutually orthogonal but it is always possible to choose states such that (1.2.38) holds and $\langle \psi_n | \psi_m \rangle = \delta_{nm}$. The mean value of A is then given by

$$\bar{A} = \langle \hat{A} \rangle = \mathrm{Tr}(\rho\hat{A}), \tag{1.2.39}$$

where Tr denotes the trace operation which is carried out by summing the diagonal matrix elements of the operator $\rho\hat{A}$ in any basis consisting of a complete orthonormal set of states. Consider, for example, using the basis $\{|\lambda_m\rangle\}$ to calculate (1.2.39) as

$$\begin{aligned}
\mathrm{Tr}(\rho\hat{A}) &= \sum_m \langle \lambda_m | \Big(\sum_n P_n |\psi_n\rangle\langle\psi_n| \hat{A} \Big) | \lambda_m \rangle \\
&= \sum_n P_n \langle \psi_n | \hat{A} \sum_m |\lambda_m\rangle\langle\lambda_m | \psi_n\rangle \\
&= \sum_n P_n \langle \psi_n | \hat{A} | \psi_n \rangle,
\end{aligned} \tag{1.2.40}$$

which has the same value as (1.2.37). Three important results follow from (1.2.40). Firstly, (1.2.40) is independent of the basis $\{|\lambda_m\rangle\}$ used to calculate $\langle \hat{A} \rangle$

and therefore the trace can be evaluated using any basis. Secondly, choosing $\hat{A}$ to be the identity operator **1** gives

$$\mathrm{Tr}(\rho) = \sum_n P_n = 1, \tag{1.2.41}$$

since the probabilities sum to unity. Finally, choosing $\hat{A}$ to be ρ, (1.2.40) becomes

$$\begin{aligned}\mathrm{Tr}(\rho^2) &= \sum_n P_n \langle \psi_n | \left(\sum_m P_m |\psi_m\rangle \langle \psi_m| \right) |\psi_n\rangle \\ &= \sum_n \sum_m P_n P_m |\langle \psi_n | \psi_m \rangle|^2,\end{aligned} \tag{1.2.42}$$

which is less than or equal to unity, as we now show. The Cauchy–Schwarz inequality

$$|\langle \phi_1 | \phi_2 \rangle|^2 \leqslant \langle \phi_1 | \phi_1 \rangle \langle \phi_2 | \phi_2 \rangle, \tag{1.2.43}$$

which follows from (1.2.3) with

$$|\psi\rangle = |\phi_2\rangle - \frac{\langle \phi_1 | \phi_2 \rangle}{\langle \phi_1 | \phi_1 \rangle} |\phi_1\rangle, \tag{1.2.44}$$

implies that $|\langle \psi_n | \psi_m \rangle|^2 \leqslant \langle \psi_n | \psi_n \rangle \langle \psi_m | \psi_m \rangle = 1$. Using this inequality in (1.2.42) gives

$$\mathrm{Tr}(\rho^2) \leqslant \sum_n \sum_m P_n P_m = 1. \tag{1.2.45}$$

It follows from the normalization of the states $|\psi_n\rangle$ that equality in (1.2.45) holds if and only if $\rho = |\psi_n\rangle\langle\psi_n|$, in which case $\rho^2 = \rho$ and the system is in a pure state which can be represented by the ket $|\psi_n\rangle$. An important property of the trace operation is that the trace of a product of operators is invariant under cyclic permutation of these operators, so that

$$\mathrm{Tr}(\rho\hat{A}\hat{B}) = \mathrm{Tr}(\hat{B}\rho\hat{A}) = \mathrm{Tr}(\hat{A}\hat{B}\rho). \tag{1.2.46}$$

Other permutations do not in general preserve the trace since

$$\mathrm{Tr}(\rho\hat{A}\hat{B}) - \mathrm{Tr}(\rho\hat{B}\hat{A}) = \mathrm{Tr}(\rho[\hat{A}, \hat{B}]). \tag{1.2.47}$$

The proof of (1.2.46) follows by evaluating the trace in the basis $\{|\lambda_n\rangle\}$ and using the resolution of the identity (1.2.26) since

$$\begin{aligned}\mathrm{Tr}(\rho\hat{A}\hat{B}) &= \sum_l \sum_m \sum_n \langle \lambda_n | \rho | \lambda_m \rangle \langle \lambda_m | \hat{A} | \lambda_l \rangle \langle \lambda_l | \hat{B} | \lambda_n \rangle \\ &= \sum_l \sum_m \sum_n \langle \lambda_l | \hat{B} | \lambda_n \rangle \langle \lambda_n | \rho | \lambda_m \rangle \langle \lambda_m | \hat{A} | \lambda_l \rangle \\ &= \mathrm{Tr}(\hat{B}\rho\hat{A}).\end{aligned} \tag{1.2.48}$$

The matrix elements ρ_{nm} of the density matrix in any basis $\{|\lambda_n\rangle\}$ are

$$\rho_{nm} = \langle\lambda_n|\,\rho\,|\lambda_m\rangle = \rho^*_{mn} \tag{1.2.49}$$

and are constrained by the inequality

$$\rho_{nm}\,\rho_{mn} \leqslant \rho_{nn}\,\rho_{mm}. \tag{1.2.50}$$

The proof of (1.2.50) follows from (1.2.43) with $|\phi_1\rangle = \rho^{1/2}|\lambda_n\rangle$ and $|\phi_2\rangle = \rho^{1/2}|\lambda_m\rangle$, where the operator $\rho^{1/2}$ is defined using (1.2.28). For a pure state, the equality holds in (1.2.50) for all n and m. Summing (1.2.50) over n and m gives the inequality (1.2.45).

For two independent systems we can write a composite state $|\psi\rangle$ as the direct product

$$|\psi\rangle = |\lambda_n\rangle\,|\phi_n\rangle, \tag{1.2.51}$$

where $|\lambda_n\rangle$ and $|\phi_n\rangle$ are states for the two systems. Not all states can be written in this way since the superposition principle (1.2.6) implies that

$$|\psi\rangle = \sum_n a_n\,|\lambda_n\rangle\,|\phi_n\rangle \tag{1.2.52}$$

is also a possible composite state. States of the type (1.2.52) which cannot be written as a direct product of the form (1.2.51) are termed entangled states. We choose, as is always possible, the states $|\lambda_n\rangle$ and $|\phi_n\rangle$ to be such that each forms a complete, orthonormal basis. The density matrix associated with the entangled state (1.2.52) is $\rho = |\psi\rangle\langle\psi|$. The expectation value of any operator $\hat{A}$ acting only on the space of the first system spanned by the basis $\{|\lambda_n\rangle\}$ is then

$$\begin{aligned}\langle\psi|\,\hat{A}\,|\psi\rangle &= \sum_m \sum_n a^*_m a_n \langle\phi_m\,|\,\phi_n\rangle\langle\lambda_m|\,\hat{A}\,|\lambda_n\rangle \\ &= \sum_n |a_n|^2 \langle\lambda_n|\,\hat{A}\,|\lambda_n\rangle.\end{aligned} \tag{1.2.53}$$

Comparing this with (1.2.38) and (1.2.40) shows that the same expectation value is obtained by using the density matrix

$$\rho_\lambda = \sum_n |a_n|^2\,|\lambda_n\rangle\langle\lambda_n|. \tag{1.2.54}$$

This is the reduced density matrix which provides a complete description of the statistical properties of the system spanned by the basis $\{|\lambda_n\rangle\}$ but contains no information on the other system. If we restrict our attention to observables associated with only one of the two entangled systems, then this leads to loss of information on the correlations between the two systems with the result that we can at best only construct the density matrix for the system under observation. Interactions between systems usually lead to entangled states so that the best description of one of the systems is provided by its reduced density matrix. We will meet examples of this in Section 3.7, and in Section 5.6 which deals with the interaction between a quantum system and its environment.

1.3 Atoms and fields

The simplest and most important model of an atom includes only two discrete, orthonormal states $|1\rangle$ and $|2\rangle$ which, in the absence of coupling, are energy eigenstates with energies $\hbar\omega_1$ and $\hbar\omega_2 > \hbar\omega_1$, respectively. A general pure state $|\psi\rangle$ is given by the superposition $|\psi\rangle = a_1|1\rangle + a_2|2\rangle$ where $|a_1|^2 + |a_2|^2 = 1$. For this simple system the projectors $|2\rangle\langle 2|$, $|2\rangle\langle 1|$, $|1\rangle\langle 2|$, and $|1\rangle\langle 1|$, and combinations of them, are the only possible operators. We will make regular use of these in the form of the four Pauli operators

$$\mathbf{1} = |2\rangle\langle 2| + |1\rangle\langle 1|, \tag{1.3.1}$$

$$\hat{\sigma}_3 = |2\rangle\langle 2| - |1\rangle\langle 1|, \tag{1.3.2}$$

$$\hat{\sigma}_+ = |2\rangle\langle 1|, \tag{1.3.3}$$

$$\hat{\sigma}_- = |1\rangle\langle 2|. \tag{1.3.4}$$

The first two of these are Hermitian operators and $\hat{\sigma}_+$ and $\hat{\sigma}_-$ are a Hermitian conjugate pair so that $\hat{\sigma}_+ = \hat{\sigma}_-^\dagger$. In place of $\hat{\sigma}_+$ and $\hat{\sigma}_-$ we also use the Hermitian operators

$$\hat{\sigma}_1 = \hat{\sigma}_+ + \hat{\sigma}_-, \tag{1.3.5}$$

$$\hat{\sigma}_2 = \mathrm{i}(\hat{\sigma}_- - \hat{\sigma}_+). \tag{1.3.6}$$

The operators $\hat{\sigma}_1$, $\hat{\sigma}_2$, and $\hat{\sigma}_3$ correspond to observables: $\hat{\sigma}_1$ and $\hat{\sigma}_2$ are the real and imaginary parts of the complex dipole moment and $\hat{\sigma}_3$ is the atomic inversion.

A convenient representation of the state $|\psi\rangle = a_1|1\rangle + a_2|2\rangle$ is as the column vector

$$\psi = \begin{pmatrix} a_2 \\ a_1 \end{pmatrix} \tag{1.3.7}$$

In this representation, the operators (1.3.1) to (1.3.6) have the forms

$$\mathbf{1} = \begin{pmatrix} 1 & 0 \\ 0 & 1 \end{pmatrix}, \tag{1.3.8}$$

$$\hat{\sigma}_3 = \begin{pmatrix} 1 & 0 \\ 0 & -1 \end{pmatrix}, \tag{1.3.9}$$

$$\hat{\sigma}_+ = \begin{pmatrix} 0 & 1 \\ 0 & 0 \end{pmatrix}, \tag{1.3.10}$$

$$\hat{\sigma}_- = \begin{pmatrix} 0 & 0 \\ 1 & 0 \end{pmatrix}, \tag{1.3.11}$$

$$\hat{\sigma}_1 = \begin{pmatrix} 0 & 1 \\ 1 & 0 \end{pmatrix}, \tag{1.3.12}$$

$$\hat{\sigma}_2 = \begin{pmatrix} 0 & -\mathrm{i} \\ \mathrm{i} & 0 \end{pmatrix}. \tag{1.3.13}$$

We can use these matrices to derive the properties of the operators. For example, $\hat{\sigma}_1\hat{\sigma}_2 = \mathrm{i}\hat{\sigma}_3$ and all operators except **1** have zero trace. A full discussion of the properties of these operators is given in Section 3.2.

The density matrix for a general state of the two-state atom is given by

$$\rho = \rho_{22}\,|2\rangle\langle 2| + \rho_{21}\,|2\rangle\langle 1| + \rho_{12}\,|1\rangle\langle 2| + \rho_{11}\,|1\rangle\langle 1| \tag{1.3.14}$$

and has a representation as the Hermitian square matrix

$$\rho = \begin{pmatrix} \rho_{22} & \rho_{21} \\ \rho_{12} & \rho_{11} \end{pmatrix}, \tag{1.3.15}$$

where ρ_{22} and ρ_{11} are real and $\rho_{12} = \rho_{21}^*$. Comparing (1.3.15) with (1.3.8), (1.3.9), (1.3.12), and (1.3.13), we find that ρ can be written in the form

$$\rho = \tfrac{1}{2}\left(\mathbf{1} + u\hat{\sigma}_1 + v\hat{\sigma}_2 + w\hat{\sigma}_3\right), \tag{1.3.16}$$

where u, v, and w are the real quantities $u = \rho_{21} + \rho_{12}$, $v = \mathrm{i}(\rho_{21} - \rho_{12})$, and $w = \rho_{22} - \rho_{11}$. From (1.2.41)

$$\rho_{22} + \rho_{11} = 1, \tag{1.3.17}$$

while (1.2.50) shows that

$$\rho_{22}\,\rho_{11} \geqslant \rho_{21}\,\rho_{12}, \tag{1.3.18}$$

with the equality holding only for pure states. These simple results are enough to obtain the equation of motion for the density matrix describing an atom undergoing spontaneous emission. The excited state probability ρ_{22} decays at the Einstein rate 2Γ so that $\rho_{22} = \rho_{22}(0)\exp(-2\Gamma t)$. It follows from the conservation of probability (1.3.17) together with (1.3.18) that $\rho_{22} \geqslant |\rho_{21}|^2$ and hence that $|\rho_{21}|$ must decay at least at rate Γ. The simplest evolution consistent with this requirement is given by the equations

$$\dot{\rho}_{22} = -2\Gamma\rho_{22} = -\dot{\rho}_{11}, \tag{1.3.19}$$

$$\dot{\rho}_{21} = -\Gamma\rho_{21}, \quad \dot{\rho}_{12} = -\Gamma\rho_{12}, \tag{1.3.20}$$

which may be combined into the single master equation for the density matrix

$$\dot{\rho} = \Gamma\left(2\hat{\sigma}_-\rho\hat{\sigma}_+ - \hat{\sigma}_+\hat{\sigma}_-\rho - \rho\hat{\sigma}_+\hat{\sigma}_-\right). \tag{1.3.21}$$

We will see in Section 5.6 that this is indeed the equation derived by considering the interaction between the atom and the electromagnetic field in the Schrödinger picture.

The most common interaction between an atom and the electromagnetic field is of the electric dipole form coupling states $|1\rangle$ and $|2\rangle$ of differing parity. This electric dipole approximation is valid when the wavelengths of the driving field are much greater than the mean separation of the electron and the nucleus. As quantum optics is primarily concerned with the interaction between atoms and optical or infrared radiation, this approximation is usually a good one. The atomic dipole operator associated with the two states $|1\rangle$ and $|2\rangle$ has the general form

$$\hat{\boldsymbol{\mu}} = \boldsymbol{\mu}^* \hat{\sigma}_+ + \boldsymbol{\mu} \hat{\sigma}_-, \tag{1.3.22}$$

so that the action of $\hat{\boldsymbol{\mu}}$ on the lower state $|1\rangle$ produces the state $\hat{\boldsymbol{\mu}}|1\rangle = \boldsymbol{\mu}^*|2\rangle$, and similarly $\hat{\boldsymbol{\mu}}|2\rangle = \boldsymbol{\mu}|1\rangle$. The Hamiltonian describing the two-state atom interacting with a classical electric field $\mathbf{E}(t)$ within the electric dipole approximation is

$$\hat{H} = \frac{\hbar}{2}(\omega_2 + \omega_1)\mathbf{1} + \frac{\hbar}{2}(\omega_2 - \omega_1)\hat{\sigma}_3 - \boldsymbol{\mu} \cdot \mathbf{E}(t)(\hat{\sigma}_+ + \hat{\sigma}_-), \tag{1.3.23}$$

where, for simplicity, we have taken $\boldsymbol{\mu}$ to be real. The first two terms in this Hamiltonian are the free Hamiltonian $\hat{H}_0$ for the atom having eigenstates $|1\rangle$ and $|2\rangle$ with eigenenergies $\hbar\omega_1$ and $\hbar\omega_2$, respectively. The evolution of the state

$$|\psi(t)\rangle = a_1(t)|1\rangle + a_2(t)|2\rangle \tag{1.3.24}$$

is determined by the amplitude equations (1.2.31) with the Hamiltonian (1.3.23). We find

$$\dot{a}_1(t) = -\mathrm{i}\omega_1 a_1(t) + \frac{\mathrm{i}}{\hbar}\boldsymbol{\mu} \cdot \mathbf{E}(t) a_2(t), \tag{1.3.25}$$

$$\dot{a}_2(t) = -\mathrm{i}\omega_2 a_2(t) + \frac{\mathrm{i}}{\hbar}\boldsymbol{\mu} \cdot \mathbf{E}(t) a_1(t). \tag{1.3.26}$$

For a monochromatic field with $\mathbf{E}(t) = \mathbf{E}_0 \cos(\omega t + \varphi)$, we obtain

$$\dot{a}_1(t) = -\mathrm{i}\omega_1 a_1(t) + \mathrm{i}V\cos(\omega t + \varphi) a_2(t), \tag{1.3.27}$$

$$\dot{a}_2(t) = -\mathrm{i}\omega_2 a_2(t) + \mathrm{i}V\cos(\omega t + \varphi) a_1(t), \tag{1.3.28}$$

where $V = \boldsymbol{\mu} \cdot \mathbf{E}_0/\hbar$. These amplitude equations are the starting point for our analysis of coherent interactions in Chapter 2.

In the quantum theory of light, the electric and magnetic fields are Hermitian operators. The simplest method of constructing these operators is to decompose the fields into modes and associate each mode with a quantum mechanical harmonic oscillator. It is most natural to describe the observables associated with such an oscillator in terms of annihilation and creation operators $\hat{a}$ and $\hat{a}^\dagger$. These non-Hermitian operators do not commute, having the commutator

$$[\hat{a}, \hat{a}^\dagger] = 1. \tag{1.3.29}$$

The Hamiltonian for the oscillator, or field mode, is

$$\hat{H} = \hbar\omega\left(\hat{a}^\dagger\hat{a} + \tfrac{1}{2}\right) = \hbar\omega(\hat{n} + \tfrac{1}{2}), \tag{1.3.30}$$

where the number operator $\hat{n} = \hat{a}^\dagger\hat{a}$ is Hermitian since, by (1.2.10) and (1.2.8), $\hat{n}^\dagger = (\hat{a}^\dagger\hat{a})^\dagger = \hat{a}^\dagger\hat{a} = \hat{n}$. The energy eigenstates are also eigenstates of $\hat{n}$ and are denoted $|n\rangle$, where $n = 0, 1, 2, \ldots$. These orthonormal states satisfy the eigenvalue equation

$$\hat{n}\,|n\rangle = n\,|n\rangle \tag{1.3.31}$$

or, equivalently,

$$\hat{H}\,|n\rangle = \hbar\omega(n + \tfrac{1}{2})|n\rangle. \tag{1.3.32}$$

The number states $|n\rangle$ form a complete set since any state of the mode can be expressed as a superposition of them and we can resolve the identity operator as

$$\mathbf{1} = \sum_{n=0}^{\infty} |n\rangle\langle n|. \tag{1.3.33}$$

Furthermore, the completeness of the set $\{|n\rangle\}$ allows the trace of any operator $\hat{A}$ to be evaluated in this basis as

$$\mathrm{Tr}(\hat{A}) = \sum_{n=0}^{\infty} \langle n|\,\hat{A}\,|n\rangle. \tag{1.3.34}$$

The commutator (1.3.29) and the form of $\hat{n}$ determine the actions of $\hat{a}$ and $\hat{a}^\dagger$ on $|n\rangle$ apart from an arbitrary phase factor. The conventional choice sets

$$\hat{a}\,|n\rangle = n^{1/2}\,|n-1\rangle, \tag{1.3.35}$$

$$\hat{a}^\dagger\,|n\rangle = (n+1)^{1/2}\,|n+1\rangle. \tag{1.3.36}$$

The action of $\hat{a}$ reduces the number of quanta by one (except for the vacuum state $|0\rangle$) while that of $\hat{a}^\dagger$ increases it by one. This, of course, is the reason for referring to $\hat{a}$ and $\hat{a}^\dagger$ as annihilation and creation operators. The number state representations of $\hat{a}$ and $\hat{a}^\dagger$ can be obtained using the resolution of the identity (1.3.33), the actions of these operators on the number states (1.3.35) and (1.3.36), and the orthonormality of the number states. We find

$$\begin{aligned}\hat{a} &= \sum_{n=0}^{\infty} |n\rangle\langle n|\,\hat{a}\sum_{m=0}^{\infty} |m\rangle\langle m| \\ &= \sum_{n=0}^{\infty} (n+1)^{1/2}\,|n\rangle\langle n+1|\end{aligned} \tag{1.3.37}$$

and, similarly,

$$\hat{a}^\dagger = \sum_{n=0}^{\infty} (n+1)^{1/2}\,|n+1\rangle\langle n|. \tag{1.3.38}$$

The commutators of $\hat{a}$ and $\hat{a}^\dagger$ with $\hat{n}$ are, from (1.2.22),

$$[\hat{a}, \hat{n}] = [\hat{a}, \hat{a}^\dagger\hat{a}] = \hat{a}, \tag{1.3.39}$$

$$[\hat{a}^\dagger, \hat{n}] = [\hat{a}^\dagger, \hat{a}^\dagger\hat{a}] = -\hat{a}^\dagger. \tag{1.3.40}$$

These, together with the free-field Hamiltonian (1.3.30) and the Heisenberg equation (1.2.36), lead to the Heisenberg equations of motion for $\hat{a}(t)$ and $\hat{a}^\dagger(t)$ given by

$$\dot{\hat{a}}(t) = -\mathrm{i}\,\omega\hat{a}(t), \tag{1.3.41}$$

$$\dot{\hat{a}}^\dagger(t) = \mathrm{i}\,\omega\hat{a}^\dagger(t), \tag{1.3.42}$$

with the simple solutions $\hat{a}(t) = \hat{a}\exp(-\mathrm{i}\,\omega t)$ and $\hat{a}^\dagger(t) = \hat{a}^\dagger\exp(\mathrm{i}\,\omega t)$, where $\hat{a}$ and $\hat{a}^\dagger$ denote the initial values $\hat{a}(0)$ and $\hat{a}^\dagger(0)$, respectively.

The electric field operator at position $\mathbf{r}$ for the plane-wave mode with wavevector $\mathbf{k}$, frequency $\omega = |\mathbf{k}|c$ and linear polarization $\boldsymbol{\varepsilon}$ is

$$\hat{\mathbf{E}}(\mathbf{r},t) = \mathrm{i}\left(\frac{\hbar\omega}{2\varepsilon_0\mathscr{V}}\right)^{1/2}\boldsymbol{\varepsilon}[\hat{a}\exp(-\mathrm{i}\,\omega t + \mathrm{i}\mathbf{k}\cdot\mathbf{r}) - \hat{a}^\dagger\exp(\mathrm{i}\,\omega t - \mathrm{i}\mathbf{k}\cdot\mathbf{r})], \tag{1.3.43}$$

where ε_0 is the permittivity of free space and $\mathscr{V}$ is the quantization volume. The full electric field is the sum over all allowed wavevectors and polarizations of single-mode fields of the form (1.3.43). For problems involving the full electromagnetic field, it is often more convenient to use a continuum description involving continuum annihilation and creation operators. We will describe this in Section 3.2 and use it to treat damping in Sections 5.5 and 5.6.

The interaction between a two-state atom and the quantized electric field is described by the electric dipole Hamiltonian (1.3.23) with $\mathbf{E}(t)$ replaced by the quantized electric field operator. For an atom placed at the origin $\mathbf{r} = 0$ interacting with a single plane-wave mode of frequency ω this Hamiltonian becomes

$$\hat{H}_1 = \frac{\hbar}{2}(\omega_2 + \omega_1)\mathbf{1} + \frac{\hbar}{2}(\omega_2 - \omega_1)\hat{\sigma}_3 - \mathrm{i}\hbar\lambda(\hat{\sigma}_+ + \hat{\sigma}_-)[\hat{a}\exp(-\mathrm{i}\,\omega t) - \hat{a}^\dagger\exp(\mathrm{i}\,\omega t)], \tag{1.3.44}$$

where $\lambda = \boldsymbol{\mu}\cdot\boldsymbol{\varepsilon}(\omega/2\hbar\varepsilon_0\mathscr{V})^{1/2}$. We will use this Hamiltonian to begin our discussion of the Jaynes–Cummings model in Section 2.4. An equivalent and time-independent form of (1.3.44) is

$$\hat{H}_2 = \frac{\hbar}{2}(\omega_2 + \omega_1)\mathbf{1} + \frac{\hbar}{2}(\omega_2 - \omega_1)\hat{\sigma}_3 + \hbar\omega\left(\hat{a}^\dagger\hat{a} + \tfrac{1}{2}\right) - \mathrm{i}\hbar\lambda(\hat{\sigma}_+ + \hat{\sigma}_-)(\hat{a} - \hat{a}^\dagger). \tag{1.3.45}$$

The appearance of the free-field Hamiltonian $\hat{H}_0 = \hbar\omega(\hat{a}^\dagger\hat{a} + \frac{1}{2})$ in (1.3.45) accounts for the evolution of $\hat{a}$ and $\hat{a}^\dagger$ in (1.3.44) since $\hat{H}_0$ alone leads to the Heisenberg equations (1.3.41) and (1.3.42). The relationship between the two Hamiltonians $\hat{H}_1$ and $\hat{H}_2$ will be discussed in Section 2.2.

2
COHERENT INTERACTIONS

2.1 Introduction

In quantum optics, we are often interested in the dynamics of atoms or molecules coupled to an electromagnetic field. Only simple models are required to describe many of the most important features of this dynamics. In these models, the atomic system is adequately described by a small number of essential states, together with a free-electron continuum in problems involving ionization. The field may be described either classically or fully quantum mechanically and may by monochromatic, have a time-dependent amplitude or phase, or be broad band. The time evolution of an initially occupied state typically produces a coherent superposition of the coupled states. This chapter discusses some of these models and the methods used to analyze them. For this, we utilize the Schrödinger picture and its related interaction pictures. We concentrate on the dynamics of a single level coupled to either a second level or to a quasi-continuum of discrete levels. The former exhibits a variety of dynamics depending upon the nature of the driving field. The latter exhibits exponential decay for short times followed by repopulation. Exponential decay without repopulation is found only for coupling to a true continuum of states; this problem is treated in detail in Chapter 5.

The final section of this chapter deals with models of coherent interaction between field modes. These are used in the study of simple non-linear optical processes and of the quantum theory of dielectric media. For this, we use the Heisenberg picture.

2.2 Interaction pictures

Quantum mechanics provides a number of equivalent descriptions, or pictures, of the dynamics of a system. These are related by unitary transformations of both the state and the operators acting on it. In the Schrödinger picture, the operators are time independent and all the dynamics is contained within the state. At the other extreme, in the Heisenberg picture, the operators evolve while the state is time independent. Intermediate between them are the interaction pictures in which part of the dynamics, usually that associated with free, uncoupled evolution, is contained within the operators, while that arising from the coupling appears in the state. Alternatively, the free evolution may be contained within the state, with the operators containing the effect of the coupling. The former are termed Schrödinger interaction pictures and the latter Heisenberg interaction pictures.

The relationship between the Schrödinger pictures is most readily appreciated by considering a simple example. In Section 1.3, we described an electric dipole interaction coupling a state $|1\rangle$ with energy $\hbar\omega_1$ to a state $|2\rangle$ with energy $\hbar\omega_2 > \hbar\omega_1$. For a monochromatic electric field of frequency ω, the Schrödinger picture amplitudes a_1 and a_2 for these states satisfy the coupled differential equations (see (1.3.27) and (1.3.28))

$$\dot{a}_1(t) = -\mathrm{i}\,\omega_1 a_1(t) + \mathrm{i}V\cos(\omega t + \varphi)a_2(t), \tag{2.2.1}$$

$$\dot{a}_2(t) = -\mathrm{i}\,\omega_2 a_2(t) + \mathrm{i}V\cos(\omega t + \varphi)a_1(t). \tag{2.2.2}$$

There are two important interaction pictures which we distinguish by labelling the amplitudes b_1 and b_2, and c_1 and c_2. In the first of these, the free evolution is removed from the differential equations (2.2.1) and (2.2.2) by writing $a_1(t) = b_1(t)\exp(-\mathrm{i}\,\omega_1 t)$ and $a_2(t) = b_2(t)\exp(-\mathrm{i}\,\omega_2 t)$ to give

$$\dot{b}_1(t) = \mathrm{i}V\cos(\omega t + \varphi)\exp[\mathrm{i}(\omega_1 - \omega_2)t]b_2(t), \tag{2.2.3}$$

$$\dot{b}_2(t) = \mathrm{i}V\cos(\omega t + \varphi)\exp[\mathrm{i}(\omega_2 - \omega_1)t]b_1(t). \tag{2.2.4}$$

The effect of this transformation is that the evolution of the amplitudes b_1 and b_2 only depends on the coupling. The second interaction picture is primarily used within the so-called rotating-wave approximation to remove the explicit time dependence from the amplitude equations. This approximation, which will be discussed in more detail in the next section, consists of neglecting the rapidly oscillating exponential term $\exp[\mathrm{i}(-\omega + \omega_1 - \omega_2)t]$ and its complex conjugate obtained by expressing cosine as a sum of complex exponentials in (2.2.3) and (2.2.4). The resulting amplitude equations are

$$\dot{b}_1(t) = \mathrm{i}\frac{V}{2}\exp(\mathrm{i}\,\varphi)\exp(-\mathrm{i}\,\Delta t)b_2(t), \tag{2.2.5}$$

$$\dot{b}_2(t) = \mathrm{i}\frac{V}{2}\exp(-\mathrm{i}\,\varphi)\exp(\mathrm{i}\,\Delta t)b_1(t), \tag{2.2.6}$$

where $\Delta = \omega_2 - \omega_1 - \omega$ is the detuning between the atomic transition frequency and that of the driving field. The explicit time dependence in (2.2.5) and (2.2.6) can be removed by the transformation $b_1(t) = c_1(t)\exp(-\mathrm{i}\,\delta t)$ and $b_2(t) = c_2(t)\exp[\mathrm{i}(\Delta - \delta)t]$, where δ is any chosen frequency. The resulting amplitude equations for $c_1(t)$ and $c_2(t)$ are

$$\dot{c}_1(t) = \mathrm{i}\,\delta c_1(t) + \mathrm{i}\frac{V}{2}\exp(\mathrm{i}\varphi)c_2(t), \tag{2.2.7}$$

$$\dot{c}_2(t) = \mathrm{i}(\delta - \Delta)c_2(t) + \mathrm{i}\frac{V}{2}\exp(-\mathrm{i}\varphi)c_1(t). \tag{2.2.8}$$

Common choices for δ are zero and $\Delta/2$.

The Schrödinger pictures are related to each other by unitary transformations. A measurable quantity is the expectation value of an operator $\hat{A}$ in a state

$|\psi\rangle$ given by $\langle\psi|\hat{A}|\psi\rangle$, and must be independent of the picture in which it is evaluated. The transformation to a Schrödinger interaction picture is achieved by the action of a unitary operator $\hat{U}$ so that the state becomes

$$|\psi_I\rangle = \hat{U}|\psi\rangle, \tag{2.2.9}$$

while $\hat{A}$ becomes $\hat{U}\hat{A}\hat{U}^\dagger$. The invariance of the expectation value $\langle\psi|\hat{A}|\psi\rangle$ then follows from $\hat{U}^\dagger = \hat{U}^{-1}$ (see (1.2.33)). The interaction picture Hamiltonian $\hat{H}_I$ can be derived from the Schrödinger equation

$$i\hbar\frac{\partial}{\partial t}|\psi\rangle = \hat{H}|\psi\rangle \tag{2.2.10}$$

together with the requirement that $|\psi_I\rangle$ satisfies the transformed Schrödinger equation

$$i\hbar\frac{\partial}{\partial t}|\psi_I\rangle = \hat{H}_I|\psi_I\rangle. \tag{2.2.11}$$

Substituting for $|\psi_I\rangle$ from (2.2.9), we find

$$i\hbar\left(\dot{\hat{U}}|\psi\rangle + \hat{U}\frac{\partial}{\partial t}|\psi\rangle\right) = i\hbar\dot{\hat{U}}|\psi\rangle + \hat{U}\hat{H}|\psi\rangle = \hat{H}_I\hat{U}|\psi\rangle, \tag{2.2.12}$$

using (2.2.10). It follows that the interaction picture Hamiltonian is

$$\hat{H}_I = i\hbar\dot{\hat{U}}\hat{U}^\dagger + \hat{U}\hat{H}\hat{U}^\dagger. \tag{2.2.13}$$

As an example we transform the Hamiltonian

$$\hat{H} = \frac{\hbar}{2}(\omega_2+\omega_1)\mathbf{1} + \frac{\hbar}{2}(\omega_2-\omega_1)\hat{\sigma}_3 - \hbar V(\hat{\sigma}_+ + \hat{\sigma}_-)\cos(\omega t+\varphi) \tag{2.2.14}$$

obtained from (1.3.23) with $\mathbf{E}(t) = \mathbf{E}_0\cos(\omega t+\varphi)$ and $V = \boldsymbol{\mu}\cdot\mathbf{E}_0/\hbar$ for which the Schrödinger equation leads to the amplitude equations (2.2.1) and (2.2.2). Consider the unitary transformation associated with the unitary operator

$$\begin{aligned}\hat{U} &= \exp\left(\frac{i}{2}[(\omega_2+\omega_1)\mathbf{1} + (\omega_2-\omega_1)\hat{\sigma}_3]t\right)\\ &= \exp(i\omega_1 t)|1\rangle\langle 1| + \exp(i\omega_2 t)|2\rangle\langle 2|,\end{aligned} \tag{2.2.15}$$

where we have used (1.3.1) and (1.3.2) together with the definition (1.2.28) of the function of an Hermitian operator. Substituting (2.2.14) and (2.2.15) into (2.2.13), we find that the corresponding interaction picture Hamiltonian is

$$\hat{H}_I = -\hbar V\{\hat{\sigma}_+\exp[i(\omega_2-\omega_1)t] + \hat{\sigma}_-\exp[i(\omega_1-\omega_2)t]\}\cos(\omega t+\varphi). \tag{2.2.16}$$

Writing $|\psi_\mathrm{I}\rangle = b_1(t)|1\rangle + b_2(t)|2\rangle$ and substituting this and $\hat{H}_\mathrm{I}$ into (2.2.11) reproduces the amplitude equations (2.2.3) and (2.2.4). As might have been expected, the interaction picture in which the evolution of the amplitudes is determined only by the coupling is associated with a Hamiltonian $\hat{H}_\mathrm{I}$ in which the free evolution terms are absent.

Making the rotating-wave approximation corresponds to replacing the Hamiltonian (2.2.14) by

$$\hat{H} = \frac{\hbar}{2}(\omega_2 + \omega_1)\mathbf{1} + \frac{\hbar}{2}(\omega_2 - \omega_1)\hat{\sigma}_3$$
$$- \frac{\hbar V}{2}\{\hat{\sigma}_+ \exp[-\mathrm{i}(\omega t + \varphi)] + \hat{\sigma}_- \exp[\mathrm{i}(\omega t + \varphi)]\}. \quad (2.2.17)$$

Consider the effect on this Hamiltonian of the unitary transformation produced by

$$\hat{U} = \exp\left(\frac{\mathrm{i}}{2}[(\omega_2 + \omega_1 + \Delta - 2\delta)\mathbf{1} + \omega\hat{\sigma}_3]t\right)$$
$$= \exp[\mathrm{i}(\omega_2 - \omega - \delta)t]|1\rangle\langle 1| + \exp[\mathrm{i}(\omega_2 - \delta)t]|2\rangle\langle 2|. \quad (2.2.18)$$

Substituting (2.2.17) and (2.2.18) into (2.2.13), we find that the corresponding interaction picture Hamiltonian is

$$\hat{H}_\mathrm{I} = \hbar\left(\frac{\Delta}{2} - \delta\right)\mathbf{1} + \frac{\hbar}{2}\Delta\hat{\sigma}_3 - \frac{\hbar V}{2}[\hat{\sigma}_+ \exp(-\mathrm{i}\varphi) + \hat{\sigma}_- \exp(\mathrm{i}\varphi)]. \quad (2.2.19)$$

Writing $|\psi_\mathrm{I}\rangle = c_1(t)|1\rangle + c_2(t)|2\rangle$ and substituting this and $\hat{H}_\mathrm{I}$ into (2.2.11) reproduces the amplitude equations (2.2.7) and (2.2.8). The absence of explicit time dependence in these equations is a consequence of the time independence of (2.2.19).

Unitary transformations can be used quite generally to transfer time dependence between the operators and the states. An important example applies to the Hamiltonians which contain the term $\hat{H}_0 = \hbar\omega(\hat{a}^\dagger\hat{a} + \frac{1}{2})$ representing the free evolution of a field mode of frequency ω. The unitary operator $\hat{U} = \exp(\mathrm{i}\hat{H}_0 t/\hbar)$ produces an interaction Hamiltonian from which $\hat{H}_0$ is absent and in which $\hat{a}$ and $\hat{a}^\dagger$ are replaced by $\hat{a}\exp(-\mathrm{i}\omega t)$ and $\hat{a}^\dagger\exp(\mathrm{i}\omega t)$, respectively, as we now show. It follows from (1.2.28) that $\hat{U}$ can be expressed in the form

$$\hat{U} = \exp[\mathrm{i}(\hat{n} + \tfrac{1}{2})\omega t] = \sum_{n=0}^{\infty} \exp[\mathrm{i}(n + \tfrac{1}{2})\omega t]|n\rangle\langle n|. \quad (2.2.20)$$

Under the action of this unitary transformation, the annihilation operator becomes

$$\hat{U}\hat{a}\hat{U}^{\dagger} = \sum_{l=0}^{\infty} \exp[\mathrm{i}(l+\tfrac{1}{2})\omega t]\,|l\rangle\langle l| \sum_{n=0}^{\infty} (n+1)^{1/2}\,|n\rangle\langle n+1|$$

$$\times \sum_{m=0}^{\infty} \exp[-\mathrm{i}(m+\tfrac{1}{2})\omega t]\,|m\rangle\langle m|$$

$$= \sum_{l=0}^{\infty}\sum_{m=0}^{\infty}\sum_{n=0}^{\infty} \exp[\mathrm{i}(l-m)\omega t](n+1)^{1/2}\,|n\rangle\langle n+1|\,\delta_{ln}\delta_{m,n+1}$$

$$= \sum_{n=0}^{\infty} \exp(-\mathrm{i}\,\omega t)(n+1)^{1/2}\,|n\rangle\langle n+1| = \hat{a}\exp(-\mathrm{i}\,\omega t), \qquad (2.2.21)$$

where we have used (1.3.37) and the orthogonality of the number states. Similarly, the creation operator $\hat{a}^{\dagger}$ becomes

$$\hat{U}\hat{a}^{\dagger}\hat{U}^{\dagger} = \hat{a}^{\dagger}\exp(\mathrm{i}\,\omega t). \qquad (2.2.22)$$

The action of $\hat{U}$ transforms the time-independent Hamiltonian (1.3.45) into the time-dependent interaction picture Hamiltonian (1.3.44) since the term $\mathrm{i}\hbar\dot{\hat{U}}\hat{U}^{\dagger}$ in (2.2.13) is $-\hat{H}_0$.

In a Heisenberg interaction picture, the equation of motion for an operator having no explicit time dependence is

$$\dot{\hat{A}} = \frac{\mathrm{i}}{\hbar}\left[\hat{H}_{\mathrm{I}}, \hat{A}\right], \qquad (2.2.23)$$

where $\hat{H}_{\mathrm{I}}$ is given by (2.2.13). The remaining time dependence is now contained within the state. We will use this interaction picture to study interactions between field modes in Section 2.6.

2.3 Two-state dynamics

Real atoms consist of a large number of discrete states and a number of free-electron continua. In most cases, the field only causes transitions between a small number of discrete states, in the simplest of which only two states are involved. The amplitude equations (2.2.3) and (2.2.4) describe the electric dipole interaction between two states $|1\rangle$ and $|2\rangle$ and a classical monochromatic field of frequency ω. The field is switched on at time $t = 0$ when state $|1\rangle$ is occupied, after which time V and φ are taken to be constant. No closed-form analytic solution of these amplitude equations is known and so it is necessary to adopt approximation methods. We begin by finding a perturbative solution for $b_2(t)$ which is valid for short times and weak coupling. To do this, we insert the free evolution of b_1, that is $b_1(t)=1$, into (2.2.4) and integrate. The resulting approximate solution for $b_2(t)$ is

$$b_2(t)=\frac{V}{2}\left[\exp(\mathrm{i}\varphi)\left(\frac{\exp[\mathrm{i}(2\omega+\Delta)t]-1}{2\omega+\Delta}\right)+\exp(-\mathrm{i}\varphi)\frac{[\exp(\mathrm{i}\,\Delta t)-1]}{\Delta}\right], \tag{2.3.1}$$

where $\Delta=\omega_2-\omega_1-\omega$ as before. The coupling V and the detuning Δ are typically many orders of magnitude smaller than the driving frequency ω and therefore the first term in (2.3.1) is many orders of magnitude smaller than the second. The physical origin of this difference is that the second term represents transitions detuned from resonance by Δ whereas, for the first term, the detuning from resonance is $2\omega+\Delta$. Neglecting terms associated with such far off-resonance effects constitutes the rotating-wave approximation. The rotating-wave forms of the amplitude equations (2.2.3) and (2.2.4) are found by dropping the rapidly oscillating terms as described in the previous section. It should be noted that transitions to other states not included in the two-state model will typically lead to larger effects than those omitted in making the rotating-wave approximation. Within this approximation, the probability for making a transition to state $|2\rangle$ is

$$P_2(t)=|b_2(t)|^2=\frac{V^2}{\Delta^2}\sin^2(\Delta t/2). \tag{2.3.2}$$

This result is enough to obtain an expression for the absorption rate produced by a broad band of driving frequencies, that is different detunings, with a spectral density $\rho(\Delta)$. The probability of transition to $|2\rangle$ is then a weighted average of the probability (2.3.2) for all possible detunings given by

$$\bar{P}_2(t)=\int \mathrm{d}\Delta\,\rho(\Delta)\frac{V^2}{\Delta^2}\sin^2(\Delta t/2)\approx\rho(0)\frac{\pi V^2}{2}t, \tag{2.3.3}$$

where we have assumed that $\rho(\Delta)$ is slowly varying compared with the rest of the integrand and have extended the range of integration from $-\infty$ to $+\infty$. The transition rate is then $\rho(0)\pi V^2/2$. A more detailed analysis of this result is given in Section 5.2.

An exact solution of the amplitude equations does exist within the rotating-wave approximation. In our second interaction picture, the amplitude equations (2.2.7) and (2.2.8) with $\delta=\Delta/2$ have the form

$$\dot{c}_1(t)=\mathrm{i}\frac{\Delta}{2}c_1(t)+\mathrm{i}\frac{V}{2}\exp(\mathrm{i}\varphi)c_2(t), \tag{2.3.4}$$

$$\dot{c}_2(t)=-\mathrm{i}\frac{\Delta}{2}c_2(t)+\mathrm{i}\frac{V}{2}\exp(-\mathrm{i}\varphi)c_1(t). \tag{2.3.5}$$

These can be solved by elimination to obtain a second-order differential equation for one of the amplitudes, or by Laplace transforms. Taking the Laplace transforms of (2.3.4) and (2.3.5) with the initial conditions $c_1(0) = 1$ and $c_2(0) = 0$, we obtain

$$s\bar{c}_1(s) - 1 = \mathrm{i}\frac{\Delta}{2}\bar{c}_1(s) + \mathrm{i}\frac{V}{2}\exp(\mathrm{i}\varphi)\bar{c}_2(s), \tag{2.3.6}$$

$$s\bar{c}_2(s) = -\mathrm{i}\frac{\Delta}{2}\bar{c}_2(s) + \mathrm{i}\frac{V}{2}\exp(-\mathrm{i}\varphi)\bar{c}_1(s), \tag{2.3.7}$$

where $\bar{c}_1(s)$ denotes the Laplace transform of $c_1(t)$ (see Appendix 9). Solving for $\bar{c}_1(s)$ gives

$$\bar{c}_1(s) = \frac{s + \mathrm{i}\Delta/2}{s^2 + \frac{1}{4}\Omega_\mathrm{R}^2}, \tag{2.3.8}$$

where

$$\Omega_\mathrm{R} = (V^2 + \Delta^2)^{1/2} \tag{2.3.9}$$

is the Rabi frequency. Inverting the transform (2.3.8) gives the amplitude

$$c_1(t) = \cos(\Omega_\mathrm{R} t/2) + \mathrm{i}\frac{\Delta}{\Omega_\mathrm{R}}\sin(\Omega_\mathrm{R} t/2) \tag{2.3.10}$$

and similarly

$$c_2(t) = \mathrm{i}\frac{V}{\Omega_\mathrm{R}}\exp(-\mathrm{i}\varphi)\sin(\Omega_\mathrm{R} t/2). \tag{2.3.11}$$

The probability for being in state $|1\rangle$ at time t is

$$P_1(t) = |c_1(t)|^2 = 1 - \frac{V^2}{2\Omega_\mathrm{R}^2}(1 - \cos\Omega_\mathrm{R} t), \tag{2.3.12}$$

and is known as the Rabi solution. This probability oscillates at the Rabi frequency Ω_R between a maximum of unity and a minimum of $(\Delta/\Omega_\mathrm{R})^2$.

It is often the case that both the amplitude and the phase of the field will vary in time. We can account for this by allowing both V and φ in (2.2.5) and (2.2.6) to be functions of time. Eliminating $b_2(t)$ between the resulting equations leads to a second-order differential equation for $b_1(t)$, with time-dependent coefficients, given by

$$\ddot{b}_1(t) - \left(\frac{\dot{V}(t)}{V(t)} + \mathrm{i}(\dot{\varphi}(t) - \Delta)\right)\dot{b}_1(t) + \frac{V^2(t)}{4}b_1(t) = 0. \tag{2.3.13}$$

The time derivative of the phase appears with Δ as an effective time-dependent detuning, reflecting the fact that the derivative of a phase is a frequency. Our ability to solve (2.3.13) analytically depends on the precise time dependences of $V(t)$ and $\varphi(t)$ and many exact solutions for b_1 are known. A general solution in terms of $V(t)$ is possible if we set $\dot{\varphi}(t) = \Delta = 0$. Then on making the change of variable from t to the pulse area up to time t given by

$$2x = \int_0^t V(t')\,\mathrm{d}t', \tag{2.3.14}$$

we find that (2.3.13) reduces to the equation

$$\frac{\mathrm{d}^2 b_1}{\mathrm{d}x^2} + b_1 = 0. \tag{2.3.15}$$

The general solution for arbitrary $V(t)$ is therefore

$$b_1(t) = A\cos\left(\tfrac{1}{2}\int_0^t V(t')\,\mathrm{d}t' + \theta\right), \tag{2.3.16}$$

where A and θ are determined by initial conditions. For $b_1(0) = 1$, we have $A = 1$ and $\theta = 0$, and the probability for being in state $|1\rangle$ is

$$P_1(t) = \tfrac{1}{2}\left[1 + \cos\left(\int_0^t V(t')\,\mathrm{d}t'\right)\right]. \tag{2.3.17}$$

We see that, in this case, the probability depends only on the pulse area up to time t. It follows that a pulse of total area $(2n+1)\pi$ will transfer the atom from state $|1\rangle$ to state $|2\rangle$ with unit probability.

For noisy fields, it is not possible to specify the time evolution of $V(t)$ and $\varphi(t)$ but only to make statistical statements about them. Rather then attempting to solve (2.2.5) and (2.2.6) for the amplitude $b_1(t)$, we obtain a differential equation for the probability $|b_1(t)|^2$. Starting from (2.2.5) and its complex conjugate, we find

$$\begin{aligned}\frac{\mathrm{d}}{\mathrm{d}t}|b_1|^2 &= b_1^*\dot{b}_1 + \dot{b}_1^* b_1\\ &= \mathrm{i}\frac{V(t)}{2}(\exp[\mathrm{i}\varphi(t)]\exp(-\mathrm{i}\,\Delta t)b_1^* b_2 - \exp[-\mathrm{i}\varphi(t)]\exp(\mathrm{i}\,\Delta t)b_2^* b_1).\end{aligned} \tag{2.3.18}$$

From the amplitude equations we also obtain

$$\frac{\mathrm{d}}{\mathrm{d}t}(b_1^* b_2) = \mathrm{i}\frac{V(t)}{2}\exp[-\mathrm{i}\varphi(t)]\exp(\mathrm{i}\,\Delta t)\left(|b_1|^2 - |b_2|^2\right) \tag{2.3.19}$$

and

$$\frac{\mathrm{d}}{\mathrm{d}t}(b_2^* b_1) = -\mathrm{i}\frac{V(t)}{2}\exp[\mathrm{i}\varphi(t)]\exp(-\mathrm{i}\,\Delta t)\left(|b_1|^2 - |b_2|^2\right). \tag{2.3.20}$$

Formally integrating (2.3.19) and (2.3.20), substituting the results into (2.3.18), and using the fact that $|b_1(t)|^2 + |b_2(t)|^2 = 1$ gives the following integro-differential equation for $|b_1(t)|^2$:

$$\frac{\mathrm{d}}{\mathrm{d}t}|b_1(t)|^2 = -\tfrac{1}{2}\int_0^t \left[2|b_1(t')|^2 - 1\right] V(t)V(t') \times \cos[\varphi(t) - \varphi(t') - \Delta(t-t')]\,\mathrm{d}t'. \qquad (2.3.21)$$

In order to proceed, we need a method for averaging (2.3.21) over the field fluctuations. The easiest way of doing this is to decorrelate the probabilities from the couplings in the integrand. We mention two situations in which this procedure leads to a good approximation. Firstly, if the coupling $V\exp(\mathrm{i}\varphi)$ is rapidly fluctuating with a zero mean value then the integrand of (2.3.21) will only give a significant value if $t' \simeq t$, in which case $|b_1(t')|^2$ can be replaced by $|b_1(t)|^2$ to give the equation

$$\frac{\mathrm{d}}{\mathrm{d}t}|b_1(t)|^2 = -\tfrac{1}{2}\left[2|b_1(t)|^2 - 1\right]\int_0^t V(t)V(t') \times \cos[\varphi(t) - \varphi(t') - \Delta(t-t')]\,\mathrm{d}t'. \qquad (2.3.22)$$

Writing $P_1(t) = \langle |b_1(t)|^2 \rangle$ and

$$K(t,t') = \langle V(t)V(t')\cos[\varphi(t) - \varphi(t') - \Delta(t-t')]\rangle, \qquad (2.3.23)$$

where the angle brackets denote a statistical average over the fluctuations, (2.3.22) becomes

$$\dot{P}_1(t) = -\tfrac{1}{2}[2P_1(t) - 1]\int_0^t K(t,t')\,\mathrm{d}t'. \qquad (2.3.24)$$

For this result to be valid, $P_1(t)$ must be approximately constant over the short time scale during which the kernel $K(t,t')$ is significant. As a simple example, consider a broad-band noisy field for which the kernel is a sharply peaked function of $t - t'$ approximated by the delta function

$$K(t,t') = K_0\,\delta(t-t'). \qquad (2.3.25)$$

Then (2.3.24) becomes the rate equation

$$\begin{aligned}\dot{P}_1(t) &= -\frac{K_0}{4}[2P_1(t) - 1] \\ &= -\frac{K_0}{4}[P_1(t) - P_2(t)],\end{aligned} \qquad (2.3.26)$$

where the additional factor of $1/2$ has arisen because the delta function is situated at the upper limit of integration (see Appendix 2), and $P_2(t) = \langle |b_2(t)|^2 \rangle$. The coefficient $K_0/4$ is the rate of absorption or of simulated emission.

The second situation in which decorrelation can be performed is if the fluctuations are small in the sense that $|b_1(t)|^2$ does not differ significantly from $P_1(t)$. We can then replace $|b_1(t')|^2$ in the integrand of (2.3.21) by $P_1(t')$ and perform a statistical average to give the approximate equation

$$\dot{P}_1(t) = -\tfrac{1}{2}\int_0^t [2P_1(t') - 1]K(t,t')\,\mathrm{d}t'. \tag{2.3.27}$$

The simplest models of noise are said to be stationary, meaning that the correlation between the values of the field at times t and t' depends only on the difference $t - t'$, so that $K(t,t') = K(t - t')$. In this case, the right-hand side of (2.3.27) is a convolution integral and the equation can be solved using Laplace transforms (see Appendix 9). If the state $|1\rangle$ is initially occupied, we obtain

$$s\bar{P}_1(s) - 1 = -\tfrac{1}{2}\left(2\bar{P}_1(s) - \frac{1}{s}\right)\bar{K}(s), \tag{2.3.28}$$

so that

$$\bar{P}_1(s) = \tfrac{1}{2}\left(\frac{1}{s} + \frac{1}{s + \bar{K}(s)}\right), \tag{2.3.29}$$

where $\bar{P}_1(s)$ and $\bar{K}(s)$ are the Laplace transforms of $P_1(t)$ and $K(t)$, respectively. If $\bar{K}(s)$ can be calculated and (2.3.29) inverted, then $P_1(t)$ can be found. Consider for example the case where $\Delta = 0$ and the kernel is given by $K(t - t') = V_0^2 \exp(-\gamma|t - t'|)$, corresponding to a field having a Lorentzian spectrum of width 2γ which for small fluctuations is typically very much less than V_0. Then $\bar{K}(s) = V_0^2(s + \gamma)^{-1}$ and

$$\bar{P}_1(s) = \tfrac{1}{2}\left(\frac{1}{s} + \frac{s + \gamma}{s^2 + \gamma s + V_0^2}\right). \tag{2.3.30}$$

The nature of the inversion of (2.3.30) depends on the relative sizes of γ and V_0. For $V_0 > \gamma/2$, we find the damped, oscillatory solution

$$P_1(t) = \tfrac{1}{2}\left[1 + \exp(-\gamma t/2)\left(\cos\Omega_\mathrm{D}t + \frac{\gamma}{2\Omega_\mathrm{D}}\sin\Omega_\mathrm{D}t\right)\right], \tag{2.3.31}$$

where $\Omega_\mathrm{D} = (V_0^2 - \gamma^2/4)^{1/2}$ is the resonant Rabi frequency V_0 modified by the non-zero bandwidth. The averaged probability tends to its steady-state value of $1/2$.

The treatment in this section has excluded the effects of spontaneous emission which is important in real systems. We investigate these effects in Sections 5.5 and 5.6.

2.4 The Jaynes–Cummings model

The field coupling the two states $|1\rangle$ and $|2\rangle$ may be either classical or quantum in nature. In this section we consider the Jaynes–Cummings model of the interaction between a two-state atom and a single-mode quantized electromagnetic field having annihilation and creation operators $\hat{a}$ and $\hat{a}^\dagger$, respectively. The time-independent Hamiltonian is

$$\hat{H} = \frac{\hbar}{2}(\omega_2 + \omega_1)\mathbf{1} + \frac{\hbar}{2}(\omega_2 - \omega_1)\hat{\sigma}_3 + \hbar\omega\left(\hat{a}^\dagger\hat{a} + \tfrac{1}{2}\right) - \mathrm{i}\hbar\lambda(\hat{\sigma}_+ + \hat{\sigma}_-)(\hat{a} - \hat{a}^\dagger), \tag{2.4.1}$$

as given in (1.3.45). Here, as before, $\hbar\omega_1$ and $\hbar\omega_2$ are the energies of the uncoupled states $|1\rangle$ and $|2\rangle$, respectively, and ω is the frequency of the field mode. Applying the unitary transformations produced by (2.2.20) and (2.2.18) with $\delta = \Delta/2$ and $\Delta = \omega_2 - \omega_1 - \omega$, we find the Hamiltonian in the interaction picture is

$$\hat{H}_\mathrm{I} = \frac{\hbar\Delta}{2}\hat{\sigma}_3 - \mathrm{i}\hbar\lambda[\hat{\sigma}_+ \exp(\mathrm{i}\omega t) + \hat{\sigma}_- \exp(-\mathrm{i}\omega t)][\hat{a}\exp(-\mathrm{i}\omega t) - \hat{a}^\dagger\exp(\mathrm{i}\omega t)]. \tag{2.4.2}$$

Making the rotating-wave approximation removes the explicit time dependence and gives the Jaynes–Cummings Hamiltonian

$$\hat{H}_\mathrm{I} = \frac{\hbar\Delta}{2}\hat{\sigma}_3 - \mathrm{i}\hbar\lambda\left(\hat{\sigma}_+\hat{a} - \hat{a}^\dagger\hat{\sigma}_-\right). \tag{2.4.3}$$

The omitted interaction terms have the forms $\hat{\sigma}_-\hat{a}$ and $\hat{\sigma}_+\hat{a}^\dagger$ and correspond to a downward transition in the atom accompanied by absorption of a photon and to an upward transition with the emission of a photon. These processes do not preserve the total number of quanta and are strongly suppressed.

We consider an initial state of the system in which the atom is in the lower state $|1\rangle$ while the field is in a superposition of photon number states. The corresponding state is

$$|\psi_\mathrm{I}(0)\rangle = \sum_{n=0}^{\infty} a_n |1\rangle|n\rangle, \tag{2.4.4}$$

where $|1\rangle|n\rangle$ is the product state with the atom in $|1\rangle$ and n photons in the field. At subsequent times, the state evolves into a superposition

$$|\psi_\mathrm{I}(t)\rangle = \sum_{n=0}^{\infty} [c_{1,n}(t)|1\rangle|n\rangle + c_{2,n}(t)|2\rangle|n\rangle]. \tag{2.4.5}$$

In general, this state is entangled in the sense that it cannot be expressed as a product of an atom state and a field state. The solution for the dynamics of this model is simplified by the fact that the operator

$$\hat{N} = \hat{a}^\dagger\hat{a} + \hat{\sigma}_+\hat{\sigma}_-, \tag{2.4.6}$$

representing the total number of quanta, commutes with the Hamiltonian (2.4.3) and hence that the number of quanta is conserved. This means that the state $|1\rangle|n\rangle$ is only coupled to $|2\rangle|n-1\rangle$ so that the state $|1\rangle|0\rangle$ is uncoupled. Each coupled pair of states evolves as a distinct two-state system. Substituting (2.4.5) into the Schrödinger equation with the Hamiltonian (2.4.3), we obtain the equations of motion

$$\dot{c}_{1,n} = \mathrm{i}\frac{\Delta}{2}c_{1,n} + \lambda n^{1/2}c_{2,n-1}, \tag{2.4.7}$$

$$\dot{c}_{2,n-1} = -\mathrm{i}\frac{\Delta}{2}c_{2,n-1} - \lambda n^{1/2}c_{1,n}, \tag{2.4.8}$$

for the probability amplitudes. The initial conditions, following from (2.4.4), are $c_{1,n}(0) = a_n$ and $c_{2,n-1}(0) = 0$. The solution of (2.4.7) and (2.4.8) with these conditions, obtained from (2.3.4), (2.3.5), (2.3.10), and (2.3.11) with $V = 2\lambda n^{1/2}$ and $\varphi = -\pi/2$, is

$$c_{1,n}(t) = a_n\left(\cos[\Omega_R(n)t/2] + \mathrm{i}\frac{\Delta}{\Omega_R(n)}\sin[\Omega_R(n)t/2]\right), \tag{2.4.9}$$

$$c_{2,n-1}(t) = -a_n\frac{2\lambda n^{1/2}}{\Omega_R(n)}\sin[\Omega_R(n)t/2], \tag{2.4.10}$$

where $\Omega_R(n) = (\Delta^2 + 4\lambda^2 n)^{1/2}$ is the photon-number-dependent Rabi frequency. For simplicity, we consider the resonant case ($\Delta = 0$) for which the probability for being in the atomic state $|1\rangle$ at time t is

$$P_1(t) = \sum_{n=0}^{\infty}|c_{1,n}(t)|^2 = \tfrac{1}{2}\sum_{n=0}^{\infty}|a_n|^2[1 + \cos(2\lambda n^{1/2}t)]. \tag{2.4.11}$$

This probability depends on the initial photon number probability distribution $|a_n|^2$, rather than simply on an initial amplitude as in the case of the interaction with a classical field.

As an example, we consider the field to be initially in a coherent state (see Section 3.6) of mean photon number $\bar{n}$ for which

$$|a_n|^2 = \exp(-\bar{n})\bar{n}^n/n!. \tag{2.4.12}$$

In Fig. 2.1, we plot $P_1(t)$ for $\bar{n} = 10$. We observe Rabi oscillations at a frequency approximately equal to $2\lambda\bar{n}^{1/2}$ under an envelope that periodically collapses and then revives. The origin of this collapse and revival phenomenon is the interference between individual oscillatory terms in (2.4.11) having the incommensurate frequencies $2\lambda n^{1/2}$. The collapses arise from the spread of the initial photon numbers in the field. To see this, we approximate $n^{1/2}$ by

$$n^{1/2} = [\bar{n} + (n - \bar{n})]^{1/2} \simeq \bar{n}^{1/2}\left(1 + \frac{n - \bar{n}}{2\bar{n}}\right) \tag{2.4.13}$$

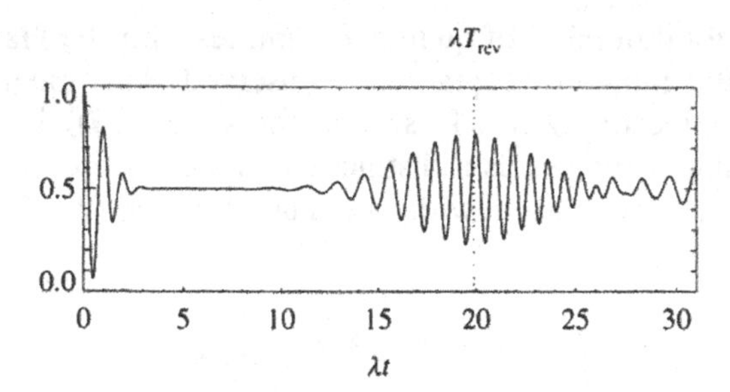

Fig. 2.1 Collapses and revivals of the excited state probability in the coherent state Jaynes–Cummings model.

and replace the Poisson distribution $|a_n|^2$ by a Gaussian probability distribution

$$\exp(-\bar{n})\bar{n}^n/n! \simeq (2\pi\bar{n})^{-1/2}\exp\left[-(n-\bar{n})^2/2\bar{n}\right]. \tag{2.4.14}$$

The Gaussian is chosen to give the correct mean and variance for the photon number and is a good approximation for large $\bar{n}$. Inserting (2.4.13) and (2.4.14) into (2.4.11), we find the approximate form

$$P_1(t) = \tfrac{1}{2} + \frac{1}{2(2\pi\bar{n})^{1/2}}\int \mathrm{d}n\,\exp\left(-\frac{(n-\bar{n})^2}{2\bar{n}}\right) \times \cos\left[2\lambda\bar{n}^{1/2}t\left(1+\frac{n-\bar{n}}{2\bar{n}}\right)\right]. \tag{2.4.15}$$

On writing the cosine factor as the real part of a complex exponential and extending the range of integration from $-\infty$ to ∞, the integral becomes the Fourier transform of a Gaussian function and can be evaluated to give

$$P_1(t) = \tfrac{1}{2}[1+\cos(2\lambda\bar{n}^{1/2}t)\exp(-\lambda^2t^2/2)]. \tag{2.4.16}$$

This result is valid at short times and accurately reproduces the first collapse which occurs in a time of the order of $1/\lambda$, as shown in Fig. 2.2. The revival occurs because the photon number is not a continuous variable but discrete.

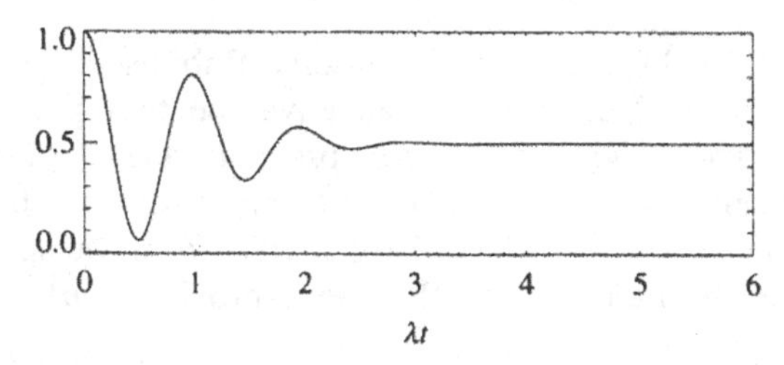

Fig. 2.2 The approximate form of the first collapse in the coherent state Jaynes–Cummings model.

The peak of the revival occurs at the time T_{rev} at which a significant number of the oscillating terms in the sum (2.4.11) are in phase. We can estimate the revival time by requiring that the terms at the peak of the photon number probability distribution are in phase. The first time this occurs is given approximately by

$$2\lambda\bar{n}^{1/2}T_{rev} - 2\lambda(\bar{n}-1)^{1/2}T_{rev} = 2\pi. \qquad (2.4.17)$$

Approximating $(\bar{n}-1)^{1/2}$ by the first two terms of its binomial expansion gives

$$T_{rev} \simeq 2\pi\bar{n}^{1/2}/\lambda. \qquad (2.4.18)$$

This revival time is shown in Fig. 2.1.

Other initial field states also produce collapses and revivals, the form of which depends on the details of their photon number probability distributions.

2.5 The Bixon–Jortner quasi-continuum

Models of coherent interactions often involve coupling between more than two atomic levels. The most important of these, with three coupled levels, will be treated in Chapter 6. In this section we analyze an exactly soluble model in which a single, initially occupied state is coupled to infinitely many others forming a quasi-continuum. We work in an interaction picture in which the energy of the single state is zero and denote its amplitude by $b(t)$. The quasi-continuum is an infinite set of states equally spaced in energy with the separation $\hbar\Delta$ between adjacent states (see Fig. 2.3). The amplitudes of the states in the quasi-continuum are denoted $c_n(t)$ with the state with amplitude $c_0(t)$ being resonantly coupled to the initially occupied single state. The amplitude equations within the rotating-wave approximation are

$$\dot{b}(t) = -\mathrm{i}\sum_n W_n c_n(t), \qquad (2.5.1)$$

$$\dot{c}_n(t) = -\mathrm{i}n\Delta c_n(t) - \mathrm{i}W_n b(t), \qquad (2.5.2)$$

where W_n, assumed to be real, is the coupling between the single state and the nth state of the quasi-continuum. We solve (2.5.1) and (2.5.2) for $b(t)$ with $b(0)=1$ using Laplace transforms. Denoting the Laplace transforms of $b(t)$ and $c_n(t)$ by $\bar{b}(s)$ and $\bar{c}_n(s)$, we obtain

$$s\bar{b}(s) - 1 = -\mathrm{i}\sum_n W_n \bar{c}_n(s), \qquad (2.5.3)$$

$$s\bar{c}_n(s) = -\mathrm{i}n\Delta\bar{c}_n(s) - \mathrm{i}W_n\bar{b}(s). \qquad (2.5.4)$$

Eliminating $\bar{c}_n(s)$ between (2.5.3) and (2.5.4), we find

$$\bar{b}(s) = \left(s + \sum_n \frac{W_n^2}{s+\mathrm{i}n\Delta}\right)^{-1} \qquad (2.5.5)$$

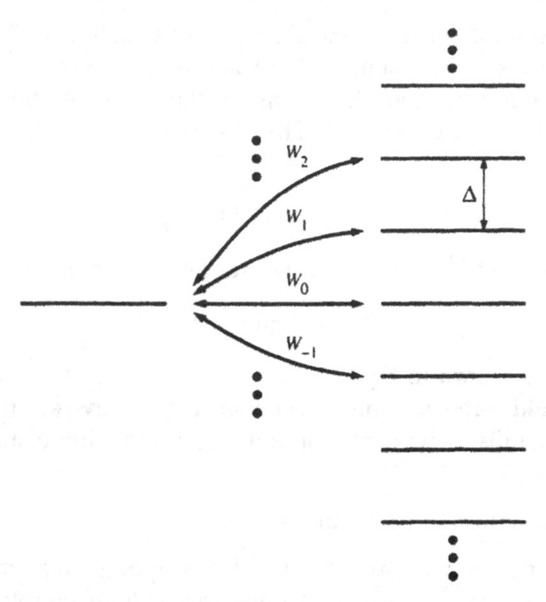

Fig. 2.3 Level diagram for a single state coupled to a quasi-continuum.

which, if we can invert, gives $b(t)$. In the Bixon–Jortner model, we choose $W_n^2 = W^2$ a constant, and, in order that the summation in (2.5.5) converges, we let n run from $-\infty$ to ∞. Using (A8.18) with $\omega = s$ and $\omega_0 = -\mathrm{i}\Delta$, this summation can be evaluated to give

$$\sum_{n=-\infty}^{\infty} \frac{W^2}{s + \mathrm{i}n\Delta} = \frac{\pi W^2}{\Delta} \coth\left(\frac{\pi s}{\Delta}\right), \tag{2.5.6}$$

so that

$$\bar{b}(s) = \left[s + \frac{\pi W^2}{\Delta} \coth\left(\frac{\pi s}{\Delta}\right) \right]^{-1}. \tag{2.5.7}$$

We show in Appendix 9 that the inversion of (2.5.7) is

$$b(T) = \exp(-\beta T)\left(1 - 2\beta \sum_{k=1}^{\infty} \exp(k\beta) \frac{(T-k)}{k} H(T-k) L_k^{(1)}[2\beta(T-k)]\right), \tag{2.5.8}$$

where $T = \Delta t/2\pi$, $\beta = 2\pi^2 W^2/\Delta^2$, H is the unit step function, and $L_k^{(1)}$ is the associated Laguerre polynomial (see Appendix 3). At each integer value of the scaled time T, an additional term is added to $b(T)$ owing to the unit step functions in the summation in (2.5.8). Each addition of a term produces a discontinuity, or kick, in the gradient of $b(T)$ and hence of $P(T) = |b(T)|^2$.

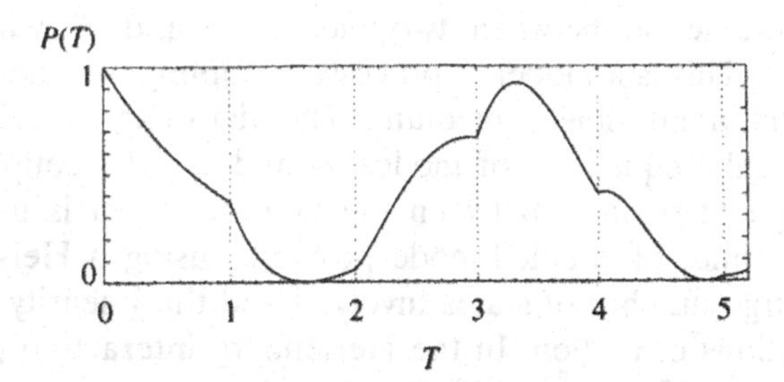

Fig. 2.4 The occupation probability for the single state coupled to a Bixon–Jortner quasi-continuum as a function of the scaled time with $\beta = 0.5$.

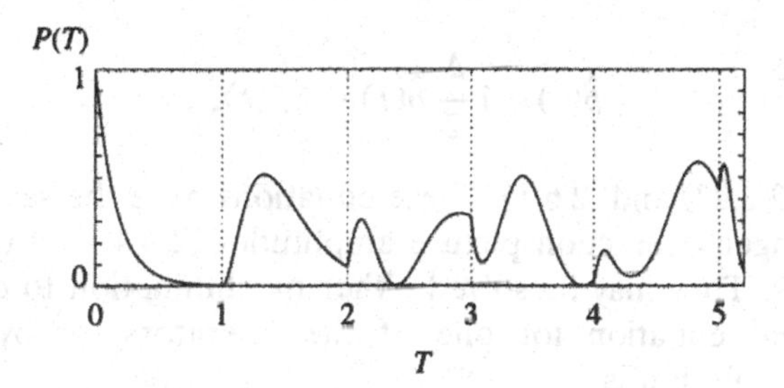

Fig. 2.5 The occupation probability for the single state coupled to a Bixon–Jortner quasi-continuum as a function of the scaled time with $\beta = 3$.

These kicks occur at intervals of $2\pi/\Delta$ in true time t and arise from the periodic nature of the quasi-continuum. In Figs 2.4 and 2.5, we plot $P(T)$ against T for $\beta = 0.5$ and $\beta = 3$, respectively. Prior to the first kick at $T = 1$ the probability $P(T)$ decays exponentially, and after some of the kicks, revivals of the population occur. In the limit where $\Delta \to 0$ and $W^2 \to 0$ such that $\pi W^2/\Delta = \Gamma$ is a constant, the true time elapsed before the first kick tends to infinity and for all finite time $b(t) = \exp(-\Gamma t)$. This result is the same as that found for a discrete state coupled to a true continuum, as we will show in Section 5.3. For non-zero Δ, the behaviour of $b(t)$ in the first interval $0 \leqslant t < 2\pi/\Delta$ is identical to this exponential decay since the structure of the quasi-continuum has not yet been resolved.

2.6 Interactions between field modes

The simplest quantum models of non-linear optical processes involve the coherent interaction between a small number of discrete field modes. Such models are used in the treatment of non-classical states of light such as the squeezed states which we shall introduce in Section 3.7.

Consider the interaction picture Hamiltonian

$$\hat{H}_{\mathrm{I}} = \frac{\hbar\Delta}{2}(\hat{a}^\dagger\hat{a} - \hat{b}^\dagger\hat{b}) - \mathrm{i}\hbar\lambda(\hat{b}^\dagger\hat{a} - \hat{a}^\dagger\hat{b}) \tag{2.6.1}$$

describing the interaction between two modes A and B with annihilation operators $\hat{a}$ and $\hat{b}$. This is a model of processes coupling two modes of different frequencies within a non-linear medium. The detuning Δ is the difference between the natural frequencies of modes A and B. The coupling causes the coherent exchange of photons between the two modes. It is usually easier to solve for the dynamics of coupled-mode problems using a Heisenberg picture because of the large number of states involved and the linearity of the coupled Heisenberg equations of motion. In the Heisenberg interaction picture the two annihilation operators $\hat{a}$ and $\hat{b}$ satisfy the equations

$$\dot{\hat{a}}(t) = -\mathrm{i}\frac{\Delta}{2}\hat{a}(t) + \lambda\hat{b}(t), \tag{2.6.2}$$

$$\dot{\hat{b}}(t) = \mathrm{i}\frac{\Delta}{2}\hat{b}(t) - \lambda\hat{a}(t), \tag{2.6.3}$$

obtained from (2.2.23) and (2.6.1). These equations have the same form as those for the Schrödinger interaction picture amplitudes (2.3.4) and (2.3.5) leading to Rabi oscillations. They may be solved either by elimination to obtain a second-order differential equation for one of the operators, or by using Laplace transforms. The solution is

$$\hat{a}(t) = \hat{a}(0)\left(\cos\Omega t - \frac{\mathrm{i}\Delta}{2\Omega}\sin\Omega t\right) + \hat{b}(0)\frac{\lambda}{\Omega}\sin\Omega t, \tag{2.6.4}$$

$$\hat{b}(t) = \hat{b}(0)\left(\cos\Omega t + \frac{\mathrm{i}\Delta}{2\Omega}\sin\Omega t\right) - \hat{a}(0)\frac{\lambda}{\Omega}\sin\Omega t, \tag{2.6.5}$$

where $\Omega^2 = \lambda^2 + \Delta^2/4$. The interaction induces a periodic exchange of energy between the two modes with a period $2\pi/\Omega$. The equal-time commutation relations between the annihilation and creation operators are required to be constants of the motion and this is easily verified from (2.6.4) and (2.6.5) since

$$\begin{aligned}[\hat{a}(t), \hat{a}^\dagger(t)] &= [\hat{a}(0), \hat{a}^\dagger(0)]\left|\cos\Omega t - \frac{\mathrm{i}\Delta}{2\Omega}\sin\Omega t\right|^2 \\ &\quad + [\hat{b}(0), \hat{b}^\dagger(0)]\frac{\lambda^2}{\Omega^2}\sin^2\Omega t \\ &= 1.\end{aligned} \tag{2.6.6}$$

Similarly $[\hat{b}(t), \hat{b}^\dagger(t)] = 1$ and

$$\begin{aligned}[\hat{a}(t), \hat{b}^\dagger(t)] &= -[\hat{a}(0), \hat{a}^\dagger(0)]\left(\cos\Omega t - \frac{\mathrm{i}\Delta}{2\Omega}\sin\Omega t\right)\frac{\lambda}{\Omega}\sin\Omega t \\ &\quad + [\hat{b}(0), \hat{b}^\dagger(0)]\frac{\lambda}{\Omega}\sin\Omega t\left(\cos\Omega t - \frac{\mathrm{i}\Delta}{2\Omega}\sin\Omega t\right) \\ &= 0.\end{aligned} \tag{2.6.7}$$

The time-evolved operators (2.6.4) and (2.6.5) and their Hermitian conjugates can now be used to express the moments of any observable in terms of the initial properties of the two modes. For example, the mean number of photons in mode A at time t is

$$\begin{aligned}\langle \hat{a}^\dagger(t)\hat{a}(t)\rangle = &\langle \hat{a}^\dagger(0)\hat{a}(0)\rangle\left(\cos^2\Omega t + \frac{\Delta^2}{4\Omega^2}\sin^2\Omega t\right)\\ &+\langle \hat{a}^\dagger(0)\hat{b}(0)\rangle\left(\cos\Omega t + \frac{\mathrm{i}\Delta}{2\Omega}\sin\Omega t\right)\frac{\lambda}{\Omega}\sin\Omega t\\ &+\langle \hat{b}^\dagger(0)\hat{a}(0)\rangle\frac{\lambda}{\Omega}\sin\Omega t\left(\cos\Omega t - \frac{\mathrm{i}\Delta}{2\Omega}\sin\Omega t\right)\\ &+\langle \hat{b}^\dagger(0)\hat{b}(0)\rangle\frac{\lambda^2}{\Omega^2}\sin^2\Omega t. \end{aligned} \tag{2.6.8}$$

Degenerate parametric oscillators produce photons in pairs in a single cavity mode of frequency ω_a. The energy is provided by a pump field of frequency ω which may be treated classically. The simplest Hamiltonian describing this process is

$$\hat{H} = \hbar\omega_a\hat{a}^\dagger\hat{a} - \frac{g}{2}[\hat{a}^2\exp(\mathrm{i}\omega t) + \hat{a}^{\dagger 2}\exp(-\mathrm{i}\omega t)], \tag{2.6.9}$$

where we have omitted the zero-point energy $\hbar\omega_a/2$ since this is a constant term and does not affect the dynamics. The Hamiltonian (2.6.9) can be transformed into a time-independent form using the unitary transformation (2.2.13) with $\hat{U} = \exp(\mathrm{i}\omega t\hat{a}^\dagger\hat{a}/2)$ to give

$$\hat{H}_\mathrm{I} = \hbar\delta\hat{a}^\dagger\hat{a} - \frac{g}{2}(\hat{a}^2 + \hat{a}^{\dagger 2}), \tag{2.6.10}$$

where $\delta = \omega_a - \omega/2$. In this interaction picture, the Heisenberg equations of motion for the annihilation operator $\hat{a}$ and creation operator $\hat{a}^\dagger$ are, from (2.2.23),

$$\dot{\hat{a}}(t) = -\mathrm{i}\delta\hat{a}(t) + \mathrm{i}g\hat{a}^\dagger(t), \tag{2.6.11}$$

$$\dot{\hat{a}}^\dagger(t) = \mathrm{i}\delta\hat{a}^\dagger(t) - \mathrm{i}g\hat{a}(t). \tag{2.6.12}$$

These are formally equivalent to (2.6.2) and (2.6.3) with $\Delta = 2\delta$, $\lambda = \mathrm{i}g$, and $\hat{b}(t)$ replaced by $\hat{a}^\dagger(t)$. From (2.6.4) and (2.6.5), therefore, the solution is

$$\hat{a}(t) = \hat{a}(0)\left(\cos\Omega t - \frac{\mathrm{i}\delta}{\Omega}\sin\Omega t\right) + \mathrm{i}\hat{a}^\dagger(0)\frac{g}{\Omega}\sin\Omega t, \tag{2.6.13}$$

$$\hat{a}^\dagger(t) = \hat{a}^\dagger(0)\left(\cos\Omega t + \frac{\mathrm{i}\delta}{\Omega}\sin\Omega t\right) - \mathrm{i}\hat{a}(0)\frac{g}{\Omega}\sin\Omega t, \tag{2.6.14}$$

where $\Omega^2 = \delta^2 - g^2$. If $g > \delta$, as is usually the case, then Ω is imaginary so that $\cos \Omega t = \cosh |\Omega| t$ and $\mathrm{i} \sin \Omega t = -\sinh |\Omega| t$. For the resonant case $\delta = 0$, $\Omega = \mathrm{i}g$ and

$$\hat{a}(t) = \hat{a}(0)\cosh gt + \mathrm{i}\hat{a}^\dagger(0)\sinh gt, \tag{2.6.15}$$

$$\hat{a}^\dagger(t) = \hat{a}^\dagger(0)\cosh gt - \mathrm{i}\hat{a}(0)\sinh gt. \tag{2.6.16}$$

The equal-time commutator $[\hat{a}(t), \hat{a}^\dagger(t)]$ is again preserved since

$$[\hat{a}(t), \hat{a}^\dagger(t)] = [\hat{a}(0), \hat{a}^\dagger(0)]\cosh^2 gt + [\hat{a}^\dagger(0), \hat{a}(0)]\sinh^2 gt = 1. \tag{2.6.17}$$

This model is fundamental to the study of squeezed states of light and we will find relationships of the type (2.6.15) and (2.6.16) in Section 3.7.

All Hamiltonians containing only terms that are either linear or quadratic in creation and annihilation operators lead to linear Heisenberg equations and therefore, in principle, are exactly soluble. As an example we consider the Hamiltonian

$$\hat{H} = \hbar\omega\hat{a}^\dagger\hat{a} + \hbar\omega\hat{b}^\dagger\hat{b} - \mathrm{i}\hbar\lambda(\hat{b}^\dagger + \hat{b})(\hat{a} - \hat{a}^\dagger) \tag{2.6.18}$$

describing the resonant interaction between a field mode and a linear polarization associated with a dielectric medium. Note that (2.6.18) contains terms proportional to $\hat{b}^\dagger\hat{a}^\dagger$ and $\hat{b}\hat{a}$ so that the total number of quanta is not conserved. Analogous counter-rotating terms were omitted in our derivation of the Jaynes–Cummings Hamiltonian (2.4.3) in making the rotating-wave approximation. If they were omitted here, the interaction term in the Hamiltonian (2.6.18) would be identical to that in (2.6.1). Comparing the dynamics associated with $\hat{H}$ in (2.6.18) with that generated by (2.6.1) enables the effect of counter-rotating terms to be demonstrated. The Heisenberg equations of motion for $\hat{a}(t)$ and $\hat{b}(t)$ for (2.6.18) are

$$\dot{\hat{a}}(t) = -\mathrm{i}\omega\hat{a}(t) + \lambda[\hat{b}^\dagger(t) + \hat{b}(t)], \tag{2.6.19}$$

$$\dot{\hat{b}}(t) = -\mathrm{i}\omega\hat{b}(t) - \lambda[\hat{a}(t) - \hat{a}^\dagger(t)]. \tag{2.6.20}$$

In order to solve these, we introduce the two normal modes with annihilation operators

$$\hat{c}(t) = \frac{1}{\sqrt{2}}[\hat{a}(t) + \mathrm{i}\hat{b}(t)], \tag{2.6.21}$$

$$\hat{d}(t) = \frac{1}{\sqrt{2}}[\hat{b}(t) + \mathrm{i}\hat{a}(t)]. \tag{2.6.22}$$

The equations of motion for these operators are then

$$\dot{\hat{c}}(t) = -\mathrm{i}(\omega + \lambda)\hat{c}(t) + \mathrm{i}\lambda\hat{c}^\dagger(t), \tag{2.6.23}$$

$$\dot{\hat{d}}(t) = -\mathrm{i}(\omega - \lambda)\hat{d}(t) + \mathrm{i}\lambda\hat{d}^\dagger(t), \tag{2.6.24}$$

which are of the same form as (2.6.11). From (2.6.13), therefore, the solution is

$$\hat{c}(t) = \hat{c}(0)\left(\cos\Omega_+ t - \frac{i(\omega+\lambda)}{\Omega_+}\sin\Omega_+ t\right) + i\hat{c}^\dagger(0)\frac{\lambda}{\Omega_+}\sin\Omega_+ t, \quad (2.6.25)$$

$$\hat{d}(t) = \hat{d}(0)\left(\cos\Omega_- t - \frac{i(\omega-\lambda)}{\Omega_-}\sin\Omega_- t\right) + i\hat{d}^\dagger(0)\frac{\lambda}{\Omega_-}\sin\Omega_- t, \quad (2.6.26)$$

where

$$\Omega_\pm^2 = \omega^2 \pm 2\omega\lambda. \quad (2.6.27)$$

Expressions for the original annihilation operators $\hat{a}(t)$ and $\hat{b}(t)$ follow from the inversions $\hat{a} = (\hat{c} - i\hat{d})/\sqrt{2}$ and $\hat{b} = (\hat{d} - i\hat{c})/\sqrt{2}$ of (2.6.21) and (2.6.22). If, as is usually the case, the coupling λ is much less than the natural frequency ω, then we can expand $\Omega_\pm$ and $\hat{c}(t)$ and $\hat{d}(t)$ in (2.6.25) and (2.6.26) in terms of λ/ω to give the approximate expressions

$$\hat{c}(t) \simeq \hat{c}(0)\exp\left[-i\left(\omega + \lambda - \frac{\lambda^2}{2\omega}\right)t\right], \quad (2.6.28)$$

$$\hat{d}(t) \simeq \hat{d}(0)\exp\left[-i\left(\omega - \lambda - \frac{\lambda^2}{2\omega}\right)t\right]. \quad (2.6.29)$$

Had we made the rotating-wave approximation by omitting the terms $i\hbar\lambda(\hat{b}^\dagger\hat{a}^\dagger - \hat{b}\hat{a})$ from (2.6.18), then the superposition mode operators $\hat{c}(t)$ and $\hat{d}(t)$ would have been $\hat{c}(0)\exp[-i(\omega+\lambda)t]$ and $\hat{d}(0)\exp[-i(\omega-\lambda)t]$; that is, as in (2.6.28) and (2.6.29) with the frequency shift $\lambda^2/2\omega$ absent. These rotating-wave approximation forms can also be obtained by substituting (2.6.4) and (2.6.5) with $\Delta = 0$ into (2.6.21) and (2.6.22) and multiplying by the factor $\exp(-i\omega t)$ to account for the change of picture. The principal effect of the counter-rotating terms is therefore to change the natural frequency of the field by $\lambda^2/2\omega$. An analogous frequency shift can also be found from a perturbative treatment of the effect of the counter-rotating terms on a driven two-state system.

3
OPERATORS AND STATES

3.1 Introduction

It is a fundamental principle of quantum mechanics that we cannot, in general, specify with arbitrary accuracies the values of a pair of observable quantities. This principle is exhibited mathematically by the fact that observable quantities are represented by operators, rather than by functions as they are in classical physics, and that these operators do not, in general, commute with each other. Solving problems in quantum theory requires an ability to manipulate these operators. The quantum model of a physical system is completed by states which, together with the operators, provide the statistical description of any desired property of the system.

This chapter is concerned with providing the necessary rules governing the manipulation of the operators and the properties of the states to enable us to model and describe some physical systems. The emphasis is on operators and states commonly occurring in quantum optics, especially those required in later chapters of this book.

3.2 Atom and field operators

Quantum optics is the study of the quantum theory of the interaction between matter and radiation, principally at low energies. It is natural therefore that the most commonly occurring operators are those associated with atom and field energy levels and transitions between them. Some of these operators, in particular the Pauli operators and the single-mode creation and annihilation operators, were introduced in Chapter 1 and used in the discussion of coherent interactions in Chapter 2. In this section, we enlarge upon the theory of these and introduce a class of related operators.

The simplest models involve transitions between only two discrete atomic levels represented by $|1\rangle$ and $|2\rangle$. All of the observable quantities of this two-state system can be conveniently represented by the Pauli operators which have the form

$$\hat{\sigma}_3 = |2\rangle\langle 2| - |1\rangle\langle 1|, \tag{3.2.1}$$

$$\hat{\sigma}_+ = |2\rangle\langle 1|, \tag{3.2.2}$$

$$\hat{\sigma}_- = |1\rangle\langle 2|. \tag{3.2.3}$$

The expectation value of the operator $\hat{\sigma}_3$ is the atomic inversion, while $\hat{\sigma}_+$ and $\hat{\sigma}_-$ induce upward and downward transitions respectively. It is sometimes convenient to work with Hermitian combinations of $\hat{\sigma}_+$ and $\hat{\sigma}_-$, in particular the operators

$$\hat{\sigma}_1 = \hat{\sigma}_+ + \hat{\sigma}_-, \tag{3.2.4}$$

$$\hat{\sigma}_2 = \mathrm{i}(\hat{\sigma}_- - \hat{\sigma}_+). \tag{3.2.5}$$

If, as is normally the case, we adopt the electric dipole approximation, then these operators represent the two quadrature components of the electric dipole associated with the two-state system. It follows directly from (3.2.1) to (3.2.5) that $\hat{\sigma}_1^2 = \hat{\sigma}_2^2 = \hat{\sigma}_3^2 = 1$ and that $\hat{\sigma}_1$, $\hat{\sigma}_2$, and $\hat{\sigma}_3$ anticommute with each other. This may be expressed in the compact form $\{\hat{\sigma}_i, \hat{\sigma}_j\} = 2\delta_{ij}$. The commutator of any two of the set of Pauli operators is proportional to one of the set. These commutators are given by

$$[\hat{\sigma}_3, \hat{\sigma}_\pm] = \pm 2\hat{\sigma}_\pm, \tag{3.2.6}$$

$$[\hat{\sigma}_+, \hat{\sigma}_-] = \hat{\sigma}_3. \tag{3.2.7}$$

Note that these commutation relations, in common with the others in this section, are valid in the Schrödinger picture, or in the Heisenberg picture if the operators are evaluated at equal times. It follows from (3.2.4) to (3.2.7) that the commutators involving $\hat{\sigma}_1$ and $\hat{\sigma}_2$ are

$$[\hat{\sigma}_3, \hat{\sigma}_1] = 2\mathrm{i}\hat{\sigma}_2, \tag{3.2.8}$$

$$[\hat{\sigma}_2, \hat{\sigma}_3] = 2\mathrm{i}\hat{\sigma}_1, \tag{3.2.9}$$

$$[\hat{\sigma}_1, \hat{\sigma}_2] = 2\mathrm{i}\hat{\sigma}_3. \tag{3.2.10}$$

These three commutators may be expressed in the compact form (see Appendix 1) $[\hat{\sigma}_j, \hat{\sigma}_k] = 2\mathrm{i}\sum_{l=1}^{3} \epsilon_{jkl}\hat{\sigma}_l$, where ϵ_{jkl} is the permutation symbol. Using (3.2.4) and (3.2.5), the set of commutators is completed by

$$[\hat{\sigma}_\pm, \hat{\sigma}_1] = \pm\hat{\sigma}_3, \tag{3.2.11}$$

$$[\hat{\sigma}_\pm, \hat{\sigma}_2] = \mathrm{i}\hat{\sigma}_3. \tag{3.2.12}$$

The Pauli operators were originally introduced to describe the angular momentum of a spin-$\frac{1}{2}$ particle, for which each of the cartesian components $\hat{J}_i$ of its angular momentum operator is one-half of the corresponding Pauli operator (in units of $\hbar$). The angular momentum raising and lowering operators are then $\hat{J}_+ = \hat{J}_1 + \mathrm{i}\hat{J}_2$ and $\hat{J}_- = \hat{J}_1 - \mathrm{i}\hat{J}_2$, so that the operators $\hat{J}_3$, $\hat{J}_+$, and $\hat{J}_-$ satisfy the commutation relations

$$\left[\hat{J}_3, \hat{J}_\pm\right] = \pm\hat{J}_\pm, \tag{3.2.13}$$

$$\left[\hat{J}_+, \hat{J}_-\right] = 2\hat{J}_3. \tag{3.2.14}$$

The angular momentum operators are not restricted to two-state systems but can be applied to any value of total angular momentum corresponding to an arbitrary number of levels. They have a wider application in quantum optics in describing the interaction between pairs of field modes, as we will illustrate in the last section of this chapter.

The electric and magnetic field operators for each mode are linear combinations of the annihilation and creation operators $\hat{a}$ and $\hat{a}^\dagger$ for the mode. The commutator of these non-Hermitian operators is

$$[\hat{a}, \hat{a}^\dagger] = \mathbf{1}. \tag{3.2.15}$$

Operators corresponding to different modes commute. The energy eigenstates for a single mode are the number states $|n\rangle$ analogous to those for the harmonic oscillator. As their names suggest, the action of the annihilation and creation operators on the number states is to decrease or increase n by unity corresponding to the annihilation or creation of one quantum of energy or photon in the mode. Explicitly,

$$\hat{a}\,|n\rangle = n^{1/2}\,|n-1\rangle, \tag{3.2.16}$$

$$\hat{a}^\dagger\,|n\rangle = (n+1)^{1/2}\,|n+1\rangle. \tag{3.2.17}$$

It follows from (3.2.17) that any number state $|n\rangle$ can be generated from the vacuum state $|0\rangle$ by repeated action of the creation operator $\hat{a}^\dagger$, so that

$$|n\rangle = (n!)^{-1/2}(\hat{a}^\dagger)^n\,|0\rangle. \tag{3.2.18}$$

Observable quantities associated with the field mode are represented by operators formed by taking Hermitian combinations of $\hat{a}$ and $\hat{a}^\dagger$. The two most important of these are the number operator

$$\hat{n} = \hat{a}^\dagger\hat{a} \tag{3.2.19}$$

and the quadrature operator

$$\hat{x}_\lambda = \frac{1}{\sqrt{2}}[\hat{a}\exp(-\mathrm{i}\lambda) + \hat{a}^\dagger\exp(\mathrm{i}\lambda)], \tag{3.2.20}$$

where λ is a real phase. The number operator $\hat{n}$ has as its eigenstates the number states since, from (3.2.16) and (3.2.17), $\hat{n}\,|n\rangle = n\,|n\rangle$. Hence the expectation value of $\hat{n}$ is the mean number of photons in the mode and the moments of the photon number probability distribution can be found by taking the expectation values of appropriate powers of $\hat{n}$. The Hamiltonian or energy operator of the non-interacting field mode of frequency ω is given by $\hbar\omega(\hat{a}^\dagger\hat{a} + \frac{1}{2}) = \hbar\omega(\hat{n} + \frac{1}{2})$. The commutators of $\hat{n}$ with $\hat{a}$ and with $\hat{a}^\dagger$ follow from the definition of $\hat{n}$ in (3.2.19) and the fundamental commutator (3.2.15). We find

$$[\hat{n}, \hat{a}] = -\hat{a}, \tag{3.2.21}$$

$$[\hat{n}, \hat{a}^\dagger] = \hat{a}^\dagger. \tag{3.2.22}$$

The quadrature operator (3.2.20) has an important role in the theory of squeezed states. The operators $\hat{x}_\lambda$ and $\hat{x}_{\lambda+\pi/2}$ are a canonically conjugate pair analogous to the position and momentum operators for a single particle. The commutator of $\hat{x}_\lambda$ and $\hat{x}_{\lambda+\pi/2}$ is, using (3.2.20) and (3.2.15),

$$[\hat{x}_\lambda, \hat{x}_{\lambda+\pi/2}] = \mathrm{i}. \tag{3.2.23}$$

It is not possible to construct normalized eigenstates $|x_\lambda\rangle$ of $\hat{x}_\lambda$ in the conventional Hilbert space of bounded, square-integrable states (see Appendix 4). We can, however, construct states $|x_\lambda\rangle$ satisfying the eigenvalue equation

$$\hat{x}_\lambda |x_\lambda\rangle = x_\lambda |x_\lambda\rangle \tag{3.2.24}$$

using the mathematics of generalized functions such as the delta function (see Appendix 2). The eigenvalues x_λ are continuous and unbounded and the eigenstates are orthogonal only in the sense that the overlap between two eigenstates $|x_\lambda\rangle$ and $|x'_\lambda\rangle$ of the quadrature operator $\hat{x}_\lambda$ is

$$\langle x_\lambda | x'_\lambda\rangle = \delta(x_\lambda - x'_\lambda). \tag{3.2.25}$$

The overlap between a pure state $|\psi\rangle$ and $|x_\lambda\rangle$ is equivalent to the familiar position representation of the states of a harmonic oscillator. In particular

$$\langle x_\lambda | n\rangle = \frac{1}{\pi^{1/4}} \frac{1}{2^{n/2}(n!)^{1/2}} \exp(-x_\lambda^2/2) H_n(x_\lambda) \exp(-\mathrm{i}n\lambda), \tag{3.2.26}$$

where H_n is the Hermite polynomial of order n. The eigenstates $|x_\lambda\rangle$ form a complete set in that

$$\int_{-\infty}^{\infty} \mathrm{d}x_\lambda\, |x_\lambda\rangle\langle x_\lambda| = 1. \tag{3.2.27}$$

To prove this, consider the effect of inserting the left-hand side of (3.2.27) between the overlap of two number states $|n\rangle$ and $|n'\rangle$. We have

$$\begin{aligned}\int_{-\infty}^{\infty} \mathrm{d}x_\lambda \langle n|x_\lambda\rangle\langle x_\lambda|n'\rangle &= \frac{1}{\pi^{1/2}} \frac{1}{2^{(n+n')/2}} \\ &\quad \times \frac{\exp[\mathrm{i}\lambda(n-n')]}{\sqrt{[(n!)(n'!)]}} \int_{-\infty}^{\infty} \mathrm{d}x_\lambda \exp(-x_\lambda^2) H_n(x_\lambda) H_{n'}(x_\lambda) \\ &= \delta_{nn'},\end{aligned} \tag{3.2.28}$$

where we have used the orthogonality relation of the Hermite polynomials (see Appendix 3). Similarly, we can establish that the trace of an operator $\hat{A}$ acting in the space of the mode is expressible as

$$\mathrm{Tr}(\hat{A}) = \int_{-\infty}^{\infty} \mathrm{d}x_\lambda \langle x_\lambda|\hat{A}|x_\lambda\rangle. \tag{3.2.29}$$

The proof of this follows from inserting the resolution (1.3.33) of the identity in terms of the number states before and after $\hat{A}$ in (3.2.29). On evaluating the resulting integral using (3.2.28), we recover the expression for the trace of $\hat{A}$ evaluated in the number state basis given in (1.3.34).

The simplest models in quantum optics utilize only a small number of discrete modes. However, this is not sufficient for all purposes. For some problems, including the treatment of dissipative processes, which we shall discuss in Chapter 5, a model incorporating the full electromagnetic field is required. This is done in one of two ways: either a full set of discrete modes for a sufficiently large but finite volume $\mathscr{V}$ of space is constructed, or the field is expanded in terms of continuum annihilation and creation operators. In the former approach, the limit $\mathscr{V} \to \infty$ is taken after the calculation of the required expectation values. In the latter, the electric and magnetic field operators are expressed as integrals over the wavevector **k** which include the continuum annihilation and creation operators $\hat{b}(\mathbf{k}, \lambda)$ and $\hat{b}^\dagger(\mathbf{k}, \lambda)$, where λ is the polarization index. These operators do not act on single discrete modes and their commutator is expressed in terms of the three-dimensional delta function (see Appendix 2) as

$$[\hat{b}(\mathbf{k}, \lambda), \hat{b}^\dagger(\mathbf{k}', \lambda')] = \delta^{(3)}(\mathbf{k} - \mathbf{k}')\delta_{\lambda\lambda'}. \tag{3.2.30}$$

For many purposes, however, it is appropriate to use a simpler form of continuum operator depending on a single parameter, for example the frequency ω. This is because in many cases the coupling between the system of interest and the field involves only one superposition of the continuum modes for each frequency. By suitably redefining this superposition to be a new continuum mode, we reduce the number of required continuum annihilation operators to one per frequency ω and denote this operator as $\hat{b}(\omega)$. The commutation relation between this annihilation operator and the corresponding creation operator is

$$[\hat{b}(\omega), \hat{b}^\dagger(\omega')] = \delta(\omega - \omega'). \tag{3.2.31}$$

We emphasize that the action of these operators is not to annihilate or create a photon in a single mode, in contrast to the single-mode operators $\hat{a}$ and $\hat{a}^\dagger$. Rather the annihilation or creation of a photon is produced by the action of a suitably weighted integral over ω of $\hat{b}(\omega)$ or $\hat{b}^\dagger(\omega)$, respectively. The operator representing any observable field quantity will be expressed as an integral over ω involving $\hat{b}(\omega)$ and $\hat{b}^\dagger(\omega)$ so that, for example, the electric dipole coupling between an atom and the field introduces into the Hamiltonian an interaction term expressed as the product of the atomic dipole operator with an integral over ω containing $\hat{b}(\omega)$ and $\hat{b}^\dagger(\omega)$. We will use interactions of this type to model dissipative phenomena in Chapters 5 and 6.

The energy of the free field is represented by the Hamiltonian operator

$$\hat{H} = \int \mathrm{d}\omega\, \hbar\omega \hat{b}^\dagger(\omega)\hat{b}(\omega), \tag{3.2.32}$$

where we have omitted the divergent zero-point energy. We can also define a continuum generalization of the number operator as the integral

$$\hat{N} = \int \mathrm{d}\omega\, \hat{b}^\dagger(\omega)\hat{b}(\omega) \tag{3.2.33}$$

which will be of use in our discussion of continuum number states in Section 3.4. Using the continuum operators requires some care since not all the properties of the single-mode operators carry over to the continuum operators, as we will see in Section 3.4.

3.3 Functions of operators and ordering theorems

In Chapter 1, we introduced the idea of a function of an Hermitian operator $\hat{A}$ with a complete, orthonormal set of eigenstates $|A_n\rangle$ and eigenvalues A_n. A suitably well-behaved function f of $\hat{A}$ is then defined by the eigenvalue equation

$$f(\hat{A})|A_n\rangle = f(A_n)|A_n\rangle, \tag{3.3.1}$$

so that

$$f(\hat{A}) = \sum_n f(A_n)|A_n\rangle\langle A_n|. \tag{3.3.2}$$

Clearly, from (3.3.2), an equivalent definition of $f(\hat{A})$ is obtained by inserting $\hat{A}$ into the power series (the Maclaurin expansion) of f. If $f(x) = \sum_{l=0}^{\infty} f^{(l)}(0)x^l/l!$ then

$$f(\hat{A}) = \sum_{l=0}^{\infty} \frac{f^{(l)}(0)}{l!}\hat{A}^l. \tag{3.3.3}$$

Expanding $\hat{A}$ in terms of its eigenstates as $\hat{A} = \sum_n A_n|A_n\rangle\langle A_n|$ and inserting this into (3.3.3), we recover (3.3.2). Our original definition of a function of an operator required the operator to have a complete, orthonormal set of eigenstates, but not all operators of interest, including $\hat{a}$ and $\hat{a}^\dagger$, have this property. Consequently it is necessary to adopt the more general definition of a function of an operator given by (3.3.3). We shall therefore employ this definition for all operators $\hat{A}$. Not all functions have Maclaurin series so that (3.3.3) cannot be employed for all functions. However, most commonly occurring functions of operators do possess Maclaurin series, as will all operator functions discussed here.

The most important function is the exponential since in the Schrödinger picture a state evolves under the action of the operator $\exp(-\mathrm{i}\hat{H}t/\hbar)$, where $\hat{H}$ is the time-independent Hamiltonian. The exponential is also important because any unitary operator $\hat{U}$ can be expressed in terms of an Hermitian operator $\hat{B}$ as $\hat{U} = \exp(\mathrm{i}\hat{B})$. From the definition (3.3.3), we have

$$\exp(\hat{A}) = \sum_{l=0}^{\infty} \frac{1}{l!}\hat{A}^l. \tag{3.3.4}$$

It follows directly from this that $\exp(\hat{A})\exp(-\hat{A}) = \mathbf{1}$ for all $\hat{A}$. As an example of (3.3.4), consider the operator $\exp(\theta\hat{n}) = \exp(\theta\hat{a}^\dagger\hat{a})$ given by the expansion

$$\exp(\theta\hat{a}^\dagger\hat{a}) = \sum_{l=0}^{\infty} \frac{1}{l!}\,\theta^l(\hat{a}^\dagger\hat{a})^l. \tag{3.3.5}$$

Since $\hat{a}^\dagger\hat{a}$ is Hermitian and has a complete, orthonormal set of eigenstates $|n\rangle$, we can also use (3.3.2) and write

$$\exp(\theta\hat{a}^\dagger\hat{a}) = \sum_{n=0}^{\infty} \exp(n\theta)\,|n\rangle\langle n|. \tag{3.3.6}$$

Any number z raised to the power of an operator $\hat{A}$ is evaluated by writing $z = \exp(\ln z)$ so that

$$z^{\hat{A}} = \exp(\hat{A}\ln z). \tag{3.3.7}$$

In particular

$$\begin{aligned} z^{\hat{a}^\dagger\hat{a}} &= \exp(\hat{a}^\dagger\hat{a}\ln z) \\ &= \sum_{n=0}^{\infty} \exp(n\ln z)\,|n\rangle\langle n| \\ &= \sum_{n=0}^{\infty} z^n\,|n\rangle\langle n|. \end{aligned} \tag{3.3.8}$$

The identification of the exponential of an operator with the power series (3.3.4) allows us to prove a very useful theorem which states that for two operators $\hat{A}$ and $\hat{B}$,

$$\begin{aligned} \exp(\hat{A})\hat{B}\exp(-\hat{A}) = \hat{B} + [\hat{A},\hat{B}] &+ \frac{1}{2!}\big[\hat{A},[\hat{A},\hat{B}]\big] \\ &+ \frac{1}{3!}\Big[\hat{A},\big[\hat{A},[\hat{A},\hat{B}]\big]\Big] + \cdots. \end{aligned} \tag{3.3.9}$$

The proof follows by inserting the expression (3.3.4) for $\exp(\hat{A})$ and $\exp(-\hat{A})$ into the left-hand side of (3.3.9) and collecting together operator products containing equal numbers of the operator $\hat{A}$. Alternatively, consider the function $\hat{F}(\theta)$ of the operators $\hat{A}$ and $\hat{B}$ defined as

$$\hat{F}(\theta) = \exp(\theta\hat{A})\hat{B}\exp(-\theta\hat{A}). \tag{3.3.10}$$

This function is defined by its power series expansion in $\hat{A}$ which, because it is a function of $\theta\hat{A}$, is simply related to its power series expansion in θ. Thus

$$\hat{F}(\theta) = \hat{F}(0) + \left.\frac{\mathrm{d}\hat{F}}{\mathrm{d}\theta}\right|_{\theta=0}\theta + \frac{1}{2!}\left.\frac{\mathrm{d}^2\hat{F}}{\mathrm{d}\theta^2}\right|_{\theta=0}\theta^2 + \cdots. \tag{3.3.11}$$

The first term is simply $\hat{F}(0)=\hat{B}$. The second- and higher-order terms follow from the fact that

$$\frac{\mathrm{d}\hat{F}}{\mathrm{d}\theta}=\hat{A}\hat{F}-\hat{F}\hat{A}=[\hat{A},\hat{F}], \tag{3.3.12}$$

so that

$$\left.\frac{\mathrm{d}\hat{F}}{\mathrm{d}\theta}\right|_{\theta=0}=[\hat{A},\hat{B}],\quad \left.\frac{\mathrm{d}^2\hat{F}}{\mathrm{d}\theta^2}\right|_{\theta=0}=\left[\hat{A},[\hat{A},\hat{B}]\right], \tag{3.3.13}$$

and so on. The result (3.3.9) follows on putting $\theta=1$ in (3.3.11). An important consequence of this result is that if $\hat{B}$ in (3.3.9) is a function of another operator $\hat{C}$, so that $\hat{B}=f(\hat{C})$, then

$$\begin{aligned}\exp(\hat{A})f(\hat{C})\exp(-\hat{A}) &= \sum_{l=0}^{\infty}\frac{f^{(l)}(0)}{l!}\exp(\hat{A})\hat{C}^l\exp(-\hat{A})\\ &= \sum_{l=0}^{\infty}\frac{f^{(l)}(0)}{l!}\left[\exp(\hat{A})\hat{C}\exp(-\hat{A})\right]^l\\ &= f\left(\exp(\hat{A})\hat{C}\exp(-\hat{A})\right),\end{aligned} \tag{3.3.14}$$

where we have used the definition (3.3.3) together with the fact that $\exp(\hat{A})\exp(-\hat{A})=\mathbf{1}$.

The number operator $\hat{n}$ is the product of two non-commuting operators. Implicit in its definition is a prescription for ordering these two operators, so that $\hat{n}=\hat{a}^\dagger\hat{a}$ but $\hat{n}\neq\hat{a}\hat{a}^\dagger$. All functions of non-commuting operators require a specification of the operator ordering. For functions of $\hat{a}$ and $\hat{a}^\dagger$ there are three special orderings commonly encountered. These are normal, symmetric, and antinormal ordering. In normal ordering, the operators are positioned so that no annihilation operator $\hat{a}$ ever appears to the left of any creation operator $\hat{a}^\dagger$. This ordering is denoted by placing pairs of dots before and after the normal ordered function. For example, the normal ordered form of $\hat{n}^l$ is

$$:\hat{n}^l:\,=\,:(\hat{a}^\dagger\hat{a})^l:\,=\hat{a}^{\dagger l}\hat{a}^l. \tag{3.3.15}$$

Antinormal ordering, as its name suggests, is the opposite of normal ordering in that no annihilation operator ever appears to the right of any creation operator. This ordering is denoted by placing triplets of dots before and after the antinormal ordered function. For example, the antinormal ordered form of $\hat{n}^l$ is

$$\vdots\,\hat{n}^l\,\vdots\,=\,\vdots\,(\hat{a}^\dagger\hat{a})^l\,\vdots\,=\hat{a}^l\hat{a}^{\dagger l}. \tag{3.3.16}$$

Symmetric ordering, denoted $S(\ )$, is the average of all possible orderings of $\hat{a}$

and $\hat{a}^{\dagger}$. For example, the first two symmetric ordered powers of the number operator are

$$S(\hat{n}) = S(\hat{a}^{\dagger}\hat{a}) = \tfrac{1}{2}(\hat{a}^{\dagger}\hat{a} + \hat{a}\hat{a}^{\dagger}), \tag{3.3.17}$$

$$S(\hat{n}^2) = S\left((\hat{a}^{\dagger}\hat{a})^2\right) = \tfrac{1}{6}\Big(\hat{a}^{\dagger 2}\hat{a}^2 + \hat{a}^{\dagger}\hat{a}\hat{a}^{\dagger}\hat{a} + \hat{a}^{\dagger}\hat{a}^2\hat{a}^{\dagger} + \hat{a}\hat{a}^{\dagger 2}\hat{a} + \hat{a}\hat{a}^{\dagger}\hat{a}\hat{a}^{\dagger} + \hat{a}^2\hat{a}^{\dagger 2}\Big). \tag{3.3.18}$$

The effect of these different orderings is clearly demonstrated by calculating the first two moments of the number operator in a number state $|n\rangle$ for each of these orderings. From (3.3.15) to (3.3.18), we find

$$\langle n|{:}\hat{n}{:}|n\rangle = n; \quad \langle n|{:}\hat{n}^2{:}|n\rangle = n(n-1), \tag{3.3.19}$$

$$\langle n|\,\vdots\,\hat{n}\,\vdots\,|n\rangle = n+1; \quad \langle n|\,\vdots\,\hat{n}^2\,\vdots\,|n\rangle = (n+1)(n+2), \tag{3.3.20}$$

$$\langle n|\,S(\hat{n})\,|n\rangle = n+\tfrac{1}{2}; \quad \langle n|\,S(\hat{n}^2)\,|n\rangle = n^2+n+\tfrac{1}{2}. \tag{3.3.21}$$

These three orderings discussed above are not the only ones used. We will consider more general operator orderings in Chapter 4. The concept of normal, symmetric, and antinormal ordering, as well as more general ordering, can be extended to apply to functions of the continuum operators $\hat{b}(\omega)$ and $\hat{b}^{\dagger}(\omega)$. This extension is a natural one in that, for example, a normal ordered function of the continuum creation and annihilation operators has all creation operators to the left of all annihilation operators.

It is often convenient to work with a particular operator ordering but the best ordering for calculational purposes is not always the one in which the problem is formulated. It is useful, therefore, to be able to relate one ordering of a function of operators to another. The rest of this section will be concerned with constructing such relationships for exponential functions of simple combinations of operators. For operators which commute there is, of course, no ordering problem, so that for two commuting operators $\hat{A}$ and $\hat{B}$ we have

$$\exp[\theta(\hat{A}+\hat{B})] = \exp(\theta\hat{A})\exp(\theta\hat{B}) = \exp(\theta\hat{B})\exp(\theta\hat{A}).$$

The simplest non-trivial ordering theorem is concerned with two operators $\hat{A}$ and $\hat{B}$ which commute with their commutator so that $[\hat{A},[\hat{A},\hat{B}]] = [\hat{B},[\hat{A},\hat{B}]] = 0$. Consider the function

$$\hat{F}_1(\theta) = \exp[\theta(\hat{A}+\hat{B})]. \tag{3.3.22}$$

This function is defined by its power series in the usual way and the expansion (3.3.4) will contain products of powers of the operators $\hat{A}$ and $\hat{B}$ with all possible orderings. It may be advantageous to express $\hat{F}_1(\theta)$ in a form in which all occurrences of $\hat{A}$ appear to the left of all those of $\hat{B}$. This ordering could be achieved by repeated use of the commutator $[\hat{A}, \hat{B}]$ to invert the order of $\hat{A}$ and $\hat{B}$ where required. Since $\hat{A}$ and $\hat{B}$ both commute with $[\hat{A}, \hat{B}]$ by assumption, the resulting ordered power series expansion will only contain the operators $\hat{A}$, $\hat{B}$, and $[\hat{A}, \hat{B}]$. It seems reasonable, therefore, to expect that $\hat{F}_1(\theta)$ can be written as a product of a function of $\hat{A}$, a function of $\hat{B}$, and a function of $[\hat{A}, \hat{B}]$. This is indeed the case as we now show. We seek to express $\hat{F}_1(\theta)$ in the form

$$\hat{F}_1(\theta) = \exp\left[p(\theta)\hat{A}\right] \exp\left[q(\theta)\hat{B}\right] \exp\{r(\theta)[\hat{A}, \hat{B}]\}, \tag{3.3.23}$$

where p, q, and r are functions to be determined. We proceed by calculating the derivative of $\hat{F}_1$ with respect to θ as given by the two forms (3.3.22) and (3.3.23). The derivative of the first is

$$\frac{\mathrm{d}\hat{F}_1(\theta)}{\mathrm{d}\theta} = (\hat{A} + \hat{B})\hat{F}_1(\theta), \tag{3.3.24}$$

while that of the second is

$$\frac{\mathrm{d}\hat{F}_1(\theta)}{\mathrm{d}\theta} = \left\{p'\hat{A} + q' \exp(p\hat{A})\hat{B} \exp(-p\hat{A}) + r'[\hat{A}, \hat{B}]\right\}\hat{F}_1(\theta), \tag{3.3.25}$$

where primes denote differentiation with respect to θ and we have again used the fact that $[\hat{A}, \hat{B}]$ commutes with both $\hat{A}$ and $\hat{B}$ and therefore with $\exp(p\hat{A})$ and $\exp(q\hat{B})$. Using (3.3.9), the coefficient of q' on the right-hand side of (3.3.25) can be written simply as

$$\exp(p\hat{A})\hat{B} \exp(-p\hat{A}) = \hat{B} + p[\hat{A}, \hat{B}], \tag{3.3.26}$$

so that (3.3.25) becomes

$$\frac{\mathrm{d}\hat{F}_1(\theta)}{\mathrm{d}\theta} = \{p'\hat{A} + q'\hat{B} + (pq' + r')[\hat{A}, \hat{B}]\}\hat{F}_1(\theta). \tag{3.3.27}$$

Comparing coefficients of $\hat{A}$, $\hat{B}$, and $[\hat{A}, \hat{B}]$ in the right-hand sides of (3.3.24) and (3.3.27), we find

$$p' = q' = 1, \quad pq' + r' = 0. \tag{3.3.28}$$

Solving these simple differential equations, with the requirements that $p(0) = q(0) = r(0) = 0$ so that $\hat{F}_1(0) = 1$, we obtain $p = q = \theta$ and $r = -\theta^2/2$. Hence the ordered form of $\hat{F}_1(\theta)$ is

$$\exp[\theta(\hat{A} + \hat{B})] = \exp(\theta\hat{A}) \exp(\theta\hat{B}) \exp\{-(\theta^2/2)[\hat{A}, \hat{B}]\}. \tag{3.3.29}$$

An important application of the theorem is the normal ordering of the symmetrical ordered unitary displacement operator

$$\hat{D}(\alpha) = \exp(\alpha\hat{a}^\dagger - \alpha^*\hat{a}), \tag{3.3.30}$$

which is important in the theory of the coherent states discussed in Section 3.6. If we let $\hat{A} = \alpha\hat{a}^\dagger$ and $\hat{B} = -\alpha^*\hat{a}$, then the commutator of $\hat{A}$ and $\hat{B}$ is the number $[\hat{A}, \hat{B}] = |\alpha|^2$ which clearly commutes with all operators. Hence from (3.3.29) with $\theta = 1$ we find

$$\begin{aligned}\hat{D}(\alpha) &= \exp(\alpha\hat{a}^\dagger)\exp(-\alpha^*\hat{a})\exp(-|\alpha|^2/2)\\ &= :\hat{D}(\alpha):\exp(-|\alpha|^2/2).\end{aligned} \tag{3.3.31}$$

The relationship between $\hat{D}(\alpha)$ and its antinormal ordered form is found in a similar way from (3.3.29) with $\hat{A} = -\alpha^*\hat{a}$, $\hat{B} = \alpha\hat{a}^\dagger$, and $\theta = 1$ to be

$$\begin{aligned}\hat{D}(\alpha) &= \exp(-\alpha^*\hat{a})\exp(\alpha\hat{a}^\dagger)\exp(|\alpha|^2/2)\\ &= \vdots\,\hat{D}(\alpha)\,\vdots\exp(|\alpha|^2/2).\end{aligned} \tag{3.3.32}$$

We can apply a similar technique to order more complicated functions of $\hat{a}$ and $\hat{a}^\dagger$. In particular, we can derive a relationship between the operator $\exp(\theta\hat{a}^\dagger\hat{a})$ and its normal ordered form. All powers of the number operator $\hat{n} = \hat{a}^\dagger\hat{a}$ commute with normal ordered powers of the number operator. This leads us to conjecture that it is possible to write

$$\hat{F}_2(\theta) = \exp(\theta\hat{a}^\dagger\hat{a}) = :\exp\left[p(\theta)\hat{a}^\dagger\hat{a}\right]:. \tag{3.3.33}$$

Proceeding as before, we calculate the derivative of each form of $\hat{F}_2(\theta)$ with respect to θ, giving

$$\frac{\mathrm{d}\hat{F}_2(\theta)}{\mathrm{d}\theta} = \hat{a}^\dagger\hat{a}\exp(\theta\hat{a}^\dagger\hat{a}) = p'\hat{a}^\dagger:\exp(p\hat{a}^\dagger\hat{a}):\hat{a}, \tag{3.3.34}$$

where the annihilation operator appears on the far right of the final expression as a result of the normal ordering. Using the supposed relationship (3.3.33), we write

$$\begin{aligned}p'\hat{a}^\dagger:\exp(p\hat{a}^\dagger\hat{a}):\hat{a} &= p'\hat{a}^\dagger\exp(\theta\hat{a}^\dagger\hat{a})\hat{a}\\ &= p'\hat{a}^\dagger\exp(\theta\hat{a}^\dagger\hat{a})\hat{a}\exp(-\theta\hat{a}^\dagger\hat{a})\exp(\theta\hat{a}^\dagger\hat{a}),\end{aligned} \tag{3.3.35}$$

where we have postmultiplied by $\exp(-\theta\hat{a}^\dagger\hat{a})\exp(\theta\hat{a}^\dagger\hat{a}) = \mathbf{1}$. The theorem (3.3.9) then gives

$$p'\hat{a}^\dagger:\exp(p\hat{a}^\dagger\hat{a}):\hat{a} = p'\exp(-\theta)\hat{a}^\dagger\hat{a}\exp(\theta\hat{a}^\dagger\hat{a}). \tag{3.3.36}$$

Comparing (3.3.34) and (3.3.36) leads to the requirement that $p'\exp(-\theta) = 1$. Integrating this and imposing the condition $p(0) = 0$, so that $\hat{F}_2(0) = \mathbf{1}$, gives $p(\theta) = \exp(\theta) - 1$. Hence

$$\exp(\theta\hat{a}^\dagger\hat{a}) = :\exp\{[\exp(\theta) - 1]\hat{a}^\dagger\hat{a}\}:. \tag{3.3.37}$$

This result will be important in our discussion of photon number moment generating functions in Section 4.2. The number states $|n\rangle$ are clearly the eigenstates of $\exp(\theta\hat{a}^\dagger\hat{a})$ in (3.3.37) with eigenvalues $\exp(\theta n)$. It follows that as $\theta \to -\infty$ all these eigenvalues tend to zero except that of $|0\rangle$ and hence that $:\exp(-\hat{a}^\dagger\hat{a}): = |0\rangle\langle 0|$.

The third theorem concerns the operator function $\hat{F}_3(\theta) = \exp[\theta(\hat{A}+\hat{B})]$ where we now assume that

$$[\hat{A},\hat{B}] = -\hat{A}, \tag{3.3.38}$$

so that $\hat{A}$ commutes with the commutator $[\hat{A},\hat{B}]$ but $\hat{B}$ does not. In this case a rearrangement of a product of powers of $\hat{A}$ and powers of $\hat{B}$ using the commutator (3.3.38) will lead to an expression which only contains powers of $\hat{A}$ and $\hat{B}$. Hence we seek a relationship of the form

$$\hat{F}_3(\theta) = \exp[\theta(\hat{A}+\hat{B})] = \exp\left[p(\theta)\hat{B}\right]\exp\left[q(\theta)\hat{A}\right]. \tag{3.3.39}$$

The method of obtaining the functions p and q is the same as that used above. The derivative of $\hat{F}_3(\theta)$ with respect to θ is

$$\begin{aligned}\frac{\mathrm{d}\hat{F}_3(\theta)}{\mathrm{d}\theta} &= (\hat{A}+\hat{B})\hat{F}_3(\theta) = \left[p'\hat{B} + q'\exp(p\hat{B})\hat{A}\exp(-p\hat{B})\right]\hat{F}_3(\theta)\\ &= \left[p'\hat{B} + q'\exp(p)\hat{A}\right]\hat{F}_3(\theta),\end{aligned} \tag{3.3.40}$$

where we have again used (3.3.9). Comparing coefficients of $\hat{A}$ and $\hat{B}$ leads to the differential equations $p' = 1$ and $q'\exp(p) = 1$. The solution of these with $p(0) = q(0) = 0$ is $p = \theta$ and $q = 1 - \exp(-\theta)$ so that

$$\exp[\theta(\hat{A}+\hat{B})] = \exp(\theta\hat{B})\exp\{[1-\exp(-\theta)]\hat{A}\}. \tag{3.3.41}$$

The alternative ordering of $\hat{F}_3(\theta)$ has the form

$$\exp[\theta(\hat{A}+\hat{B})] = \exp\{[\exp(\theta)-1]\hat{A}\}\exp(\theta\hat{B}), \tag{3.3.42}$$

as may be shown either by the same procedure or by postmultiplying (3.3.41) by $\exp(-\theta\hat{B})\exp(\theta\hat{B}) = \mathbf{1}$ and using (3.3.9) and (3.3.14).

Ordering theorems involving larger numbers of operators can also be derived although only at the expense of greater complexity. We shall now consider two examples involving three non-commuting operators. The first concerns the system of angular momentum operators $\hat{J}_3$, $\hat{J}_+$, and $\hat{J}_-$, the commutators of which are given by (3.2.13) and (3.2.14). Consider the exponential function of $\hat{J}_+$ and $\hat{J}_-$ in the form $\hat{F}_4(\theta) = \exp[\mathrm{i}\theta(\hat{J}_+ + \hat{J}_-)]$. The fact that the commutator of any two angular momentum operators is proportional to one of the operators suggests a relationship of the form

$$\hat{F}_4(\theta) = \exp\left[\mathrm{i}\theta\left(\hat{J}_+ + \hat{J}_-\right)\right]$$
$$= \exp\left[\mathrm{i}p(\theta)\hat{J}_+\right]\exp\left[\mathrm{i}q(\theta)\hat{J}_3\right]\exp\left[\mathrm{i}r(\theta)\hat{J}_-\right]. \qquad (3.3.43)$$

Employing the same method as before, we find

$$\frac{\mathrm{d}\hat{F}_4(\theta)}{\mathrm{d}\theta} = \mathrm{i}\left(\hat{J}_+ + \hat{J}_-\right)\hat{F}_4(\theta)$$
$$= \mathrm{i}\left[p'\hat{J}_+ + q'\exp\left(\mathrm{i}p\hat{J}_+\right)\hat{J}_3\exp\left(-\mathrm{i}p\hat{J}_+\right)\right.$$
$$\left. + r'\exp\left(\mathrm{i}p\hat{J}_+\right)\exp\left(\mathrm{i}q\hat{J}_3\right)\hat{J}_-\exp\left(-\mathrm{i}q\hat{J}_3\right)\exp\left(-\mathrm{i}p\hat{J}_+\right)\right]\hat{F}_4(\theta). \qquad (3.3.44)$$

The last form can be simplified by using the commutators (3.2.13) and (3.2.14) and applying (3.3.9) to the expressions

$$\exp\left(\mathrm{i}p\hat{J}_+\right)\hat{J}_3\exp\left(-\mathrm{i}p\hat{J}_+\right) = \hat{J}_3 - \mathrm{i}p\hat{J}_+ \qquad (3.3.45)$$

and

$$\exp\left(\mathrm{i}p\hat{J}_+\right)\exp\left(\mathrm{i}q\hat{J}_3\right)\hat{J}_-\exp\left(-\mathrm{i}q\hat{J}_3\right)\exp\left(-\mathrm{i}p\hat{J}_+\right)$$
$$= \exp\left(\mathrm{i}p\hat{J}_+\right)\left(\hat{J}_- - \mathrm{i}q\hat{J}_- - \tfrac{1}{2}q^2\hat{J}_- + \cdots\right)\exp\left(-\mathrm{i}p\hat{J}_+\right)$$
$$= \exp(-\mathrm{i}q)\exp\left(\mathrm{i}p\hat{J}_+\right)\hat{J}_-\exp\left(-\mathrm{i}p\hat{J}_+\right)$$
$$= \exp(-\mathrm{i}q)\left(\hat{J}_- + 2\mathrm{i}p\hat{J}_3 + p^2\hat{J}_+\right). \qquad (3.3.46)$$

Substituting (3.3.45) and (3.3.46) into (3.3.44) and comparing coefficients of $\hat{J}_+$, $\hat{J}_-$, and $\hat{J}_3$, we obtain the differential equations

$$p' - \mathrm{i}pq' + r'\exp(-\mathrm{i}q)p^2 = 1, \qquad (3.3.47)$$
$$r'\exp(-\mathrm{i}q) = 1, \qquad (3.3.48)$$
$$q' + 2\mathrm{i}r'\exp(-\mathrm{i}q)p = 0. \qquad (3.3.49)$$

The solution of these with $p(0) = q(0) = r(0) = 0$ is $p = r = \tan\theta$ and $q = \mathrm{i}\ln(\cos^2\theta)$, so that

$$\exp\left[\mathrm{i}\theta\left(\hat{J}_+ + \hat{J}_-\right)\right] = \exp\left[\mathrm{i}(\tan\theta)\hat{J}_+\right]\exp\left[-\ln(\cos^2\theta)\hat{J}_3\right]\exp\left[\mathrm{i}(\tan\theta)\hat{J}_-\right]. \qquad (3.3.50)$$

In the study of squeezed states, we will use operators $\hat{K}_+$, $\hat{K}_-$, and $\hat{K}_3$ similar to the angular momentum operators but having the slightly different commutation relations

$$\left[\hat{K}_3, \hat{K}_\pm\right] = \pm\hat{K}_\pm, \qquad (3.3.51)$$
$$\left[\hat{K}_+, \hat{K}_-\right] = -2\hat{K}_3. \qquad (3.3.52)$$

For these operators we can use the same method to prove the ordering theorem

$$\exp\left[i\theta\left(\hat{K}_+ + \hat{K}_-\right)\right] = \exp\left[i(\tanh\theta)\hat{K}_+\right]\exp\left[-\ln(\cosh^2\theta)\hat{K}_3\right]\exp\left[i(\tanh\theta)\hat{K}_-\right]. \tag{3.3.53}$$

It is possible to use the above method to order the exponential of any linear combination of $\hat{J}_+$, $\hat{J}_-$, and $\hat{J}_3$ or of $\hat{K}_+$, $\hat{K}_-$, and $\hat{K}_3$. The general form of these ordering theorems is included in Appendix 5, together with a summary of other ordering theorems.

The above theorems establish the equivalence of differently ordered operator functions without reference to states. If we have a complete or overcomplete set of states spanning the space on which the operators act then there are other ways of proving the equivalence of two operators. Two operators are equivalent if their matrix elements between any two basis states are equal, for all possible pairs of basis states. We illustrate this in Appendix 5. It is sometimes more convenient to reorder exponential functions of operators using the definition (3.3.4). The simplest example of this arises when we have the exponential of a sum of Pauli operators in the form

$$\exp\left[i\left(a_1\hat{\sigma}_1 + a_2\hat{\sigma}_2 + a_3\hat{\sigma}_3\right)\right] = 1 + i\left(a_1\hat{\sigma}_1 + a_2\hat{\sigma}_2 + a_3\hat{\sigma}_3\right) - \tfrac{1}{2}\left(a_1\hat{\sigma}_1 + a_2\hat{\sigma}_2 + a_3\hat{\sigma}_3\right)^2 + \cdots, \tag{3.3.54}$$

where a_1, a_2, and a_3 are arbitrary complex numbers. It follows from the anticommutator $\{\hat{\sigma}_i, \hat{\sigma}_j\} = 2\delta_{ij}$ that

$$\left(a_1\hat{\sigma}_1 + a_2\hat{\sigma}_2 + a_3\hat{\sigma}_3\right)^2 = a_1^2 + a_2^2 + a_3^2 = a^2, \tag{3.3.55}$$

and hence that

$$\exp\left[i\left(a_1\hat{\sigma}_1 + a_2\hat{\sigma}_2 + a_3\hat{\sigma}_3\right)\right] = \cos a + i\left(a_1\hat{\sigma}_1 + a_2\hat{\sigma}_2 + a_3\hat{\sigma}_3\right)\frac{\sin a}{a}. \tag{3.3.56}$$

Note that this is simply a linear combination of the Pauli operators.

Ordering theorems can also be derived for the exponential functional of some combinations of the continuum operators $\hat{b}(\omega)$ and $\hat{b}^\dagger(\omega)$. We emphasize, however, that not all orderings are meaningful. In particular $\hat{b}^\dagger(\omega)\hat{b}(\omega)$ and $\hat{b}(\omega)\hat{b}^\dagger(\omega)$ cannot both be well defined owing to the form of the commutator (3.2.31). The normal ordered form $\hat{b}^\dagger(\omega)\hat{b}(\omega)$ is well behaved in that a weighted integral over ω of this acting on a state produces another state of the field. Examples of this are the free-field Hamiltonian (3.2.32) and the continuum number operator (3.2.33). We conclude this section with two continuum operator ordering theorems. The continuum analogue of the displacement operator (3.3.30) is

$$\hat{D}[\alpha(\omega)] = \exp\left(\int d\omega[\alpha(\omega)\hat{b}^\dagger(\omega) - \alpha^*(\omega)\hat{b}(\omega)]\right). \tag{3.3.57}$$

This can be related to normal and antinormal ordered forms similar to (3.3.31) and (3.3.32) for a single-mode displacement operator. Making the identifications

$$\hat{A} = \int \mathrm{d}\omega\, \alpha(\omega)\hat{b}^\dagger(\omega), \tag{3.3.58}$$

$$\hat{B} = -\int \mathrm{d}\omega\, \alpha^*(\omega)\hat{b}(\omega), \tag{3.3.59}$$

we find that the commutator is

$$[\hat{A}, \hat{B}] = \int \mathrm{d}\omega |\alpha(\omega)|^2, \tag{3.3.60}$$

where we have used the continuum operator commutation relation (3.2.31). Clearly $\hat{A}$ and $\hat{B}$ both commute with $[\hat{A}, \hat{B}]$ and therefore we can apply (3.3.29) to obtain the normal ordered form

$$\begin{aligned}\hat{D}[\alpha(\omega)] &= \exp\left(\int \mathrm{d}\omega\, \alpha(\omega)\hat{b}^\dagger(\omega)\right) \exp\left(-\int \mathrm{d}\omega\, \alpha^*(\omega)\hat{b}(\omega)\right) \\ &\quad \times \exp\left(-\tfrac{1}{2}\int \mathrm{d}\omega |\alpha(\omega)|^2\right) \\ &= :\hat{D}[\alpha(\omega)]: \exp\left(-\tfrac{1}{2}\int \mathrm{d}\omega |\alpha(\omega)|^2\right). \end{aligned} \tag{3.3.61}$$

Similarly, the antinormal ordered form is related to $\hat{D}[\alpha(\omega)]$ by

$$\begin{aligned}\hat{D}[\alpha(\omega)] &= \exp\left(-\int \mathrm{d}\omega\, \alpha^*(\omega)\hat{b}(\omega)\right) \exp\left(\int \mathrm{d}\omega\, \alpha(\omega)\hat{b}^\dagger(\omega)\right) \\ &\quad \times \exp\left(\tfrac{1}{2}\int \mathrm{d}\omega |\alpha(\omega)|^2\right) \\ &= \,\vdots\, \hat{D}[\alpha(\omega)] \,\vdots\, \exp\left(\tfrac{1}{2}\int \mathrm{d}\omega |\alpha(\omega)|^2\right). \end{aligned} \tag{3.3.62}$$

The method of proof based on differentiating operator functions can be extended to functionals of continuum operators using the technique of functional differentiation. The functional derivative of a functional $F[g(x)]$ of a function $g(x)$ is defined by the limit

$$\frac{\delta F[g(x)]}{\delta g(y)} = \lim_{\epsilon \to 0} \frac{F[g(x) + \epsilon\delta(x-y)] - F[g(x)]}{\epsilon}, \tag{3.3.63}$$

where this limit is well behaved. In particular, the functional derivative of the integral of a function of $g(x)$ is

$$\frac{\delta \int f(g(x))\,\mathrm{d}x}{\delta g(y)} = \frac{\partial f(g(y))}{\partial g(y)} \tag{3.3.64}$$

and that of the exponential of the integral of a function $g(x)$ is

$$\frac{\delta \exp[\int f(g(x))\,\mathrm{d}x]}{\delta g(y)} = \frac{\partial f(g(y))}{\partial g(y)} \exp(\int f(g(x))\,\mathrm{d}x). \qquad (3.3.65)$$

This technique can be used to provide an alternative proof of (3.3.61) and (3.3.62). Consider the functional

$$\hat{F}_1[\theta(\omega)] = \exp\left(\int \mathrm{d}\omega\, \theta(\omega)[\alpha(\omega)\hat{b}^\dagger(\omega) - \alpha^*(\omega)\hat{b}(\omega)]\right). \qquad (3.3.66)$$

By analogy with the derivation of the single-mode result (3.3.31), we seek an equivalent expression for $\hat{F}_1$ in the form

$$\hat{F}_1[\theta(\omega)] = \exp\left(\int \mathrm{d}\omega\, p(\theta(\omega))\hat{b}^\dagger(\omega)\right) \exp\left(\int \mathrm{d}\omega\, q(\theta(\omega))\hat{b}(\omega)\right) \times \exp\left(\int \mathrm{d}\omega\, r(\theta(\omega))\right). \qquad (3.3.67)$$

Evaluating and equating the functional derivatives of (3.3.66) and (3.3.67) and imposing the requirement that $\hat{F}_1 = \mathbf{1}$ for $\theta(\omega) = 0$, we find

$$p(\theta(\omega)) = \theta(\omega)\alpha(\omega) = -q^*(\theta(\omega)), \qquad (3.3.68)$$

$$r(\theta(\omega)) = -\tfrac{1}{2}\theta^2(\omega)|\alpha(\omega)|^2. \qquad (3.3.69)$$

The result (3.3.61) follows on setting $\theta(\omega) = 1$. Finally we obtain the continuum generalization of the normal ordering theorem (3.3.37). Consider the functional

$$\hat{F}_2[\theta(\omega)] = \exp\left(\int \mathrm{d}\omega\, \theta(\omega)\hat{b}^\dagger(\omega)\hat{b}(\omega)\right). \qquad (3.3.70)$$

Once again, by analogy with the single-mode result (3.3.37), we conjecture that

$$\hat{F}_2[\theta(\omega)] = :\exp\left(\int \mathrm{d}\omega\, p(\theta(\omega))\hat{b}^\dagger(\omega)\hat{b}(\omega)\right):. \qquad (3.3.71)$$

Equating the functional derivatives with respect to $\theta(\omega')$ of (3.3.70) and (3.3.71) gives

$$\begin{aligned}\frac{\delta \hat{F}_2[\theta(\omega)]}{\delta\theta(\omega')} &= \hat{b}^\dagger(\omega')\hat{b}(\omega')\hat{F}_2[\theta(\omega)] \\ &= \hat{b}^\dagger(\omega')\hat{F}_2[\theta(\omega)]\hat{b}(\omega')\frac{\partial p(\theta(\omega'))}{\partial\theta(\omega')} \\ &= \hat{b}^\dagger(\omega')\hat{b}(\omega')\hat{F}_2[\theta(\omega)]\exp[-\theta(\omega')]\frac{\partial p(\theta(\omega'))}{\partial\theta(\omega')},\end{aligned} \qquad (3.3.72)$$

where we have again used (3.3.9) and the commutation relation (3.2.31). Comparing the coefficients of $\hat{b}^\dagger(\omega')\hat{b}(\omega')$ gives

$$\exp[-\theta(\omega')]\frac{\partial p(\theta(\omega'))}{\partial\theta(\omega')} = 1, \tag{3.3.73}$$

so that with $\hat{F}_2 = \mathbf{1}$ for $\theta(\omega) = 0$ we obtain $p(\theta(\omega)) = \exp[\theta(\omega)] - 1$. Hence

$$\exp\left(\int \mathrm{d}\omega\,\theta(\omega)\hat{b}^\dagger(\omega)\hat{b}(\omega)\right) = :\exp\left(\int \mathrm{d}\omega[\exp[\theta(\omega)] - 1]\hat{b}^\dagger(\omega)\hat{b}(\omega)\right):. \tag{3.3.74}$$

3.4 Number states

So far in this chapter we have been primarily concerned with operators representing observable properties of atoms and fields. In the remainder of this chapter, we will introduce important classes of states, principally of the electromagnetic field, and use the operators and ordering theorems of the preceding two sections to derive their properties. We introduced the number states of a single field mode in Chapter 1, and in this section we derive some of their properties and those of their continuum generalizations.

The single-mode number states $|n\rangle$ are the eigenstates of the number operator $\hat{n}$ with eigenvalues n $(=0,1,2,\ldots)$. As such, they are the energy eigenstates of the free-field Hamiltonian. They are orthonormal, so that $\langle n\,|\,n'\rangle = \delta_{nn'}$, and complete, so that they can be used to form the resolution (1.3.33) of the identity operator. The orthonormality of the number states together with the action of $\hat{a}$ and $\hat{a}^\dagger$ on these states, given by (3.2.16) and (3.2.17), ensures that the expectation value in a number state of any operator formed as a product of powers of $\hat{a}$ and of $\hat{a}^\dagger$ is zero unless the number of occurrences of $\hat{a}$ is the same as that of $\hat{a}^\dagger$. In particular, the normal ordered moments are

$$\langle n|\,\hat{a}^{\dagger l}\hat{a}^m\,|n\rangle = \frac{n!}{(n-l)!}\,\delta_{lm}, \tag{3.4.1}$$

for $l \leqslant n$ and zero otherwise, while the antinormal ordered moments are

$$\langle n|\,\hat{a}^m\hat{a}^{\dagger l}\,|n\rangle = \frac{(n+l)!}{n!}\,\delta_{lm}. \tag{3.4.2}$$

The symmetrical ordered moment is more complicated to evaluate.

For two or more modes, we can define a number operator for each mode but it is also useful to introduce a total number operator which is simply the sum of the single-mode operators. For example, the total number operator for two modes A and B with annihilation operators $\hat{a}$ and $\hat{b}$ is

$$\hat{N}_2 = \hat{n}_A + \hat{n}_B, \tag{3.4.3}$$

where $\hat{n}_A = \hat{a}^\dagger \hat{a}$ and $\hat{n}_B = \hat{b}^\dagger \hat{b}$. The set of eigenstates of $\hat{N}_2$ includes those formed as products of the single-mode eigenstates of $\hat{n}_A$ and $\hat{n}_B$, which we denote as

$$|n_A; n'_B\rangle = |n\rangle_A |n'\rangle_B. \tag{3.4.4}$$

In addition, it also includes superpositions of these product states corresponding to the same total photon number so that, for example,

$$\hat{N}_2 \sum_{m=0}^{n} c_m |m_A; (n-m)_B\rangle = n \sum_{m=0}^{n} c_m |m_A; (n-m)_B\rangle. \tag{3.4.5}$$

Hence the set of eigenstates of $\hat{N}_2$ is overcomplete.

It is possible to define states of the continuum field each having a precisely determined number of quanta. These states are the generalization of the eigenstates of the total photon number operator described above and are eigenstates of the continuum number operator $\hat{N}$ given by (3.2.33). We begin by defining the vacuum state $|0\rangle$ of the *continuum field* by the two properties that it is normalized, so that $\langle 0 | 0 \rangle = 1$, and that

$$\hat{b}(\omega)|0\rangle = 0, \tag{3.4.6}$$

for all ω. This is analogous to the action of the single-mode annihilation operator $\hat{a}$ on the vacuum state given by (1.3.35) with $n = 0$. For single modes, the number states are generated by repeated application of the creation operator as in (3.2.18). However, the application of the continuum creation operator $\hat{b}^\dagger(\omega)$ to the continuum vacuum state $|0\rangle$ does not produce a state. This is because the norm of $\hat{b}^\dagger(\omega)|0\rangle$, being the expectation value of the operator $\hat{b}(\omega)\hat{b}^\dagger(\omega)$, is not defined since this operator cannot itself be defined, as discussed in Section 3.3. We can, however, introduce suitable operators for the construction of continuum number states by taking a weighted integral of $\hat{b}^\dagger(\omega)$ over frequency. Consider the operator

$$\hat{b}_f^\dagger = \int \mathrm{d}\omega\, f(\omega)\hat{b}^\dagger(\omega), \tag{3.4.7}$$

where $f(\omega)$ is any complex function satisfying the normalization condition

$$\int \mathrm{d}\omega |f(\omega)|^2 = 1. \tag{3.4.8}$$

The operator $\hat{b}_f^\dagger$ and its Hermitian conjugate $\hat{b}_f$ have the commutation relation

$$\begin{aligned}\left[\hat{b}_f, \hat{b}_f^\dagger\right] &= \int \mathrm{d}\omega' f^*(\omega') \int \mathrm{d}\omega\, f(\omega)[\hat{b}(\omega'), \hat{b}^\dagger(\omega)] \\ &= \int \mathrm{d}\omega' \int \mathrm{d}\omega\, f^*(\omega') f(\omega)\, \delta(\omega - \omega') = 1,\end{aligned} \tag{3.4.9}$$

using (3.4.8) and the commutator (3.2.31). The form of the commutator (3.4.9), together with the fact that $\hat{b}_f|0\rangle = 0$, implies that $\hat{b}_f$ and $\hat{b}_f^\dagger$ are boson annihilation and creation operators. The action of $\hat{b}_f^\dagger$ on the vacuum state $|0\rangle$ produces a well-behaved one-quantum continuum state the form of which is determined by the function $f(\omega)$. The normalization of $\hat{b}_f^\dagger|0\rangle$ follows directly from the commutator (3.4.9) and the property (3.4.6) of $|0\rangle$. The state $\hat{b}_f^\dagger|0\rangle$ is also an eigenstate of $\hat{N}$ with eigenvalue unity since

$$\hat{N}\hat{b}_f^\dagger|0\rangle = \int d\omega' \hat{b}^\dagger(\omega')\hat{b}(\omega') \int d\omega f(\omega)\hat{b}^\dagger(\omega)|0\rangle$$

$$= \int d\omega' \int d\omega f(\omega)\hat{b}^\dagger(\omega')[\delta(\omega-\omega') + \hat{b}^\dagger(\omega)\hat{b}(\omega')]|0\rangle$$

$$= \int d\omega f(\omega)\hat{b}^\dagger(\omega)|0\rangle = \hat{b}_f^\dagger|0\rangle, \tag{3.4.10}$$

using the commutator (3.2.31). Further, it is straightforward to show that the action of any two continuum annihilation operators on $\hat{b}_f^\dagger|0\rangle$ gives zero. This justifies the assertion that $\hat{b}_f^\dagger|0\rangle$ is a one-quantum state. This state is not, however, an eigenstate of the free-field Hamiltonian (3.2.32) and hence is not a stationary state of the field. The expectation value of the Hamiltonian in this one-quantum state is $\int d\omega\, \hbar\omega\, |f(\omega)|^2$.

The general two-quantum state is produced by the action of a second creation operator $\hat{b}_g^\dagger$ acting on the one-quantum state $\hat{b}_f^\dagger|0\rangle$. The norm of the state $\hat{b}_g^\dagger\hat{b}_f^\dagger|0\rangle$ is

$$\langle 0|\hat{b}_f\hat{b}_g\hat{b}_g^\dagger\hat{b}_f^\dagger|0\rangle = \langle 0|\hat{b}_f\hat{b}_f^\dagger|0\rangle + \langle 0|\hat{b}_f\hat{b}_g^\dagger\hat{b}_g\hat{b}_f^\dagger|0\rangle$$

$$= 1 + \langle 0|\left\{\left[\hat{b}_f, \hat{b}_g^\dagger\right] + \hat{b}_g^\dagger\hat{b}_f\right\}\left\{\left[\hat{b}_g, \hat{b}_f^\dagger\right] + \hat{b}_f^\dagger\hat{b}_g\right\}|0\rangle$$

$$= 1 + \left|\left[\hat{b}_f, \hat{b}_g^\dagger\right]\right|^2, \tag{3.4.11}$$

where we have used the boson commutator (3.4.9) for $\hat{b}_g$ and $\hat{b}_g^\dagger$. In terms of the functions $f(\omega)$ and $g(\omega)$, this norm is given by

$$1 + \left|\left[\hat{b}_f, \hat{b}_g^\dagger\right]\right|^2 = 1 + \left|\int d\omega f^*(\omega)g(\omega)\right|^2, \tag{3.4.12}$$

which can take any value from 1 to 2 depending on the degree of overlap of the functions $f^*(\omega)$ and $g(\omega)$. If the overlap is zero then the state $\hat{b}_g^\dagger\hat{b}_f^\dagger|0\rangle$ is already normalized. This is analogous to the state of two discrete modes each containing precisely one photon. At the other extreme, if $f^*(\omega)$ and $g(\omega)$ differ only by an ω-independent phase factor, then the norm has value 2. This is the counterpart of two photons excited in a single discrete mode. A normalized two-quantum state can be obtained by dividing $\hat{b}_g^\dagger\hat{b}_f^\dagger|0\rangle$ by the square root of the norm (3.4.11).

An n-quantum continuum field state is constructed by the action of n boson creation operators $\hat{b}_{f_1}^\dagger, \hat{b}_{f_2}^\dagger, \ldots, \hat{b}_{f_n}^\dagger$ on $|0\rangle$. The norm of the state $\hat{b}_{f_1}^\dagger \hat{b}_{f_2}^\dagger \cdots \hat{b}_{f_n}^\dagger |0\rangle$ depends on the overlap integrals of all possible pairs of functions $f_i(\omega)$. All states of this form are eigenstates of the continuum number operator $\hat{N}$ with eigenvalue n. Moreover, the action of any $n+1$ annihilation operators on this state gives zero, as required for an n-quantum state.

3.5 Thermal states

It is often the case that we do not have enough information to specify completely the state of a system and hence cannot form its wavefunction. In such a situation, we can only describe the system as a mixed state represented by a density matrix ρ. The thermal states are those about which we have a minimum of information, knowing only the mean value of the energy. They arise frequently in problems, such as those considered in Chapter 5, involving the coupling of a system to its environment, the detailed structure of which is unknown. This is precisely the situation when the environment is in thermodynamic equilibrium.

In thermal equilibrium, the state of a system with Hamiltonian $\hat{H}$ is represented by the density matrix

$$\rho = \frac{\exp(-\beta \hat{H})}{\mathrm{Tr}\left[\exp(-\beta \hat{H})\right]}, \tag{3.5.1}$$

where $\beta = (k_B T)^{-1}$, k_B being Boltzmann's constant and T being the absolute temperature. For a two-state system with energy eigenstates $|1\rangle$ and $|2\rangle$, $|2\rangle$ having energy $\hbar\omega$ greater than $|1\rangle$, (3.5.1) gives the density matrix

$$\rho = [1 + \exp(-\beta\hbar\omega)]^{-1}[|1\rangle\langle 1| + \exp(-\beta\hbar\omega)|2\rangle\langle 2|]. \tag{3.5.2}$$

The probabilities for finding the system in either of its energy eigenstates are those associated with Fermi–Dirac statistics. If we restrict our model of an atom to only two states, then the density matrix (3.5.2) represents the steady state of the atom when in thermal equilibrium with a black-body radiation field at temperature T. In this state, the expectation values of $\hat{\sigma}_1$ and $\hat{\sigma}_2$, the quadrature components of the electric dipole, are both zero, showing that there is no preferred dipole orientation. The expectation value of $\hat{\sigma}_3$, the atomic inversion, is given by

$$\langle \hat{\sigma}_3 \rangle = \frac{\exp(-\beta\hbar\omega) - 1}{\exp(-\beta\hbar\omega) + 1}, \tag{3.5.3}$$

the value of which is always negative.

For a single field mode with frequency ω in a thermal state corresponding to temperature T, the density matrix (3.5.1) is

$$\rho = \frac{\exp(-\beta\hbar\omega\hat{n})}{\mathrm{Tr}[\exp(-\beta\hbar\omega\hat{n})]}. \tag{3.5.4}$$

Since this is a function only of the number operator $\hat{n}$, this density matrix is diagonal in the number state basis and can be written as

$$\rho = \sum_{n=0}^{\infty} P(n)|n\rangle\langle n|, \tag{3.5.5}$$

where the probability $P(n)$ is given by the Bose–Einstein form

$$P(n) = [1 - \exp(-\beta\hbar\omega)]\exp(-\beta n\hbar\omega). \tag{3.5.6}$$

The mean photon number $\bar{n}$ is the expectation value of the number operator $\hat{n}$ given by

$$\bar{n} = \langle\hat{n}\rangle = \mathrm{Tr}(\rho\hat{n}). \tag{3.5.7}$$

Inserting the forms (3.5.5) and (3.5.6), we obtain

$$\begin{aligned}\bar{n} &= [1 - \exp(-\beta\hbar\omega)] \sum_{n=0}^{\infty} n\exp(-\beta n\hbar\omega) \\ &= [1 - \exp(-\beta\hbar\omega)]\left(-\frac{1}{\hbar\omega}\frac{\mathrm{d}}{\mathrm{d}\beta}\right)\sum_{n=0}^{\infty}\exp(-\beta n\hbar\omega) \\ &= [\exp(\beta\hbar\omega) - 1]^{-1}.\end{aligned} \tag{3.5.8}$$

We can invert (3.5.8) to express $\exp(-\beta\hbar\omega)$ in terms of $\bar{n}$ and find

$$\exp(-\beta\hbar\omega) = \frac{\bar{n}}{1+\bar{n}}. \tag{3.5.9}$$

Hence from (3.5.5) and (3.5.6), the density matrix can be written in the form

$$\rho = \sum_{n=0}^{\infty} \frac{\bar{n}^n}{(1+\bar{n})^{n+1}}|n\rangle\langle n|. \tag{3.5.10}$$

Alternative forms of this density matrix which follow from (3.3.2), (3.5.5), and (3.5.10) are

$$\rho = [1 - \exp(-\beta\hbar\omega)]\exp(-\beta\hbar\omega\hat{n}) \tag{3.5.11}$$

and

$$\rho = \frac{1}{1+\bar{n}}\left(\frac{\bar{n}}{1+\bar{n}}\right)^{\hat{n}}. \tag{3.5.12}$$

The moments of the number operator $\hat{n}$ can be calculated by differentiation, using (3.5.11), since

$$\begin{aligned}\langle\hat{n}^l\rangle &= [1 - \exp(-\beta\hbar\omega)]\left(-\frac{1}{\hbar\omega}\frac{\mathrm{d}}{\mathrm{d}\beta}\right)^l \mathrm{Tr}[\exp(-\beta\hbar\omega\hat{n})] \\ &= [1 - \exp(-\beta\hbar\omega)]\left(-\frac{1}{\hbar\omega}\frac{\mathrm{d}}{\mathrm{d}\beta}\right)^l [1 - \exp(-\beta\hbar\omega)]^{-1}.\end{aligned} \tag{3.5.13}$$

The first two of these moments are

$$\langle \hat{n} \rangle = [\exp(\beta\hbar\omega) - 1]^{-1} = \bar{n}, \tag{3.5.14}$$

as in (3.5.8), and

$$\langle \hat{n}^2 \rangle = \frac{\exp(\beta\hbar\omega) + 1}{[\exp(\beta\hbar\omega) - 1]^2} = 2\bar{n}(\bar{n} + \tfrac{1}{2}), \tag{3.5.15}$$

so that the variance in the photon number is $\Delta n^2 = \bar{n}(\bar{n} + 1)$. The expectation value of $:\hat{n}^l:$ (see (3.3.15)) can be written

$$\langle :\hat{n}^l: \rangle = \langle \hat{a}^{\dagger l}\hat{a}^l \rangle = \mathrm{Tr}(\rho\hat{a}^{\dagger l}\hat{a}^l) = \mathrm{Tr}(\hat{a}^l\rho\hat{a}^{\dagger l}), \tag{3.5.16}$$

using the cyclic property of the trace operation. Hence if we can obtain an antinormal ordered form of the density matrix ρ, then the required expectation value will be the trace of an antinormal ordered operator. We first rewrite (3.5.11) in the form

$$\begin{aligned}\rho &= [1 - \exp(-\beta\hbar\omega)]\exp(-\beta\hbar\omega\hat{a}^\dagger\hat{a}) \\ &= [\exp(\beta\hbar\omega) - 1]\exp(-\beta\hbar\omega\hat{a}\hat{a}^\dagger),\end{aligned} \tag{3.5.17}$$

Using the ordering theorem (A5.8) and (3.5.8), the antinormal ordered form of ρ is

$$\begin{aligned}\rho &= [\exp(\beta\hbar\omega) - 1] ⋮ \exp\left[-[\exp(\beta\hbar\omega) - 1]\hat{a}^\dagger\hat{a}\right] ⋮ \\ &= \frac{1}{\bar{n}} ⋮ \exp\left(-\frac{1}{\bar{n}}\hat{a}^\dagger\hat{a}\right) ⋮ .\end{aligned} \tag{3.5.18}$$

Substituting (3.5.18) into (3.5.16), we see that

$$\begin{aligned}\langle :\hat{n}^l: \rangle &= \frac{1}{\bar{n}}\left(-\frac{\mathrm{d}}{\mathrm{d}(1/\bar{n})}\right)^l \mathrm{Tr}\left[⋮ \exp\left(-\frac{1}{\bar{n}}\hat{a}^\dagger\hat{a}\right) ⋮ \right] \\ &= \frac{1}{\bar{n}}\left(-\frac{\mathrm{d}}{\mathrm{d}(1/\bar{n})}\right)^l \bar{n} = l!\,\bar{n}^l,\end{aligned} \tag{3.5.19}$$

where the trace has been evaluated by considering the case $l = 0$. In Section 4.2, we rederive (3.5.19) more easily using the moment generating function. As with the number states, the expectation value in a thermal state of any operator formed as a product of unequal numbers of $\hat{a}$ and $\hat{a}^\dagger$ is zero.

For a continuum in a thermal state it is more natural, rather than trying to define a density matrix, to express the thermal properties in terms of correlation functions of the continuum operators. By analogy with the single-mode treatment, we assume that the only non-zero moments are those which contain an equal number of continuum creation and annihilation operators. The first normal ordered moment is

$$\langle \hat{b}^\dagger(\omega_1)\hat{b}(\omega_2)\rangle = \bar{n}(\omega_1)\delta(\omega_1-\omega_2), \tag{3.5.20}$$

where

$$\bar{n}(\omega) = [\exp(\beta\hbar\omega)-1]^{-1}. \tag{3.5.21}$$

Note that the delta function in (3.5.20) implies that the expectation value of $\hat{b}^\dagger(\omega)\hat{b}(\omega)$ is not defined and neither, therefore, is the expectation value of the continuum number operator (3.2.33). The second normal ordered moment is

$$\begin{aligned}&\left\langle \hat{b}^\dagger(\omega_1)\hat{b}^\dagger(\omega_2)\hat{b}(\omega_3)\hat{b}(\omega_4)\right\rangle\\ &\quad = \bar{n}(\omega_1)\bar{n}(\omega_2)[\delta(\omega_1-\omega_3)\,\delta(\omega_2-\omega_4)+\delta(\omega_1-\omega_4)\,\delta(\omega_2-\omega_3)].\end{aligned} \tag{3.5.22}$$

More generally, all the non-zero normal ordered moments have the form

$$\begin{aligned}&\left\langle \hat{b}^\dagger(\omega_1)\hat{b}^\dagger(\omega_2)\ldots\hat{b}^\dagger(\omega_l)\hat{b}(\omega_{l+1})\ldots\hat{b}(\omega_{2l})\right\rangle\\ &\qquad = \bar{n}(\omega_1)\bar{n}(\omega_2)\ldots\bar{n}(\omega_l)\sum_{j=1}^{l!}\prod_{i=1}^{l}\delta(\omega_i-\omega_{\{i,j\}}),\end{aligned} \tag{3.5.23}$$

where $\{i,j\}$ is the ith element of the jth permutation of the integers $\{l+1, l+2,\ldots,2l\}$. We justify this by considering the properties of the boson operators $\hat{b}_f$ and $\hat{b}_f^\dagger$ introduced in the preceding section (see (3.4.7)). The expectation value of any operator formed as a product of powers of $\hat{b}_f$ and $\hat{b}_f^\dagger$ is zero unless the number of occurrences of $\hat{b}_f$ is the same as that of $\hat{b}_f^\dagger$. In particular

$$\begin{aligned}\langle \hat{b}_f^\dagger\hat{b}_f\rangle &= \int d\omega_1\, f(\omega_1)\int d\omega_2\, f^*(\omega_2)\bar{n}(\omega_1)\,\delta(\omega_1-\omega_2)\\ &= \int d\omega_1\, |f(\omega_1)|^2\bar{n}(\omega_1)\end{aligned} \tag{3.5.24}$$

and

$$\begin{aligned}\langle \hat{b}_f^{\dagger 2}\hat{b}_f^2\rangle &= \int d\omega_1\, f(\omega_1)\int d\omega_2\, f(\omega_2)\int d\omega_3\, f^*(\omega_3)\int d\omega_4\, f^*(\omega_4)\\ &\quad\times\bar{n}(\omega_1)\bar{n}(\omega_2)[\delta(\omega_1-\omega_3)\,\delta(\omega_2-\omega_4)+\delta(\omega_1-\omega_4)\,\delta(\omega_2-\omega_3)]\\ &= 2\int d\omega_1\, |f(\omega_1)|^2\bar{n}(\omega_1)\int d\omega_2\, |f(\omega_2)|^2\,\bar{n}(\omega_2)\\ &= 2\langle \hat{b}_f^\dagger\hat{b}_f\rangle^2.\end{aligned} \tag{3.5.25}$$

In general, the non-zero moments of normal ordered operators are

$$\left\langle \hat{b}_f^{\dagger l}\hat{b}_f^l\right\rangle = l!\,\langle \hat{b}_f^\dagger\hat{b}_f\rangle^l, \tag{3.5.26}$$

which is precisely the form obtained in (3.5.19) for the single-mode case. Hence the photon statistics associated with quanta annihilated by the operator $\hat{b}_f$ are of the Bose–Einstein form. If $|f(\omega)|^2$ is sharply peaked compared with $\bar{n}(\omega)$ at $\omega = \omega_0$, say, then (3.5.24) reduces to

$$\langle \hat{b}_f^\dagger \hat{b}_f \rangle \simeq \bar{n}(\omega_0), \tag{3.5.27}$$

which corresponds to the result for a single mode at frequency ω_0. This justifies the assertion that the factors $\bar{n}(\omega)$ in the correlation functions (3.5.20) and (3.5.23) are given by (3.5.22).

3.6 Coherent states

Of all states of the radiation field, the coherent states are the most important and arise frequently in quantum optics. Not only can they be an accurate representation of the field produced by a stabilized laser operating well above threshold, but also many of the techniques for studying the properties of fields rely on the mathematics of the coherent states. The single-mode coherent states are generated by the action on the vacuum state $|0\rangle$ of the Glauber displacement operator

$$\hat{D}(\alpha) = \exp(\alpha \hat{a}^\dagger - \alpha^* \hat{a}), \tag{3.6.1}$$

where $\alpha = |\alpha| \exp(i\theta)$ is any complex number with real and imaginary parts α_r and α_i, respectively. This operator is unitary since

$$\hat{D}^\dagger(\alpha) = \exp(-\alpha \hat{a}^\dagger + \alpha^* \hat{a}) = \hat{D}(-\alpha), \tag{3.6.2}$$

which is the inverse of $\hat{D}(\alpha)$. This follows from the fact that $\exp(\hat{A})\exp(-\hat{A}) = 1$ for all operators $\hat{A}$, as shown in Section 3.3. It is useful to be able to write $\hat{D}(\alpha)$ in normal and antinormal ordered forms. These are found using the ordering theorem (3.3.29) and explicit relationships between $\hat{D}(\alpha)$ and its normal and antinormal ordered forms are given by (3.3.31) and (3.3.32). The single-mode coherent state $|\alpha\rangle$ is given by

$$|\alpha\rangle = \hat{D}(\alpha)|0\rangle \tag{3.6.3}$$

and it then follows from (3.3.31) that

$$\begin{aligned} |\alpha\rangle &= \exp(-|\alpha|^2/2) \exp(\alpha \hat{a}^\dagger) \exp(-\alpha^* \hat{a}) |0\rangle \\ &= \exp(-|\alpha|^2/2) \exp(\alpha \hat{a}^\dagger) |0\rangle \\ &= \exp(-|\alpha|^2/2) \sum_{n=0}^{\infty} \frac{\alpha^n}{\sqrt{(n!)}} |n\rangle, \end{aligned} \tag{3.6.4}$$

where we have used the definition (3.3.4) of the exponential function of an operator together with (3.2.16) and (3.2.17). The state $|\alpha\rangle$ is normalized since $\langle\alpha|\alpha\rangle = \langle 0|\hat{D}(-\alpha)\hat{D}(\alpha)|0\rangle = 1$, using the unitarity of $\hat{D}(\alpha)$ together with the fact that $\langle 0|0\rangle = 1$.

One of the reasons for the significance of the coherent states is that $|\alpha\rangle$ is a *right eigenstate* of $\hat{a}$ with eigenvalue α. This is easily seen using (3.6.4) since

$$\begin{aligned}\hat{a}|\alpha\rangle &= \exp(-|\alpha|^2/2)\sum_{n=0}^{\infty}\frac{\alpha^n}{\sqrt{(n!)}}\hat{a}|n\rangle \\ &= \exp(-|\alpha|^2/2)\sum_{n=1}^{\infty}\frac{\alpha^n}{\sqrt{(n!)}}n^{1/2}|n-1\rangle \\ &= \exp(-|\alpha|^2/2)\sum_{m=0}^{\infty}\frac{\alpha^{m+1}}{\sqrt{(m!)}}|m\rangle = \alpha|\alpha\rangle.\end{aligned} \tag{3.6.5}$$

It is straightforward to see that $|\alpha\rangle$ is *not* a right eigenstate of $\hat{a}^\dagger$. This is not surprising since $\hat{a}$ and $\hat{a}^\dagger$ do not commute. However, it follows from (3.6.5) that the coherent state $|\alpha\rangle$ is a *left eigenstate* of $\hat{a}^\dagger$ with eigenvalue α^*, so that

$$\langle\alpha|\hat{a}^\dagger = \alpha^*\langle\alpha|, \tag{3.6.6}$$

but that it is not a left eigenstate of $\hat{a}$. From the above, we conclude that any expectation value in a coherent state $|\alpha\rangle$ of a normal ordered function of $\hat{a}$ and $\hat{a}^\dagger$ is simply found by replacing $\hat{a}$ by α and $\hat{a}^\dagger$ by α^* in the function so that, for example, $\langle\alpha|\hat{a}^{\dagger l}\hat{a}^m|\alpha\rangle = \alpha^{*l}\alpha^m$.

The photon number probability distribution $P(n)$ for the coherent state $|\alpha\rangle$ is, from (3.6.4),

$$P(n) = |\langle n|\alpha\rangle|^2 = \exp(-|\alpha|^2)|\alpha|^{2n}/n! \tag{3.6.7}$$

which is the Poisson distribution. The expectation value of the photon number operator $\hat{n}$ is

$$\begin{aligned}\langle\hat{n}\rangle &= \sum_{n=0}^{\infty} nP(n) \\ &= \exp(-|\alpha|^2)\sum_{n=0}^{\infty} n|\alpha|^{2n}/n! = |\alpha|^2.\end{aligned} \tag{3.6.8}$$

Alternatively, we can use the fact that $|\alpha\rangle$ is a right eigenstate of $\hat{a}$ to write $\langle\hat{n}\rangle = \langle\alpha|\hat{a}^\dagger\hat{a}|\alpha\rangle = \alpha^*\alpha\langle\alpha|\alpha\rangle = |\alpha|^2$. The variance of $\hat{n}$ is

$$\begin{aligned}\Delta n^2 &= \langle\alpha|\hat{n}^2|\alpha\rangle - (\langle\alpha|\hat{n}|\alpha\rangle)^2 \\ &= \langle\alpha|\hat{n}^2|\alpha\rangle - |\alpha|^4.\end{aligned} \tag{3.6.9}$$

The expectation value of $\hat{n}^2$ is easily calculated using (3.6.5) and (3.6.6) since

$$\begin{aligned}\langle\alpha|\,\hat{n}^2\,|\alpha\rangle &= \langle\alpha|\,\hat{a}^\dagger\hat{a}\hat{a}^\dagger\hat{a}\,|\alpha\rangle\\ &= \langle\alpha|\,\alpha^*\hat{a}\hat{a}^\dagger\alpha\,|\alpha\rangle\\ &= |\alpha|^2\,\langle\alpha|(1+\hat{a}^\dagger\hat{a})\,|\alpha\rangle\\ &= |\alpha|^2(1+|\alpha|^2).\end{aligned} \tag{3.6.10}$$

Hence, from (3.6.9), the variance of $\hat{n}$ is given by

$$\Delta n^2 = |\alpha|^2, \tag{3.6.11}$$

so that the variance is equal to the mean as it should be for a Poisson distribution. It is often useful to calculate the normal ordered variance which for an Hermitian operator $\hat{A}$ is defined to be

$$:\Delta A^2: = \langle:\hat{A}^2:\rangle - \langle:\hat{A}:\rangle^2. \tag{3.6.12}$$

For the number operator $\hat{n}$, the normal ordered variance in a coherent state is

$$:\Delta n^2: = \langle\alpha|\,\hat{a}^{\dagger 2}\hat{a}^2\,|\alpha\rangle - (\langle\alpha|\,\hat{a}^\dagger\hat{a}\,|\alpha\rangle)^2 = 0. \tag{3.6.13}$$

This implies that the expectation value of $:\hat{n}^2:$ in any coherent state is equal to the square of the mean photon number $|\alpha|^2$. More generally, the expectation value of $:\hat{n}^l:$ is $|\alpha|^{2l}$. States for which the normal ordered variance (3.6.13) is positive are termed super-Poissonian while those for which it is negative are termed sub-Poissonian. The former include the thermal states for which $:\Delta n^2: = \bar{n}^2$, while the latter include the number states $|n\rangle$, other than $|0\rangle$, for which $:\Delta n^2: = -n$.

The expectation values in the coherent state $|\alpha\rangle$ of the annihilation operator $\hat{a}$ and the creation operator $\hat{a}^\dagger$ are α and α^*, respectively. This follows directly from the eigenvalue equations (3.6.5) and (3.6.6) together with $\langle\alpha\,|\,\alpha\rangle = 1$. Hence the quadrature operator $\hat{x}_\lambda$ given by (3.2.20) has the expectation value

$$\langle\alpha|\,\hat{x}_\lambda\,|\alpha\rangle = \frac{1}{\sqrt{2}}[\alpha\exp(-\mathrm{i}\lambda) + \alpha^*\exp(\mathrm{i}\lambda)], \tag{3.6.14}$$

which, being the expectation value of an Hermitian operator, is real. The expectation value of $\hat{x}_\lambda^2$ is

$$\begin{aligned}\langle\alpha|\,\hat{x}_\lambda^2\,|\alpha\rangle &= \tfrac{1}{2}\langle\alpha|\big[\hat{a}^2\exp(-2\mathrm{i}\lambda) + \hat{a}^{\dagger 2}\exp(2\mathrm{i}\lambda) + 2\hat{a}^\dagger\hat{a} + 1\big]|\alpha\rangle\\ &= \tfrac{1}{2}\big[\alpha^2\exp(-2\mathrm{i}\lambda) + \alpha^{*2}\exp(2\mathrm{i}\lambda) + 2\,|\alpha|^2 + 1\big].\end{aligned} \tag{3.6.15}$$

Hence for all the coherent states, the variance of $\hat{x}_\lambda$ is

$$\Delta x_\lambda^2 = \tfrac{1}{2}. \tag{3.6.16}$$

For all states, the product of the variances of the two quadrature components $\hat{x}_\lambda$ and $\hat{x}_{\lambda+\pi/2}$ is bounded by the Heisenberg uncertainty principle which is expressed as the inequality

$$\Delta x_\lambda^2 \, \Delta x_{\lambda+\pi/2}^2 \geqslant \tfrac{1}{4}\left|\left\langle\left[\hat{x}_\lambda, \hat{x}_{\lambda+\pi/2}\right]\right\rangle\right|^2 = \tfrac{1}{4}, \tag{3.6.17}$$

where we have used the commutator (3.2.23). It follows immediately that the coherent states are minimum uncertainty states having the same uncertainty associated with each quadrature. As with the number operator, it is often useful to calculate the normal ordered variance (3.6.12) of $\hat{x}_\lambda$, which for $|\alpha\rangle$ is

$$:\Delta x_\lambda^2: = \tfrac{1}{2}\langle\alpha|\left[\hat{a}^2\exp(-2\mathrm{i}\lambda) + \hat{a}^{\dagger 2}\exp(2\mathrm{i}\lambda) + 2\hat{a}^\dagger\hat{a}\right]|\alpha\rangle - (\langle\alpha|\,\hat{x}_\lambda\,|\alpha\rangle)^2 = 0. \tag{3.6.18}$$

States for which the normal ordered variance in one of the quadratures is negative are termed squeezed states. We shall consider examples of these in Section 3.7.

Rather than using the number state expansion (3.6.4) of the coherent state $|\alpha\rangle$ to derive the eigenvalue equations (3.6.5) and (3.6.6), we can work directly from the definition (3.6.3). Consider the action of the annihilation operator $\hat{a}$ on the state $|\alpha\rangle$ given by

$$\hat{a}\,|\alpha\rangle = \hat{a}\hat{D}(\alpha)\,|0\rangle = \hat{D}(\alpha)\hat{D}(-\alpha)\hat{a}\hat{D}(\alpha)\,|0\rangle, \tag{3.6.19}$$

where we have used the fact that $\hat{D}(\alpha)\hat{D}(-\alpha) = \mathbf{1}$. A simple expression for the operator $\hat{D}(-\alpha)\hat{a}\hat{D}(\alpha)$ can be found using the operator theorem (3.3.9). Identifying $\hat{A}$ and $\hat{B}$ in (3.3.9) with $-\alpha\hat{a}^\dagger + \alpha^*\hat{a}$ and $\hat{a}$, respectively, we find

$$\begin{aligned}\hat{D}(-\alpha)\hat{a}\hat{D}(\alpha) &= \hat{a} + [-\alpha\hat{a}^\dagger + \alpha^*\hat{a}, \hat{a}] + \cdots \\ &= \hat{a} + \alpha.\end{aligned} \tag{3.6.20}$$

Hence (3.6.19) becomes

$$\hat{a}\,|\alpha\rangle = \hat{D}(\alpha)(\hat{a} + \alpha)\,|0\rangle = \alpha\hat{D}(\alpha)\,|0\rangle = \alpha\,|\alpha\rangle. \tag{3.6.21}$$

Similarly, we can prove that $\langle\alpha|$ is a left eigenstate of $\hat{a}^\dagger$ with eigenvalue α^* using the fact that $\langle\alpha| = \langle 0|\,\hat{D}^\dagger(\alpha) = \langle 0|\,\hat{D}(-\alpha)$ and the transformation

$$\hat{D}(-\alpha)\hat{a}^\dagger\hat{D}(\alpha) = \hat{a}^\dagger + \alpha^*, \tag{3.6.22}$$

which is the Hermitian conjugate of (3.6.20). The transformations (3.6.20) and (3.6.22) can be used to obtain expectation values of operator functions of $\hat{a}$ and $\hat{a}^\dagger$ in a coherent state. For example,

$$\begin{aligned}\langle\alpha|\,\hat{n}\,|\alpha\rangle &= \langle 0|\,\hat{D}(-\alpha)\hat{n}\hat{D}(\alpha)\,|0\rangle \\ &= \langle 0|\,\hat{D}(-\alpha)\hat{a}^\dagger\hat{D}(\alpha)\hat{D}(-\alpha)\hat{a}\hat{D}(\alpha)\,|0\rangle \\ &= \langle 0|(\hat{a}^\dagger + \alpha^*)(\hat{a} + \alpha)\,|0\rangle = |\alpha|^2,\end{aligned} \tag{3.6.23}$$

in agreement with the earlier result (3.6.8). More generally, the expectation value in a coherent state $|\alpha\rangle$ of any function of $\hat{a}$ and $\hat{a}^\dagger$ equals the expectation value of the same function of $\hat{a}+\alpha$ and $\hat{a}^\dagger+\alpha^*$ in the state $|0\rangle$. In this sense, we can picture a coherent state as a superposition of a classical field mode and a quantum field mode in its vacuum state.

In contrast to the number states, the coherent states are not mutually orthogonal. The overlap between two coherent states $|\alpha\rangle$ and $|\alpha'\rangle$ is

$$\begin{aligned}\langle\alpha|\alpha'\rangle &= \exp(-|\alpha|^2/2)\exp(-|\alpha'|^2/2)\sum_{n=0}^{\infty}\sum_{m=0}^{\infty}\frac{(\alpha')^n}{\surd(n!)}\frac{(\alpha^*)^m}{\surd(m!)}\,\delta_{nm}\\ &= \exp\left[-\tfrac{1}{2}(|\alpha|^2+|\alpha'|^2)\right]\sum_{n=0}^{\infty}\frac{(\alpha'\alpha^*)^n}{n!}\\ &= \exp\left[-\tfrac{1}{2}(|\alpha|^2+|\alpha'|^2-2\alpha'\alpha^*)\right],\end{aligned} \tag{3.6.24}$$

and hence

$$|\langle\alpha|\alpha'\rangle|^2 = \exp(-|\alpha-\alpha'|^2). \tag{3.6.25}$$

Despite the lack of orthogonality of the coherent states, they form an overcomplete set in that the identity can be resolved in terms of the coherent states as

$$\frac{1}{\pi}\int_{-\infty}^{\infty}\mathrm{d}^2\alpha\,|\alpha\rangle\langle\alpha| = 1, \tag{3.6.26}$$

where $\int_{-\infty}^{\infty}\mathrm{d}^2\alpha = \int_{-\infty}^{\infty}\int_{-\infty}^{\infty}\mathrm{d}\alpha_r\,\mathrm{d}\alpha_i$ denotes a double integral over the whole complex α-plane. We prove (3.6.26) by using the number state expansion (3.6.4) of $|\alpha\rangle$ so that writing $\alpha=|\alpha|\exp(\mathrm{i}\theta)$ and performing the double integral in polar coordinates gives

$$\begin{aligned}\frac{1}{\pi}\int_{-\infty}^{\infty}\mathrm{d}^2\alpha\,|\alpha\rangle\langle\alpha| &= \frac{1}{\pi}\int_0^{2\pi}\mathrm{d}\theta\int_0^{\infty}|\alpha|\,\mathrm{d}|\alpha|\exp(-|\alpha|^2)\\ &\quad\times\sum_{n=0}^{\infty}\sum_{m=0}^{\infty}\frac{|\alpha|^{n+m}}{\surd(n!m!)}\exp[\mathrm{i}(n-m)\theta]\,|n\rangle\langle m|\\ &= \sum_{n=0}^{\infty}|n\rangle\langle n|\int_0^{\infty}\mathrm{d}|\alpha|^2\exp(-|\alpha|^2)\,|\alpha|^{2n}/n!\\ &= \sum_{n=0}^{\infty}|n\rangle\langle n| = 1,\end{aligned} \tag{3.6.27}$$

from the resolution of the identity in terms of the number states. The overcompleteness of the coherent states allows us to express the trace of an operator $\hat{A}$

as an integral over the complex α-plane. The trace of $\hat{A}$ evaluated in the number state basis is

$$\begin{aligned}\mathrm{Tr}(\hat{A}) &= \sum_{n=0}^{\infty} \langle n|\hat{A}|n\rangle \\ &= \sum_{n=0}^{\infty} \langle n|\hat{A}\frac{1}{\pi}\int_{-\infty}^{\infty} \mathrm{d}^2\alpha\, |\alpha\rangle\langle\alpha|n\rangle, \end{aligned} \tag{3.6.28}$$

where we have used the resolution of the identity (3.6.27). Rearranging (3.6.28) gives the form of the trace of $\hat{A}$ evaluated in the coherent state basis since

$$\begin{aligned}\mathrm{Tr}(\hat{A}) &= \frac{1}{\pi}\int_{-\infty}^{\infty} \mathrm{d}^2\alpha \sum_{n=0}^{\infty} \langle\alpha|n\rangle\langle n|\hat{A}|\alpha\rangle \\ &= \frac{1}{\pi}\int_{-\infty}^{\infty} \mathrm{d}^2\alpha\, \langle\alpha|\hat{A}|\alpha\rangle. \end{aligned} \tag{3.6.29}$$

Although the coherent states form an overcomplete set of states, we can construct a complete orthonormal basis from any coherent state $|\alpha\rangle = \hat{D}(\alpha)|0\rangle$ and the states generated by the action of $\hat{D}(\alpha)$ on the remaining number states. The completeness and orthonormality of the displaced number states $\hat{D}(\alpha)|n\rangle$ follow from the unitarity of $\hat{D}(\alpha)$ and the orthonormality of the number states.

It is important to note that two displacement operators $\hat{D}(\alpha)$ and $\hat{D}(\alpha')$ with different arguments do not, in general, commute. However, the product $\hat{D}(\alpha)\hat{D}(\alpha')$ is simply $\hat{D}(\alpha+\alpha')$ multiplied by a phase factor. This phase factor can be found using (3.3.29) with $\hat{A} = \alpha\hat{a}^\dagger - \alpha^*\hat{a}$, $\hat{B} = \alpha'\hat{a}^\dagger - \alpha'^*\hat{a}$ and $\theta = 1$, so that

$$\begin{aligned}\hat{D}(\alpha)\hat{D}(\alpha') &= \hat{D}(\alpha+\alpha')\exp\left\{\tfrac{1}{2}[\alpha\hat{a}^\dagger - \alpha^*\hat{a},\, \alpha'\hat{a}^\dagger - \alpha'^*\hat{a}]\right\} \\ &= \hat{D}(\alpha+\alpha')\exp[\tfrac{1}{2}(\alpha\alpha'^* - \alpha^*\alpha')]. \end{aligned} \tag{3.6.30}$$

It follows that the action of the displacement operator on a coherent state produces another coherent state, apart from a phase factor, since

$$\hat{D}(\alpha)|\alpha'\rangle = \hat{D}(\alpha)\hat{D}(\alpha')|0\rangle = \exp[\tfrac{1}{2}(\alpha\alpha'^* - \alpha^*\alpha')]|\alpha+\alpha'\rangle. \tag{3.6.31}$$

The action of the operator formed by raising a complex number z to the power $\hat{n}$ on a coherent state $|\alpha\rangle$ produces the coherent state $|\alpha z\rangle$ multiplied by a factor. Using (3.3.8) together with the number state expansion (3.6.4) of $|\alpha\rangle$, we find

$$\begin{aligned}z^{\hat{n}}|\alpha\rangle &= \exp(-|\alpha|^2/2)\sum_{n=0}^{\infty}\frac{(\alpha z)^n}{\sqrt{(n!)}}|n\rangle \\ &= \exp\left[\tfrac{1}{2}|\alpha|^2(|z|^2-1)\right]|\alpha z\rangle. \end{aligned} \tag{3.6.32}$$

In particular, if $|z| = 1$ then writing $z = \exp(i\varphi)$ we obtain

$$\exp(i\varphi\hat{n})|\alpha\rangle = |\alpha\exp(i\varphi)\rangle. \tag{3.6.33}$$

In Section 4.5, we will need a special case of this result with $\varphi = \pi$, so that

$$(-1)^{\hat{n}}|\alpha\rangle = |-\alpha\rangle. \tag{3.6.34}$$

It is possible to construct operators as functions of the displacement operator $\hat{D}(\alpha)$. One example required in Section 4.4 is the operator $\hat{T}(p)$ given by

$$\hat{T}(p) = \frac{1}{\pi^2}\int_{-\infty}^{\infty} d^2\xi\, \hat{D}(\xi)\exp(p|\xi|^2/2), \tag{3.6.35}$$

where $p < 1$. The action of this operator on a coherent state $|\alpha\rangle$ can be found by using (3.6.30) so that

$$\begin{aligned}\hat{T}(p)|\alpha\rangle &= \frac{1}{\pi^2}\int_{-\infty}^{\infty} d^2\xi \exp(p|\xi|^2/2)\hat{D}(\xi)\hat{D}(\alpha)|0\rangle \\ &= \frac{1}{\pi^2}\int_{-\infty}^{\infty} d^2\xi \exp(p|\xi|^2/2)\exp[\tfrac{1}{2}(\xi\alpha^* - \xi^*\alpha)]|\alpha + \xi\rangle. \end{aligned}\tag{3.6.36}$$

Surprisingly, this is also a coherent state multiplied by a factor. We can show this by first normal ordering the operator $\hat{T}(p)$ using the ordering theorem (3.3.31). We have

$$\begin{aligned}\hat{T}(p) &= \frac{1}{\pi^2}\int_{-\infty}^{\infty} d^2\xi \exp(\xi\hat{a}^\dagger - \xi^*\hat{a})\exp(p|\xi|^2/2) \\ &= \frac{1}{\pi^2}\int_{-\infty}^{\infty} d^2\xi \exp(\xi\hat{a}^\dagger)\exp(-\xi^*\hat{a})\exp[(p-1)|\xi|^2/2]. \end{aligned}\tag{3.6.37}$$

Substituting the expression (3.3.4) for each of the exponential operators and performing the integral over the complex ξ-plane using polar coordinates with $\xi = |\xi|\exp(i\varphi)$ gives

$$\begin{aligned}\hat{T}(p) &= \frac{1}{\pi^2}\sum_{n=0}^{\infty}\sum_{m=0}^{\infty}\frac{(-1)^m}{n!\,m!}\hat{a}^{\dagger n}\hat{a}^m\int_0^{2\pi} d\varphi\int_0^{\infty}|\xi|d|\xi||\xi|^{n+m} \\ &\quad\times\exp[i\varphi(n-m)]\exp[(p-1)|\xi|^2/2] \\ &= \frac{1}{\pi}\sum_{n=0}^{\infty}\frac{(-1)^n}{n!}\hat{a}^{\dagger n}\hat{a}^n\left(\frac{2}{1-p}\right)^{n+1} \\ &= \frac{2}{\pi(1-p)}:\exp\left(-\frac{2}{1-p}\hat{a}^\dagger\hat{a}\right):. \end{aligned}\tag{3.6.38}$$

Alternatively, we can use (3.3.37) to write $\hat{T}(p)$ in the form

$$\hat{T}(p) = \frac{2}{\pi(1-p)}\left(\frac{p+1}{p-1}\right)^{\hat{n}}. \tag{3.6.39}$$

It follows from (3.6.38) that $\hat{T}(-1) = \pi^{-1}\,|0\rangle\langle 0|$, and from (3.6.32) and (3.6.39) that $\hat{T}(p)|\alpha\rangle$ is a coherent state multiplied by a factor depending on p and α so that, for example,

$$\hat{T}(0)|\alpha\rangle = \frac{2}{\pi}|-\alpha\rangle. \tag{3.6.40}$$

The quadrature representation $\psi_\alpha(x_\lambda)$ of the coherent state $|\alpha\rangle$ is defined to be the overlap between the quadrature eigenstate $|x_\lambda\rangle$ and $|\alpha\rangle$, that is

$$\psi_\alpha(x_\lambda) = \langle x_\lambda\,|\,\alpha\rangle. \tag{3.6.41}$$

To obtain $\psi_\alpha(x_\lambda)$ we first use the eigenvalue equation (3.6.5) to write

$$\langle x_\lambda|\,\hat{a}\,|\alpha\rangle = \alpha\psi_\alpha(x_\lambda). \tag{3.6.42}$$

From (3.2.20), the annihilation operator $\hat{a}$ can be expressed in terms of the quadrature operators $\hat{x}_\lambda$ and $\hat{x}_{\lambda+\pi/2}$ as

$$\hat{a} = \frac{\exp(\mathrm{i}\lambda)}{\sqrt{2}}\left(\hat{x}_\lambda + \mathrm{i}\hat{x}_{\lambda+\pi/2}\right). \tag{3.6.43}$$

This, together with the quadrature representation of $\hat{x}_\lambda$ and $\hat{x}_{\lambda+\pi/2}$ given in Appendix 4, enables us to rewrite (3.6.42) as the differential equation

$$\frac{\exp(\mathrm{i}\lambda)}{\sqrt{2}}\left(x_\lambda + \frac{\mathrm{d}}{\mathrm{d}x_\lambda}\right)\psi_\alpha(x_\lambda) = \alpha\psi_\alpha(x_\lambda). \tag{3.6.44}$$

The normalized solution to this equation, up to an x_λ-independent factor κ of unit modulus (see Appendix 4), is

$$\psi_\alpha(x_\lambda) = \kappa\pi^{-1/4}\exp\left[\mathrm{i}\langle\hat{x}_{\lambda+\pi/2}\rangle x_\lambda\right]\exp\left[-\tfrac{1}{2}(x_\lambda - \langle\hat{x}_\lambda\rangle)^2\right], \tag{3.6.45}$$

where the expectation values of $\hat{x}_\lambda$ and $\hat{x}_{\lambda+\pi/2}$ are given by (3.6.14). This has the form of a Gaussian in x_λ multiplied by an x_λ-dependent phase factor and will lead to a Gaussian probability distribution for x_λ given by $|\psi_\alpha(x_\lambda)|^2$. It is a characteristic of such Gaussian states that they minimize the uncertainty product (3.6.17).

Coherent states of more than one field mode are simply formed as products of single-mode coherent states. For two modes A and B with annihilation operators $\hat{a}$ and $\hat{b}$, the two-mode coherent state $|\alpha;\beta\rangle$ is generated by the action of two displacement operators $\hat{D}_A(\alpha)$ and $\hat{D}_B(\beta)$ on the two-mode vacuum state $|0_A;0_B\rangle$, so that

$$
\begin{aligned}
|\alpha;\beta\rangle &= \hat{D}_A(\alpha)\hat{D}_B(\beta)|0_A;0_B\rangle \\
&= \exp(\alpha\hat{a}^\dagger - \alpha^*\hat{a})\exp\left(\beta\hat{b}^\dagger - \beta^*\hat{b}\right)|0_A;0_B\rangle. \qquad (3.6.46)
\end{aligned}
$$

The properties of this state follow from those described above of the single-mode state. In particular, $|\alpha;\beta\rangle$ is a right eigenstate of both $\hat{a}$ and $\hat{b}$ with eigenvalues α and β, respectively, and therefore expectation values of normal ordered products of $\hat{a}$, $\hat{a}^\dagger$, $\hat{b}$, and $\hat{b}^\dagger$ can be found by replacing these operators by α, α^*, β, and β^*, respectively, in a given product. For example,

$$
\langle\alpha;\beta|\hat{a}^{\dagger k}\hat{b}^{\dagger l}\hat{b}^m\hat{a}^n|\alpha;\beta\rangle = \alpha^{*k}\beta^{*l}\beta^m\alpha^n. \qquad (3.6.47)
$$

The generalization of coherent states to a continuum of modes is achieved by acting on the vacuum state $|0\rangle$ of the continuum field with the displacement operator $\hat{D}[\alpha(\omega)]$ given in (3.3.57), producing

$$
|\{\alpha(\omega)\}\rangle = \hat{D}[\alpha(\omega)]|0\rangle = \exp\left(\int d\omega[\alpha(\omega)\hat{b}^\dagger(\omega) - \alpha^*(\omega)\hat{b}(\omega)]\right)|0\rangle. \qquad (3.6.48)
$$

This state is normalized and is a right eigenstate of the continuum annihilation operator $\hat{b}(\omega')$ with eigenvalue $\alpha(\omega')$. In order to show this, we need the continuum analogue of (3.6.20) given by

$$
\begin{aligned}
&\hat{D}[-\alpha(\omega)]\hat{b}(\omega')\hat{D}[\alpha(\omega)] \\
&\quad = \hat{b}(\omega') + \left[\int d\omega[-\alpha(\omega)\hat{b}^\dagger(\omega) + \alpha^*(\omega)\hat{b}(\omega)], \hat{b}(\omega')\right] + \cdots \\
&\quad = \hat{b}(\omega') + \alpha(\omega'), \qquad (3.6.49)
\end{aligned}
$$

where we have used the continuum operator commutation relation (3.2.31), and (3.3.9). The action of $\hat{b}(\omega')$ on $|\{\alpha(\omega)\}\rangle$ gives

$$
\begin{aligned}
\hat{b}(\omega')|\{\alpha(\omega)\}\rangle &= \hat{D}[\alpha(\omega)]\hat{D}[-\alpha(\omega)]\hat{b}(\omega')\hat{D}[\alpha(\omega)]|0\rangle \\
&= \hat{D}[\alpha(\omega)][\hat{b}(\omega') + \alpha(\omega')]|0\rangle \\
&= \alpha(\omega')|\{\alpha(\omega)\}\rangle, \qquad (3.6.50)
\end{aligned}
$$

using (3.6.49), (3.4.6), and the fact that $\hat{D}[-\alpha(\omega)]$ is the inverse of $\hat{D}[\alpha(\omega)]$. It follows from (3.6.50) that the continuum coherent state is a left eigenstate of $\hat{b}^\dagger(\omega')$ with eigenvalue $\alpha^*(\omega')$, so that

$$
\langle\{\alpha(\omega)\}|\hat{b}^\dagger(\omega') = \alpha^*(\omega')\langle\{\alpha(\omega)\}|. \qquad (3.6.51)
$$

The expectation value in the coherent state $|\{\alpha(\omega)\}\rangle$ of a normal ordered function of continuum annihilation and creation operators is simply found by replacing $\hat{b}(\omega')$ by $\alpha(\omega')$ and $\hat{b}^\dagger(\omega')$ by $\alpha^*(\omega')$ in the function.

The expectation value $\langle \hat{N}\rangle$ of the continuum number operator (3.2.33) in the coherent state $|\{\alpha(\omega)\}\rangle$ follows from the eigenvalue equation (3.6.50) since

$$\begin{aligned}\langle \hat{N}\rangle &= \langle\{\alpha(\omega)\}|\hat{N}|\{\alpha(\omega)\}\rangle \\ &= \int \mathrm{d}\omega' \langle\{\alpha(\omega)\}|\hat{b}^\dagger(\omega')\hat{b}(\omega')|\{\alpha(\omega)\}\rangle \\ &= \int \mathrm{d}\omega' |\alpha(\omega')|^2. \end{aligned} \tag{3.6.52}$$

The normal ordered form of $\hat{N}^l$ has all annihilation operators positioned to the right of every creation operator so that

$$:\hat{N}^l: = \int \mathrm{d}\omega_1 \int \mathrm{d}\omega_2 \ldots \int \mathrm{d}\omega_l \hat{b}^\dagger(\omega_1)\hat{b}^\dagger(\omega_2)\ldots\hat{b}^\dagger(\omega_l)\hat{b}(\omega_l)\ldots\hat{b}(\omega_2)\hat{b}(\omega_1). \tag{3.6.53}$$

The expectation value of this operator in the state $|\{\alpha(\omega)\}\rangle$ is found to be

$$\begin{aligned}\langle :\hat{N}^l:\rangle &= \langle\{\alpha(\omega)\}|:\hat{N}^l:|\{\alpha(\omega)\}\rangle \\ &= \left(\int \mathrm{d}\omega' |\alpha(\omega')|^2\right)^l = \langle \hat{N}\rangle^l. \end{aligned} \tag{3.6.54}$$

It follows that the continuum coherent states are Poissonian in that $:\Delta N^2: = 0$.

As with the single-mode coherent states, the continuum coherent states are not mutually orthogonal. The overlap between two states $|\{\alpha(\omega)\}\rangle$ and $|\{\alpha'(\omega)\}\rangle$ is

$$\begin{aligned}\langle\{\alpha(\omega)\}|\{\alpha'(\omega)\}\rangle &= \langle 0|\hat{D}[-\alpha(\omega)]\hat{D}[\alpha'(\omega)]|0\rangle \\ &= \langle 0|\hat{D}[\alpha'(\omega)-\alpha(\omega)]|0\rangle \\ &\quad\times \exp\left(\tfrac{1}{2}\int \mathrm{d}\omega[-\alpha(\omega)\alpha'^*(\omega)+\alpha^*(\omega)\alpha'(\omega)]\right) \\ &= \exp\left(-\tfrac{1}{2}\int \mathrm{d}\omega\left[|\alpha(\omega)|^2+|\alpha'(\omega)|^2-2\alpha'(\omega)\alpha^*(\omega)\right]\right), \end{aligned} \tag{3.6.55}$$

where we have used the ordering theorems (A5.5) and (A5.4).

In Section 3.4, we introduced boson annihilation and creation operators $\hat{b}_f$ and $\hat{b}_f^\dagger$ defined by (3.4.7) and its Hermitian conjugate. The action of these operators on a continuum field state is to remove or to add one quantum. The continuum coherent state $|\{\alpha(\omega)\}\rangle$ is a right eigenstate of

$$\hat{b}_f = \int \mathrm{d}\omega' f^*(\omega')\hat{b}(\omega') \tag{3.6.56}$$

with eigenvalue

$$\beta_f = \int \mathrm{d}\omega' f^*(\omega')\alpha(\omega'). \tag{3.6.57}$$

This follows directly from the eigenvalue equation (3.6.50). The properties of these operators are similar to those of $\hat{a}$ and $\hat{a}^\dagger$ with the single-mode coherent state. In particular, the expectation value of the normal ordered product $\hat{b}_f^{\dagger l}\hat{b}_f^m$ is

$$\langle\{\alpha(\omega)\}|\hat{b}_f^{\dagger l}\hat{b}_f^m|\{\alpha(\omega)\}\rangle = \beta_f^{*l}\beta_f^m. \tag{3.6.58}$$

For $l = m$,

$$\langle\hat{b}_f^{\dagger l}\hat{b}_f^l\rangle = \langle\hat{b}_f^\dagger\hat{b}_f\rangle^l, \tag{3.6.59}$$

showing that the photon statistics associated with quanta annihilated by the operator $\hat{b}_f$ are Poissonian.

3.7 Squeezed states

Squeezed states arise in simple quantum models of a number of non-linear optical processes including optical parametric oscillation and four-wave mixing. These states are characterized by the property that the variance of the quadrature operator $\hat{x}_\lambda$ is less than the value 1/2 associated with the vacuum and the coherent states (see (3.6.16)), for a range of values of λ. The Heisenberg uncertainty relation (3.6.17) then implies that the variance of the quadrature operator $\hat{x}_{\lambda+\pi/2}$ exceeds 1/2. It follows that the uncertainty in a quadrature depends on its phase λ. The simplest single-mode squeezed states are generated by the action on the vacuum state $|0\rangle$ of the squeezing operator

$$\hat{S}(\zeta) = \exp\left(-\frac{\zeta}{2}\hat{a}^{\dagger 2} + \frac{\zeta^*}{2}\hat{a}^2\right), \tag{3.7.1}$$

where $\zeta = r\exp(\mathrm{i}\varphi)$ is any complex number with modulus r and argument φ. The squeezing operator is similar in form to $\hat{D}(\alpha)$ given in (3.6.1) with $\hat{a}$ replaced by $\hat{a}^2$ and for this reason squeezed states are sometimes referred to as two-photon coherent states. The operator $\hat{S}(\zeta)$ is unitary since

$$\hat{S}^\dagger(\zeta) = \exp\left(\frac{\zeta}{2}\hat{a}^{\dagger 2} - \frac{\zeta^*}{2}\hat{a}^2\right) = \hat{S}(-\zeta), \tag{3.7.2}$$

which is the inverse of $\hat{S}(\zeta)$. It is useful to write $\hat{S}(\zeta)$ in an ordered form which, using the operator ordering theorem (A5.18), is

$$\hat{S}(\zeta)=\exp\left[-\tfrac{1}{2}\hat{a}^{\dagger 2}\exp(\mathrm{i}\varphi)\tanh r\right]\exp\left[-\tfrac{1}{2}(\hat{a}^{\dagger}\hat{a}+\hat{a}\hat{a}^{\dagger})\ln(\cosh r)\right]$$
$$\times\exp\left[\tfrac{1}{2}\hat{a}^{2}\exp(-\mathrm{i}\varphi)\tanh r\right]. \qquad (3.7.3)$$

The single-mode squeezed state $|\zeta\rangle$ is given by

$$|\zeta\rangle=\hat{S}(\zeta)|0\rangle \qquad (3.7.4)$$

and it then follows from (3.7.3) that

$$|\zeta\rangle=\surd(\operatorname{sech} r)\sum_{n=0}^{\infty}\frac{\surd[(2n)!]}{n!}\left[-\tfrac{1}{2}\exp(\mathrm{i}\varphi)\tanh r\right]^{n}|2n\rangle, \qquad (3.7.5)$$

where we have used the definition (3.3.4) of the exponential function of an operator, together with (3.2.16) and (3.2.17) and the fact that $(\hat{a}^{\dagger}\hat{a}+\hat{a}\hat{a}^{\dagger})|0\rangle=|0\rangle$. Note that $|\zeta\rangle$ is a superposition only of *even* photon number states. The squeezed state $|\zeta\rangle$ is normalized since $\langle\zeta|\zeta\rangle=\langle 0|\hat{S}(-\zeta)\hat{S}(\zeta)|0\rangle=1$, using the unitarity of $\hat{S}(\zeta)$ together with the fact that $\langle 0|0\rangle=1$.

The squeezed state $|\zeta\rangle$ is a right eigenstate of a ζ-dependent linear combination of $\hat{a}$ and $\hat{a}^{\dagger}$, with eigenvalue 0. Using the number state expansion (3.7.5) together with the properties of $\hat{a}$ and $\hat{a}^{\dagger}$, it is straightforward to show that

$$[\hat{a}\cosh r+\hat{a}^{\dagger}\exp(\mathrm{i}\varphi)\sinh r]|\zeta\rangle=0. \qquad (3.7.6)$$

Similarly, $\langle\zeta|$ is the zero-eigenvalue left eigenstate of $\hat{a}^{\dagger}\cosh r+\hat{a}\exp(-\mathrm{i}\varphi)\sinh r$. For these reasons, $|\zeta\rangle$ is termed the squeezed vacuum. These properties also follow from the form of the operators $\hat{S}(\zeta)\hat{a}\hat{S}(-\zeta)$ and $\hat{S}(\zeta)\hat{a}^{\dagger}\hat{S}(-\zeta)$. Using the form of the squeezing operator given in (3.7.1) and the theorem (3.3.9), we have

$$\hat{S}(\zeta)\hat{a}\hat{S}(-\zeta)=\hat{a}+\zeta\hat{a}^{\dagger}+\frac{1}{2!}|\zeta|^{2}\hat{a}+\frac{1}{3!}\zeta|\zeta|^{2}\hat{a}^{\dagger}+\cdots$$
$$=\hat{a}\cosh r+\hat{a}^{\dagger}\exp(\mathrm{i}\varphi)\sinh r. \qquad (3.7.7)$$

The transformation of $\hat{a}^{\dagger}$ is the Hermitian conjugate of (3.7.7) given by

$$\hat{S}(\zeta)\hat{a}^{\dagger}\hat{S}(-\zeta)=\hat{a}^{\dagger}\cosh r+\hat{a}\exp(-\mathrm{i}\varphi)\sinh r. \qquad (3.7.8)$$

The eigenvalue equation (3.7.6) can then be derived from (3.7.7) since

$$(\hat{a}\cosh r+\hat{a}^{\dagger}\exp(\mathrm{i}\varphi)\sinh r)|\zeta\rangle=\hat{S}(\zeta)\hat{a}\hat{S}(-\zeta)\hat{S}(\zeta)|0\rangle$$
$$=\hat{S}(\zeta)\hat{a}|0\rangle=0. \qquad (3.7.9)$$

In this section, we will also need the inverses of the transformations (3.7.7) and (3.7.8) obtained by changing the sign of ζ and hence of r. We find

$$\hat{S}(-\zeta)\hat{a}\hat{S}(\zeta) = \hat{a}\cosh r - \hat{a}^\dagger \exp(\mathrm{i}\varphi)\sinh r, \tag{3.7.10}$$

$$\hat{S}(-\zeta)\hat{a}^\dagger\hat{S}(\zeta) = \hat{a}^\dagger\cosh r - \hat{a}\exp(-\mathrm{i}\varphi)\sinh r. \tag{3.7.11}$$

The photon number probability distribution $P(n)$ for the squeezed state is, from (3.7.5),

$$P(2n) = \operatorname{sech} r \frac{(2n)!}{(n!)^2 2^{2n}}(\tanh r)^{2n}, \tag{3.7.12}$$

$$P(2n+1) = 0. \tag{3.7.13}$$

These probabilities sum to unity, as required, since the n-dependent factors in (3.7.12) are the terms of the binomial expansion of $(1-\tanh^2 r)^{-1/2} = \cosh r$. The expectation value of the number operator $\hat{n}$ can be found directly from the above photon number probability distribution since

$$\begin{aligned}
\bar{n} = \langle \hat{n} \rangle &= \operatorname{sech} r \sum_{n=0}^{\infty} \frac{(2n)!}{(n!)^2 2^{2n}}(\tanh r)^{2n} 2n \\
&= \operatorname{sech} r \cdot \tanh r \frac{\mathrm{d}}{\mathrm{d}\tanh r} \sum_{n=0}^{\infty} \frac{(2n)!}{(n!)^2 2^{2n}}(\tanh r)^{2n} \\
&= \operatorname{sech} r \cdot \tanh r \frac{\mathrm{d}}{\mathrm{d}\tanh r}(1-\tanh^2 r)^{-1/2} \\
&= \sinh^2 r.
\end{aligned} \tag{3.7.14}$$

Alternatively, we can use the transformations of (3.7.10) and (3.7.11) to rewrite $\langle \zeta | \hat{n} | \zeta \rangle$ as a vacuum expectation value since

$$\begin{aligned}
\bar{n} = \langle \zeta | \hat{n} | \zeta \rangle &= \langle 0 | \hat{S}(-\zeta)\hat{a}^\dagger\hat{a}\hat{S}(\zeta) | 0 \rangle \\
&= \langle 0 | \hat{S}(-\zeta)\hat{a}^\dagger\hat{S}(\zeta)\hat{S}(-\zeta)\hat{a}\hat{S}(\zeta) | 0 \rangle \\
&= \langle 0 | [\hat{a}^\dagger\cosh r - \hat{a}\exp(-\mathrm{i}\varphi)\sinh r][\hat{a}\cosh r - \hat{a}^\dagger\exp(\mathrm{i}\varphi)\sinh r] | 0 \rangle \\
&= \sinh^2 r.
\end{aligned} \tag{3.7.15}$$

This procedure is the same as that used for the coherent state $|\alpha\rangle$ and the operator $\hat{D}(\alpha)$ in (3.6.23). It can be used to calculate the expectation value in the squeezed state $|\zeta\rangle$ of any function of $\hat{a}$ and $\hat{a}^\dagger$ since this expectation value equals the vacuum expectation value of the same function with $\hat{a}$ replaced by

(3.7.10) and $\hat{a}^\dagger$ replaced by (3.7.11). In particular the expectation value of $\hat{n}^2$ is

$$\begin{aligned}\langle\zeta|\hat{n}^2|\zeta\rangle &= \langle\zeta|\hat{a}^\dagger\hat{a}\hat{a}^\dagger\hat{a}|\zeta\rangle \\ &= \langle 0|[\hat{a}^\dagger\cosh r - \hat{a}\exp(-\mathrm{i}\varphi)\sinh r][\hat{a}\cosh r - \hat{a}^\dagger\exp(\mathrm{i}\varphi)\sinh r] \\ &\quad\times[\hat{a}^\dagger\cosh r - \hat{a}\exp(-\mathrm{i}\varphi)\sinh r][\hat{a}\cosh r - \hat{a}^\dagger\exp(\mathrm{i}\varphi)\sinh r]|0\rangle \\ &= \sinh^2 r(2\cosh^2 r + \sinh^2 r) \\ &= \bar{n}(3\bar{n}+2), \end{aligned} \tag{3.7.16}$$

so that the variance in the photon number is $\Delta n^2 = 2\bar{n}(\bar{n}+1)$. This, being twice that for a thermal state of the same $\bar{n}$, represents a large uncertainty.

The expectation values in the squeezed state $|\zeta\rangle$ of $\hat{a}$ and $\hat{a}^\dagger$ are zero since, from (3.7.5), $\langle\zeta|$ is a superposition of even photon number states whereas $\hat{a}|\zeta\rangle$ contains only odd photon number states. Hence the quadrature operator $\hat{x}_\lambda$ given by (3.2.20) has the expectation value

$$\langle\zeta|\hat{x}_\lambda|\zeta\rangle = 0. \tag{3.7.17}$$

We can calculate the expectation value of $\hat{x}_\lambda^2$ using the transformation method described above together with (3.7.10) and (3.7.11). We have

$$\begin{aligned}\langle\zeta|\hat{x}_\lambda^2|\zeta\rangle &= \tfrac{1}{2}\langle 0|\hat{S}(-\zeta)[\hat{a}\exp(-\mathrm{i}\lambda) + \hat{a}^\dagger\exp(\mathrm{i}\lambda)]^2\hat{S}(\zeta)|0\rangle \\ &= \tfrac{1}{2}\langle 0|\{[\hat{a}\cosh r - \hat{a}^\dagger\exp(\mathrm{i}\varphi)\sinh r]\exp(-\mathrm{i}\lambda) \\ &\quad + [\hat{a}^\dagger\cosh r - \hat{a}\exp(-\mathrm{i}\varphi)\sinh r]\exp(\mathrm{i}\lambda)\}^2|0\rangle \\ &= \tfrac{1}{2}[\exp(2r)\sin^2(\lambda-\varphi/2) + \exp(-2r)\cos^2(\lambda-\varphi/2)], \end{aligned} \tag{3.7.18}$$

which is also equal to the variance Δx_λ^2 since $\langle\hat{x}_\lambda\rangle = 0$. As φ is varied, the variance ranges from a minimum of $\exp(-2r)/2$ to a maximum of $\exp(2r)/2$ and will be less than 1/2 if $\sin^2(\lambda-\varphi/2) < [\exp(2r)+1]^{-1}$. If $\Delta x_\lambda^2 < 1/2$, then we say that the quadrature $\hat{x}_\lambda$ is squeezed. For all states, the product of the variances of the two quadrature components $\hat{x}_\lambda$ and $\hat{x}_{\lambda+\pi/2}$ is bounded by the Heisenberg uncertainty principle (3.6.17). For the squeezed state $|\zeta\rangle$, this product is given by

$$\begin{aligned}\Delta x_\lambda^2\,\Delta x_{\lambda+\pi/2}^2 &= \tfrac{1}{4}[\sin^4(\lambda-\varphi/2) + \cos^4(\lambda-\varphi/2) \\ &\quad + 2\sin^2(\lambda-\varphi/2)\cos^2(\lambda-\varphi/2)\cosh 4r], \end{aligned} \tag{3.7.19}$$

where we have used (3.7.18) and the corresponding result for $\hat{x}_{\lambda+\pi/2}$ obtained by replacing λ by $\lambda+\pi/2$. The product of the variances in (3.7.19) attains its minimum value of 1/4 when $r=0$, or when either $\cos(\lambda-\varphi/2)$ or $\sin(\lambda-\varphi/2)$ is zero. The first of these conditions corresponds to the true vacuum state $|0\rangle$, while the others will hold if $2\lambda = \varphi + n\pi$, in which case the variance of one of the quadrature components is $\exp(-2r)/2$ while that of the other is $\exp(2r)/2$. It is often more convenient to express the condition for squeezing in terms of the normal ordered variance $:\Delta x_\lambda^2:$ defined by (3.6.12). The quadrature $\hat{x}_\lambda$ is squeezed if $:\Delta x_\lambda^2: < 0$, zero being its value for a coherent state (see (3.6.18)). Using the commutator of $\hat{a}$ and $\hat{a}^\dagger$, and the definition (3.2.20) of $\hat{x}_\lambda$, it is easy to show that, for all states,

$$:\Delta x_\lambda^2: = \Delta x_\lambda^2 - \tfrac{1}{2}. \qquad (3.7.20)$$

Hence the two conditions $\Delta x_\lambda^2 < \frac{1}{2}$ and $:\Delta x_\lambda^2: < 0$ for the quadrature to be squeezed are equivalent.

As with the coherent states, the squeezed states are not mutually orthogonal. The overlap between two squeezed states $|\zeta\rangle$ and $|\zeta'\rangle$ is

$$\begin{aligned}\langle\zeta|\zeta'\rangle &= \langle 0|\hat{S}(-\zeta)\hat{S}(\zeta')|0\rangle \\ &= \surd(\operatorname{sech} r \operatorname{sech} r')\langle 0|\exp\left[-\tfrac{1}{2}\hat{a}^2\exp(-\mathrm{i}\varphi)\tanh r\right] \\ &\quad\times \exp\left[-\tfrac{1}{2}\hat{a}^{\dagger 2}\exp(\mathrm{i}\varphi')\tanh r'\right]|0\rangle \\ &= \surd(\operatorname{sech} r \operatorname{sech} r')\sum_{n=0}^{\infty}\frac{(2n)!}{(n!)^2 2^{2n}}\{\exp[\mathrm{i}(\varphi'-\varphi)]\tanh r \tanh r'\}^n \\ &= \left(\frac{\operatorname{sech} r \operatorname{sech} r'}{1-\exp[\mathrm{i}(\varphi'-\varphi)]\tanh r \tanh r'}\right)^{1/2}, \end{aligned} \qquad (3.7.21)$$

However, in contrast to the coherent states, the squeezed states do not form a complete set since they are all orthogonal to the odd photon number states, that is $\langle 2n+1|\zeta\rangle = 0$ for all n and ζ. The overlap between the coherent state $|\alpha\rangle$ and the squeezed state $|\zeta\rangle$ is

$$\begin{aligned}\langle\alpha|\zeta\rangle &= \langle\alpha|\hat{S}(\zeta)|0\rangle \\ &= \surd(\operatorname{sech} r)\langle\alpha|\exp\left[-\tfrac{1}{2}\hat{a}^{\dagger 2}\exp(\mathrm{i}\varphi)\tanh r\right]|0\rangle \\ &= \surd(\operatorname{sech} r)\exp\left[-\tfrac{1}{2}\alpha^{*2}\exp(\mathrm{i}\varphi)\tanh r\right]\exp\left(-|\alpha|^2/2\right), \end{aligned} \qquad (3.7.22)$$

where we have used the eigenvalue equation (3.6.6) and the number state expansion (3.6.4) of $|\alpha\rangle$. The overlap (3.7.22) is non-zero for all α and ζ and hence the coherent and squeezed states are never orthogonal. The square of the modulus of $\langle\alpha|\zeta\rangle$ is a two-dimensional Gaussian in the real part α_{r} and imaginary part α_{i} of α. For $\varphi = 0$, we find

$$\begin{aligned}|\langle\alpha|\zeta\rangle|^2 &= \operatorname{sech} r\cdot\exp\left\{-\left[\alpha_{\mathrm{r}}^2(1+\tanh r)+\alpha_{\mathrm{i}}^2(1-\tanh r)\right]\right\} \\ &= \operatorname{sech} r\cdot\exp\left[-2\left(\frac{\alpha_{\mathrm{r}}^2}{\exp(-2r)+1}+\frac{\alpha_{\mathrm{i}}^2}{\exp(2r)+1}\right)\right]. \end{aligned} \qquad (3.7.23)$$

The quadrature representation $\psi_\zeta(x_\lambda)$ of the squeezed state $|\zeta\rangle$ is the overlap between the quadrature eigenstate $|x_\lambda\rangle$ and $|\zeta\rangle$, that is

$$\psi_\zeta(x_\lambda) = \langle x_\lambda | \zeta \rangle. \tag{3.7.24}$$

To obtain $\psi_\zeta(x_\lambda)$, we first use the eigenvalue equation (3.7.6) to write

$$\langle x_\lambda | [\hat{a} \cosh r + \hat{a}^\dagger \exp(\mathrm{i}\varphi) \sinh r] | \zeta \rangle = 0. \tag{3.7.25}$$

Using (3.6.43) and its Hermitian conjugate together with the quadrature representations of $\hat{x}_\lambda$ and $\hat{x}_{\lambda+\pi/2}$ given in Appendix 4, we rewrite (3.7.25) as the differential equation

$$\begin{aligned} &[\exp(-r)\cos(\lambda - \varphi/2) + \mathrm{i}\exp(r)\sin(\lambda - \varphi/2)]\frac{\mathrm{d}}{\mathrm{d}x_\lambda}\psi_\zeta(x_\lambda) \\ &\quad + [\exp(r)\cos(\lambda - \varphi/2) + \mathrm{i}\exp(-r)\sin(\lambda - \varphi/2)]x_\lambda \psi_\zeta(x_\lambda) = 0. \end{aligned} \tag{3.7.26}$$

The normalized solution of this equation, up to an x_λ-independent factor κ of unit modulus, is

$$\psi_\zeta(x_\lambda) = \kappa(2\pi\Delta x_\lambda^2)^{-1/4} \exp\left(-\frac{x_\lambda^2}{4\Delta x_\lambda^2}[1 - \mathrm{i}\sin(2\lambda - \varphi)\sinh 2r] \right), \tag{3.7.27}$$

where Δx_λ^2, the variance of $\hat{x}_\lambda$, is given by (3.7.18). This will be a simple Gaussian in x_λ if $2\lambda = \varphi + n\pi$ which is, as noted earlier, the condition for minimizing the product of the variances of $\hat{x}_\lambda$ and $\hat{x}_{\lambda+\pi/2}$ in (3.7.19). The probability distribution for the quadrature $\hat{x}_\lambda$ is

$$P_\zeta(x_\lambda) = (2\pi\Delta x_\lambda^2)^{-1/2} \exp\left(-\frac{x_\lambda^2}{2\Delta x_\lambda^2} \right), \tag{3.7.28}$$

which is a Gaussian, centred on $x_\lambda = 0$, with a width proportional to the square root of the variance.

The combined action of $\hat{S}(\zeta)$ followed by $\hat{D}(\alpha)$ on the vacuum state $|0\rangle$ produces a class of squeezed states characterized by a non-zero expectation value of $\hat{a}$. The state $|\alpha, \zeta\rangle$ defined by

$$|\alpha, \zeta\rangle = \hat{D}(\alpha)\hat{S}(\zeta)|0\rangle, \tag{3.7.29}$$

where ζ and α are complex numbers, is termed the coherent squeezed state. It is normalized since $\langle \alpha, \zeta | \alpha, \zeta \rangle = \langle 0 | \hat{S}(-\zeta)\hat{D}(-\alpha)\hat{D}(\alpha)\hat{S}(\zeta)|0\rangle = 1$. It is important to realize that the operators $\hat{D}(\alpha)$ and $\hat{S}(\zeta)$ do not commute and hence $\hat{S}(\zeta)\hat{D}(\alpha)|0\rangle \neq |\alpha, \zeta\rangle$. Consider

$$\begin{aligned} \hat{S}(\zeta)\hat{D}(\alpha)|0\rangle &= \hat{S}(\zeta)\hat{D}(\alpha)\hat{S}(-\zeta)\hat{S}(\zeta)|0\rangle \\ &= \exp\{\alpha[\hat{a}^\dagger \cosh r + \hat{a}\exp(-\mathrm{i}\varphi)\sinh r] \\ &\qquad - \alpha^*[\hat{a}\cosh r + \hat{a}^\dagger \exp(\mathrm{i}\varphi)\sinh r]\}\hat{S}(\zeta)|0\rangle \\ &= \hat{D}(\alpha\cosh r - \alpha^* \exp(\mathrm{i}\varphi)\sinh r)\hat{S}(\zeta)|0\rangle \\ &= |\alpha\cosh r - \alpha^* \exp(\mathrm{i}\varphi)\sinh r, \zeta\rangle, \end{aligned} \tag{3.7.30}$$

where we have used the transformations (3.7.7) and (3.7.8) together with the definition (3.6.1) of $\hat{D}(\alpha)$. Hence $\hat{S}(\zeta)\hat{D}(\alpha)|0\rangle$ is a coherent squeezed state having the same squeezing parameter ζ as $|\alpha,\zeta\rangle$ but a different coherent amplitude. It follows that $|\alpha,\zeta\rangle$ may be expressed in the alternative form

$$|\alpha,\zeta\rangle = \hat{S}(\zeta)\hat{D}(\alpha\cosh r + \alpha^*\exp(\mathrm{i}\varphi)\sinh r)|0\rangle. \tag{3.7.31}$$

It is convenient, when deriving the properties of $|\alpha,\zeta\rangle$, to transform expectation values in a coherent squeezed state into vacuum expectation values using the unitary transformations

$$\begin{aligned}\hat{S}(-\zeta)\hat{D}(-\alpha)\,\hat{a}\,\hat{D}(\alpha)\hat{S}(\zeta) &= \hat{S}(-\zeta)(\hat{a}+\alpha)\hat{S}(\zeta)\\ &= \hat{a}\cosh r - \hat{a}^\dagger\exp(\mathrm{i}\varphi)\sinh r + \alpha,\end{aligned} \tag{3.7.32}$$

$$\hat{S}(-\zeta)\hat{D}(-\alpha)\hat{a}^\dagger\hat{D}(\alpha)\hat{S}(\zeta) = \hat{a}^\dagger\sinh r - \hat{a}\exp(-\mathrm{i}\varphi)\sinh r + \alpha^*, \tag{3.7.33}$$

obtained from the unitary transformations (3.6.20), (3.6.22), (3.7.10), and (3.7.11). The expectation values in the coherent squeezed state $|\alpha,\zeta\rangle$ of $\hat{a}$ and $\hat{a}^\dagger$ are α and α^*, respectively, since

$$\begin{aligned}\langle\alpha,\zeta|\,\hat{a}\,|\alpha,\zeta\rangle &= \langle 0|\,\hat{S}(-\zeta)\hat{D}(-\alpha)\hat{a}\hat{D}(\alpha)\hat{S}(\zeta)\,|0\rangle\\ &= \langle 0|(\hat{a}\cosh r - \hat{a}^\dagger\exp(\mathrm{i}\varphi)\sinh r + \alpha)\,|0\rangle\\ &= \alpha.\end{aligned} \tag{3.7.34}$$

Hence the quadrature operator $\hat{x}_\lambda$ given by (3.2.20) has the expectation value

$$\langle\alpha,\zeta|\,\hat{x}_\lambda\,|\alpha,\zeta\rangle = \frac{1}{\sqrt{2}}[\alpha\exp(-\mathrm{i}\lambda) + \alpha^*\exp(\mathrm{i}\lambda)], \tag{3.7.35}$$

which is the value (3.6.14) found for the coherent state $|\alpha\rangle$. The expectation value of $\hat{x}_\lambda^2$ is

$$\begin{aligned}\langle\alpha,\zeta|\,\hat{x}_\lambda^2\,|\alpha,\zeta\rangle &= \tfrac{1}{2}\langle 0|\{[\hat{a}\cosh r - \hat{a}^\dagger\exp(\mathrm{i}\varphi)\sinh r + \alpha]\exp(-\mathrm{i}\lambda)\\ &\qquad + [\hat{a}^\dagger\cosh r - \hat{a}\exp(-\mathrm{i}\varphi)\sinh r + \alpha^*]\exp(\mathrm{i}\lambda)\}^2\,|0\rangle\\ &= \tfrac{1}{2}[\exp(2r)\sin^2(\lambda-\varphi/2) + \exp(-2r)\cos^2(\lambda-\varphi/2)]\\ &\qquad + \tfrac{1}{2}[\alpha\exp(-\mathrm{i}\lambda) + \alpha^*\exp(\mathrm{i}\lambda)]^2,\end{aligned} \tag{3.7.36}$$

and hence the variance of $\hat{x}_\lambda$ is

$$\Delta x_\lambda^2 = \tfrac{1}{2}[\exp(2r)\sin^2(\lambda-\varphi/2) + \exp(-2r)\cos^2(\lambda-\varphi/2)], \tag{3.7.37}$$

which is the value found in (3.7.18) for the squeezed state $|\zeta\rangle$. We see from (3.7.35) and (3.7.37) that the expectation value of $\hat{x}_\lambda$ is determined solely by the coherent amplitude α while the variance, and hence the squeezing, depends on the squeezing parameter ζ only.

The expectation value of the number operator $\hat{n}$ can be found by similar use of the unitary transformations (3.7.32) and (3.7.33). We find

$$\begin{aligned}\langle\alpha,\zeta|\hat{n}|\alpha,\zeta\rangle &= \langle 0|[\hat{a}^\dagger\cosh r - \hat{a}\exp(-\mathrm{i}\varphi)\sinh r + \alpha^*]\\ &\quad\times[\hat{a}\cosh r - \hat{a}^\dagger\exp(\mathrm{i}\varphi)\sinh r + \alpha]|0\rangle\\ &= \sinh^2 r + |\alpha|^2,\end{aligned} \tag{3.7.38}$$

which is the sum of the mean photon numbers for the coherent state $|\alpha\rangle$, found in (3.6.8), and for the squeezed state $|\zeta\rangle$, found in (3.7.14). The normal ordered variance of $\hat{n}$ is

$$\begin{aligned}:\Delta n^2: &= \langle :\hat{n}^2:\rangle - \langle\hat{n}\rangle^2\\ &= |\alpha|^2[\exp(2r)\sin^2(\theta-\varphi/2) + \exp(-2r)\cos^2(\theta-\varphi/2) - 1]\\ &\quad + \sinh^2 r(\cosh^2 r + \sinh^2 r),\end{aligned} \tag{3.7.39}$$

where θ is the argument of α. This expression can be negative corresponding to sub-Poissonian statistics. In particular, if $|\alpha|^2$ is sufficiently large for the first term of (3.7.39) to dominate then the minimum value of $:\Delta n^2:$ is $-|\alpha|^2[1-\exp(-2r)]$ and

$$:\Delta n^2: \simeq 2|\alpha|^2 :\Delta x_\theta^2:, \tag{3.7.40}$$

where we have used (3.7.20) and (3.7.37). This proportionality between $:\Delta n^2:$ and $:\Delta x_\theta^2:$ is the basis for the method of homodyne detection used to measure squeezing.

The coherent squeezed states form an overcomplete set in that the identity can be resolved as

$$\frac{1}{\pi}\int_{-\infty}^{\infty} \mathrm{d}^2\alpha\,|\alpha,\zeta\rangle\langle\alpha,\zeta| = \mathbf{1}. \tag{3.7.41}$$

We prove this by using the form (3.7.31) to write the left-hand side of (3.7.41) as

$$\begin{aligned}&\hat{S}(\zeta)\frac{1}{\pi}\int_{-\infty}^{\infty} \mathrm{d}^2\alpha\,\hat{D}(\beta)|0\rangle\langle 0|\hat{D}(-\beta)\hat{S}(-\zeta)\\ &= \hat{S}(\zeta)\frac{1}{\pi}\int_{-\infty}^{\infty} \mathrm{d}^2\alpha\,|\beta\rangle\langle\beta|\hat{S}(-\zeta),\end{aligned} \tag{3.7.42}$$

where $\beta = \alpha\cosh r + \alpha^*\exp(\mathrm{i}\varphi)\sinh r$. The Jacobian for the change of variable from α to β is unity and therefore

$$\frac{1}{\pi}\int_{-\infty}^{\infty} \mathrm{d}^2\alpha\,|\alpha,\zeta\rangle\langle\alpha,\zeta| = \hat{S}(\zeta)\frac{1}{\pi}\int_{-\infty}^{\infty} \mathrm{d}^2\beta\,|\beta\rangle\langle\beta|\hat{S}(-\zeta) = \mathbf{1}, \tag{3.7.43}$$

using (3.6.26).

The coherent squeezed state $|\alpha,\zeta\rangle$ is a right eigenstate of the transformed annihilation operator $\hat{D}(\alpha)\hat{S}(\zeta)\hat{a}\hat{S}(-\zeta)\hat{D}(-\alpha)$ with eigenvalue 0 since

$$\begin{aligned}\hat{D}(\alpha)\hat{S}(\zeta)\hat{a}\hat{S}(-\zeta)\hat{D}(-\alpha)|\alpha,\zeta\rangle \\ &= \hat{D}(\alpha)\hat{S}(\zeta)\hat{a}\hat{S}(-\zeta)\hat{D}(-\alpha)\hat{D}(\alpha)\hat{S}(\zeta)|0\rangle \\ &= \hat{D}(\alpha)\hat{S}(\zeta)\hat{a}|0\rangle = 0. \end{aligned} \tag{3.7.44}$$

The transformed annihilation operator is

$$\begin{aligned}\hat{D}(\alpha)\hat{S}(\zeta)\hat{a}\hat{S}(-\zeta)\hat{D}(-\alpha) \\ &= (\hat{a}-\alpha)\cosh r + (\hat{a}^\dagger - \alpha^*)\exp(\mathrm{i}\varphi)\sinh r, \end{aligned} \tag{3.7.45}$$

where we have used the transformations (3.7.7), and (3.6.20) and (3.6.22) with α replaced by $-\alpha$. We can calculate the quadrature representation $\psi_{\alpha,\zeta}(x_\lambda) = \langle x_\lambda|\alpha,\zeta\rangle$ by writing the equation

$$\langle x_\lambda|[(\hat{a}-\alpha)\cosh r + (\hat{a}^\dagger - \alpha^*)\exp(\mathrm{i}\varphi)\sinh r]|\alpha,\zeta\rangle = 0 \tag{3.7.46}$$

as a differential equation for $\psi_{\alpha,\zeta}(x_\lambda)$ by the method used to derive (3.7.26). The normalized solution of the resulting equation is

$$\begin{aligned}\psi_{\alpha,\zeta}(x_\lambda) &= \kappa(2\pi\,\Delta x_\lambda^2)^{-1/4}\exp\big(\mathrm{i}\langle\hat{x}_{\lambda+\pi/2}\rangle x_\lambda\big) \\ &\quad\times\exp\left(-\frac{(x_\lambda - \langle\hat{x}_\lambda\rangle)^2[1-\mathrm{i}\sin(2\lambda-\varphi)\sinh 2r]}{4\,\Delta x_\lambda^2}\right),\end{aligned} \tag{3.7.47}$$

where κ is a unit-modulus factor and, as found earlier, $\langle\hat{x}_\lambda\rangle$ and $\langle\hat{x}_{\lambda+\pi/2}\rangle$ are the same as those for the coherent state $|\alpha\rangle$ given by (3.6.14). The probability distribution for the quadrature $\hat{x}_\lambda$ is

$$P_{\alpha,\zeta}(x_\lambda) = (2\pi\,\Delta x_\lambda^2)^{-1/2}\exp\left(-\frac{(x_\lambda - \langle\hat{x}_\lambda\rangle)^2}{2\,\Delta x_\lambda^2}\right), \tag{3.7.48}$$

which is a Gaussian, similar to that given in (3.7.28) for the state $|\zeta\rangle$ but with the centre shifted to $x_\lambda = \langle\hat{x}_\lambda\rangle$.

For two modes A and B with annihilation operators $\hat{a}$ and $\hat{b}$, the two-mode squeezed state $|\zeta_{AB}\rangle$ is generated by the action on the two-mode vacuum state $|0_A;0_B\rangle$ of the two-mode squeezing operator

$$\hat{S}_{AB}(\zeta) = \exp\big(-\zeta\hat{a}^\dagger\hat{b}^\dagger + \zeta^*\hat{b}\hat{a}\big), \tag{3.7.49}$$

where, as before, $\zeta = r\exp(\mathrm{i}\varphi)$ is any complex number. It should be noted that $\hat{S}_{AB}(\zeta)$ is not simply the product of two single-mode squeezing operators for the modes A and B. The form of (3.7.49) is similar to that of the single-mode operator $\hat{S}(\zeta)$ given in (3.7.1) but with $\hat{a}^{\dagger 2}/2$ replaced by $\hat{a}^{\dagger}\hat{b}^{\dagger}$. Both single-mode and two-mode squeezing arise in models of two-photon non-linear optics being associated with degenerate and non-degenerate processes, respectively. In the first of these, pairs of photons are generated in a single mode by the action of $\hat{a}^{\dagger 2}$ while in the second, pairs of photons are generated, one in each of the modes A and B, by the action of $\hat{a}^{\dagger}\hat{b}^{\dagger}$. The operator $\hat{S}_{AB}(\zeta)$ is unitary since $\hat{S}_{AB}^{\dagger}(\zeta) = \hat{S}_{AB}(-\zeta)$ which is the inverse of $\hat{S}_{AB}(\zeta)$. It is useful to write $\hat{S}_{AB}(\zeta)$ in an ordered form which, using the operator ordering theorem (A5.18), is

$$\hat{S}_{AB}(\zeta) = \exp\left[-\hat{a}^{\dagger}\hat{b}^{\dagger}\exp(\mathrm{i}\varphi)\tanh r\right]$$
$$\times \exp\left[-(\hat{a}^{\dagger}\hat{a} + \hat{b}\hat{b}^{\dagger})\ln(\cosh r)\right]\exp\left[\hat{b}\hat{a}\exp(-\mathrm{i}\varphi)\tanh r\right]. \quad (3.7.50)$$

The two-mode squeezed state $|\zeta_{AB}\rangle$ is given by

$$|\zeta_{AB}\rangle = \hat{S}_{AB}(\zeta)|0_A;0_B\rangle \quad (3.7.51)$$

and it then follows from (3.7.50) that

$$|\zeta_{AB}\rangle = \operatorname{sech} r \sum_{n=0}^{\infty} [-\exp(\mathrm{i}\varphi)\tanh r]^n |n_A;n_B\rangle, \quad (3.7.52)$$

where we have again used the definition (3.3.4) of the exponential function of an operator. The state $|\zeta_{AB}\rangle$ is a superposition only of states in which the two modes contain the same number of photons and is normalized since $\langle\zeta_{AB}|\zeta_{AB}\rangle = \langle 0_A;0_B|\hat{S}_{AB}(-\zeta)\hat{S}_{AB}(\zeta)|0_A;0_B\rangle = 1$. Note that $|\zeta_{AB}\rangle$ is an entangled state in that it is not the product of a state for mode A and one for mode B. This leads to correlations between the properties of the modes.

The single-mode properties of each of the two modes in the state $|\zeta_{AB}\rangle$ are precisely those of a single-mode thermal state. Consider, for example, the expectation value of any operator $\hat{A}$ associated with mode A alone and given by

$$\langle\zeta_{AB}|\hat{A}|\zeta_{AB}\rangle = \operatorname{sech}^2 r \sum_{n=0}^{\infty} (\tanh r)^{2n} \langle n_A|\hat{A}|n_A\rangle, \quad (3.7.53)$$

from (3.7.52). This expectation value contains only diagonal elements in the number state basis, the off-diagonal ones vanishing owing to the orthonormality of the number states for mode B. An equivalent result to (3.7.53) can be obtained using the thermal state density matrix (3.5.5) and (3.5.6) if we identify $\tanh^2 r$ with $\exp(-\beta\hbar\omega)$, so that $\bar{n} = \sinh^2 r$ in (3.5.9). As this equivalence of expectation values holds for any single-mode operator $\hat{A}$, we can infer that the state of mode A is represented by a thermal density matrix. Hence the moments of the number operator $\hat{a}^{\dagger}\hat{a}$ are the same as those given in (3.5.14), (3.5.15), and

(3.5.19), and the expectation value and variance of the quadrature operator $\hat{x}_\lambda$ in (3.2.20) are

$$\langle \zeta_{AB} | \hat{x}_\lambda | \zeta_{AB} \rangle = 0, \tag{3.7.54}$$

$$\Delta x_\lambda^2 = \tfrac{1}{2} \cosh 2r = \bar{n} + \tfrac{1}{2}. \tag{3.7.55}$$

The symmetry between modes A and B apparent in (3.7.52) means that mode B can be represented by a thermal density matrix with the same mean photon number as that of mode A. These properties can be derived directly from the form of the state (3.7.52) but it is easier to transform expectation values into two-mode vacuum expectation values using

$$\hat{S}_{AB}(-\zeta)\hat{a}\hat{S}_{AB}(\zeta) = \hat{a}\cosh r - \hat{b}^\dagger \exp(\mathrm{i}\varphi)\sinh r, \tag{3.7.56}$$

$$\hat{S}_{AB}(-\zeta)\hat{b}\hat{S}_{AB}(\zeta) = \hat{b}\cosh r - \hat{a}^\dagger \exp(\mathrm{i}\varphi)\sinh r, \tag{3.7.57}$$

and their Hermitian conjugates

$$\hat{S}_{AB}(-\zeta)\hat{a}^\dagger\hat{S}_{AB}(\zeta) = \hat{a}^\dagger\cosh r - \hat{b} \exp(-\mathrm{i}\varphi)\sinh r, \tag{3.7.58}$$

$$\hat{S}_{AB}(-\zeta)\hat{b}^\dagger\hat{S}_{AB}(\zeta) = \hat{b}^\dagger\cosh r - \hat{a} \exp(-\mathrm{i}\varphi)\sinh r. \tag{3.7.59}$$

In this way we can re-express single-mode thermal averages as two-mode vacuum expectation values. The thermal state expectation value of any operator function of $\hat{a}$ and $\hat{a}^\dagger$ is equal to the two-mode vacuum expectation value with $\hat{a}$ replaced by (3.7.56) and $\hat{a}^\dagger$ replaced by (3.7.58), with $\sinh^2 r = \bar{n}$ and φ arbitrary.

We can express the two-mode squeezed state as a product of two single-mode squeezed states by introducing superposition modes C and D, as in Section 2.6, with annihilation operators $\hat{c}$ and $\hat{d}$ given by

$$\hat{c} = \frac{1}{\sqrt{2}}\left[\hat{a} + \exp(\mathrm{i}\delta)\hat{b}\right], \tag{3.7.60}$$

$$\hat{d} = \frac{1}{\sqrt{2}}\left[\hat{b} - \exp(-\mathrm{i}\delta)\hat{a}\right], \tag{3.7.61}$$

where δ is a real phase. Such superpositions model the action of a beam-splitter as we will discuss in the last section of this chapter. The operators $\hat{c}$ and $\hat{d}$ are independent boson annihilation operators since $[\hat{c}, \hat{c}^\dagger] = 1 = [\hat{d}, \hat{d}^\dagger]$ and $\hat{c}$ and $\hat{c}^\dagger$ commute with $\hat{d}$ and $\hat{d}^\dagger$. The inverse relations to (3.7.60) and (3.7.61) are

$$\hat{a} = \frac{1}{\sqrt{2}}\left[\hat{c} - \exp(\mathrm{i}\delta)\hat{d}\right], \tag{3.7.62}$$

$$\hat{b} = \frac{1}{\sqrt{2}}\left[\hat{d} + \exp(-\mathrm{i}\delta)\hat{c}\right]. \tag{3.7.63}$$

If we express the two-mode squeezing operator (3.7.49) in terms of $\hat{c}$ and $\hat{d}$, we find

$$\hat{S}_{AB}(\zeta) = \exp\Big\{-\tfrac{1}{2}\zeta\left[\hat{c}^{\dagger 2}\exp(\mathrm{i}\delta) - \hat{d}^{\dagger 2}\exp(-\mathrm{i}\delta)\right] + \tfrac{1}{2}\zeta^*\left[\hat{c}^2\exp(-\mathrm{i}\delta) - \hat{d}^2\exp(\mathrm{i}\delta)\right]\Big\}$$

$$= \hat{S}_C(\zeta\exp(\mathrm{i}\delta))\hat{S}_D(-\zeta\exp(-\mathrm{i}\delta)), \qquad (3.7.64)$$

where $\hat{S}_C$ and $\hat{S}_D$ are single-mode squeezing operators acting on the superposition modes C and D. The two-mode squeezed state $|\zeta_{AB}\rangle$ can then be written as the product

$$|\zeta_{AB}\rangle = |\zeta\exp(\mathrm{i}\delta)\rangle\!\rangle_C \otimes |-\zeta\exp(-\mathrm{i}\delta)\rangle\!\rangle_D \qquad (3.7.65)$$

of squeezed states for the superposition modes C and D. The properties of the modes C and D are simply those associated with the single-mode squeezed states described earlier in this section. The squeezing of the superposition mode quadratures is a consequence of the correlation between modes A and B. This correlation is also responsible for the zero variance in the difference between the photon numbers $\hat{n}_A$ and $\hat{n}_B$ since

$$\Delta(n_A - n_B)^2 = \Delta n_A^2 + \Delta n_B^2 - 2\big(\langle\hat{n}_A\hat{n}_B\rangle - \langle\hat{n}_A\rangle\langle\hat{n}_B\rangle\big)$$

$$= 0. \qquad (3.7.66)$$

The variances in both $\hat{n}_A$ and $\hat{n}_B$ are $\bar{n}(\bar{n}+1)$ but the correlation between the photon numbers ensures that $\langle\hat{n}_A\hat{n}_B\rangle \neq \langle\hat{n}_A\rangle\langle\hat{n}_B\rangle$ and leads to the zero variance in $\hat{n}_A - \hat{n}_B$.

The action of the displacement operators $\hat{D}_A(\alpha)$ and $\hat{D}_B(\beta)$ given by (3.6.46) on $|\zeta_{AB}\rangle$ produces a two-mode coherent squeezed state. As with the single-mode coherent squeezed state, the expectation values of the quadratures are determined by α and β while the squeezing properties depend on ζ only.

The generalization of the squeezed vacuum states to a continuum of modes is achieved by acting on the vacuum state $|0\rangle$ of the continuum field with the squeezing operator

$$\hat{S}[\zeta(\omega)] = \exp\left(-\tfrac{1}{2}\int \mathrm{d}\omega\left[\zeta(\omega)\hat{b}^\dagger(\omega)\hat{b}^\dagger(2\Omega-\omega) - \zeta^*(\omega)\hat{b}(2\Omega-\omega)\hat{b}(\omega)\right]\right), \qquad (3.7.67)$$

where $\zeta(\omega) = r(\omega)\exp[\mathrm{i}\varphi(\omega)]$. This operator is the natural generalization of the two-mode squeezing operator (3.7.49) in that it involves products of continuum operators associated with pairs of frequencies summing to 2Ω. We then say that the field is squeezed at frequency Ω. In (3.7.67) the range of integration is $0 \leqslant \omega \leqslant 2\Omega$ and we note that $\zeta(\omega)$ and $\zeta(2\Omega - \omega)$ multiply the same pair of operators. Therefore we may, without loss of generality, set $\zeta(\omega)=\zeta(2\Omega-\omega)$. The factor $\frac{1}{2}$ in the exponent of (3.7.67) is a consequence of the fact that the continuum operators corresponding to each frequency occur twice in the integrand. The continuum squeezed vacuum state is

$$|\{\zeta(\omega)\}\rangle = \hat{S}[\zeta(\omega)]|0\rangle. \tag{3.7.68}$$

This state is important in models of coherent two-photon processes such as parametric amplification in which excitation at frequency 2Ω results in the production of pairs of photons. The statistical properties of the state $|\{\zeta(\omega)\}\rangle$ can be found with the aid of the transformation

$$\hat{S}[-\zeta(\omega)]\hat{b}(\omega')\hat{S}[\zeta(\omega)] = \hat{b}(\omega')\cosh r(\omega') - \hat{b}^{\dagger}(2\Omega-\omega') \times \exp[\mathrm{i}\varphi(\omega')]\sinh r(\omega') \tag{3.7.69}$$

and its Hermitian conjugate, where we have used the continuum operator commutation relation (3.2.31), and (3.3.9). In particular we find that the lowest-order non-vanishing moments are

$$\langle\hat{b}^{\dagger}(\omega)\hat{b}(\omega')\rangle = \sinh^2 r(\omega)\,\delta(\omega-\omega'), \tag{3.7.70}$$

$$\langle\hat{b}(\omega)\hat{b}^{\dagger}(\omega')\rangle = \cosh^2 r(\omega)\,\delta(\omega-\omega'), \tag{3.7.71}$$

$$\begin{aligned}\langle\hat{b}(\omega)\hat{b}(\omega')\rangle &= \langle\hat{b}^{\dagger}(\omega)\hat{b}^{\dagger}(\omega')\rangle^{*} \\ &= -\exp[\mathrm{i}\varphi(\omega)]\sinh r(\omega) \\ &\quad\times\cosh r(\omega)\,\delta(\omega+\omega'-2\Omega).\end{aligned} \tag{3.7.72}$$

We will use these moments in Section 5.5 in our discussion of dissipation arising from coupling to a squeezed environment. In order to demonstrate that the state $|\{\zeta(\omega)\}\rangle$ is squeezed, we consider the continuum generalization of the quadrature operator defined as

$$\hat{X}_{\lambda} = \frac{1}{\sqrt{2}}\int \mathrm{d}\omega\left[\hat{b}(\omega)\exp(-\mathrm{i}\lambda) + \hat{b}^{\dagger}(\omega)\exp(\mathrm{i}\lambda)\right]. \tag{3.7.73}$$

It follows from the moments (3.7.70) and (3.7.72) that the normal ordered variance of $\hat{X}_{\lambda}$ is

$$\begin{aligned}:\Delta X_{\lambda}^2: &= \langle:\hat{X}_{\lambda}^2:\rangle \\ &= \int \mathrm{d}\omega\{\sinh^2 r(\omega) - \cos[\varphi(\omega)-2\lambda]\sinh r(\omega)\cosh r(\omega)\}.\end{aligned} \tag{3.7.74}$$

For $\varphi(\omega)=2\lambda$ the integrand of (3.7.74) is negative and hence the field is squeezed.

We conclude this section by emphasizing that not all squeezed states are generated by unitary transformation of the vacuum state. Consider for example the single-mode state $c_0|0\rangle + c_1|1\rangle$ where for simplicity we choose c_0 and c_1 to be real. The variance of $\hat{x}_\lambda$ is

$$\Delta x_\lambda^2 = \tfrac{1}{2} + c_1^2 - 2c_0^2 c_1^2 \cos^2\lambda, \tag{3.7.75}$$

which is clearly minimized if $\lambda = 0$. The quadrature $\hat{x}_0$ is squeezed if $2c_0^2 > 1$, that is if $c_0^2 > c_1^2$. The minimum variance for this state is $3/8$ which occurs when $c_1 = 1/2$ and $c_0 = \sqrt{3}/2$. Note that this squeezed state is not a minimum uncertainty state.

3.8 Angular momentum coherent states

Angular momentum coherent states, also known as spin coherent states and atomic coherent states, are important in many models in quantum optics, notably in the coherent excitation of atoms and in the action of a beam-splitter. Although interest in these states is not limited to rotating systems, it is simplest to define these states in terms of the angular momentum eigenstates $|j,m\rangle$ where j takes any integer or half-integer value and $m = -j, -j+1, \ldots, j$. The actions of the angular momentum operators $\hat{J}_3$, $\hat{J}_+$, and $\hat{J}_-$ on the state $|j,m\rangle$ are given by

$$\hat{J}_3\,|j,m\rangle = m\,|j,m\rangle, \tag{3.8.1}$$

$$\hat{J}_\pm\,|j,m\rangle = [(j \mp m)(j \pm m + 1)]^{1/2}\,|j, m \pm 1\rangle, \tag{3.8.2}$$

so that $\hat{J}_+\,|j,j\rangle = 0$ and $\hat{J}_-\,|j,-j\rangle = 0$. It follows from (3.8.2) that we can write $|j,m\rangle$ as

$$|j,m\rangle = \frac{1}{(j+m)!}\left(C_{j+m}^{2j}\right)^{-1/2} \hat{J}_+^{j+m}\,|j,-j\rangle. \tag{3.8.3}$$

The angular momentum coherent state is defined by the action of the rotation operator

$$\hat{R}(\theta,\phi) = \exp\left\{\tfrac{1}{2}\theta\left[\exp(-\mathrm{i}\phi)\hat{J}_+ - \exp(\mathrm{i}\phi)\hat{J}_-\right]\right\} \tag{3.8.4}$$

on $|j,-j\rangle$. This operator is unitary since $\hat{R}^\dagger(C,\phi) = \hat{R}(-\theta,\phi) = \hat{R}^{-1}(\theta,\phi)$ and rotates the angular momentum vector through an angle θ about the axis $\hat{\mathbf{\imath}}\sin\phi - \hat{\mathbf{\jmath}}\cos\phi$. It is useful to write the operator (3.8.4) in the ordered form

$$\begin{aligned}\hat{R}(\theta,\phi) = {} & \exp\left[\exp(-\mathrm{i}\phi)(\tan\tfrac{1}{2}\theta)\hat{J}_+\right]\exp\left[-\ln\left(\cos^2\tfrac{1}{2}\theta\right)\hat{J}_3\right] \\ & \times \exp\left[-\exp(\mathrm{i}\phi)(\tan\tfrac{1}{2}\theta)\hat{J}_-\right],\end{aligned} \tag{3.8.5}$$

using the ordering theorem (A5.11). The angular momentum coherent state $|\theta, \phi\rangle$ is given by

$$|\theta, \phi\rangle = \hat{R}(\theta, \phi)|j, -j\rangle \tag{3.8.6}$$

and it then follows from (3.8.5) that

$$|\theta, \phi\rangle = (\cos\tfrac{1}{2}\theta)^{2j} \sum_{m=-j}^{j} \left(C^{2j}_{j+m}\right)^{1/2} [\exp(-\mathrm{i}\phi)\tan\tfrac{1}{2}\theta]^{j+m}\,|j, m\rangle, \tag{3.8.7}$$

where we have used the definition (3.3.4) of the exponential function of an operator together with (3.8.3). The probability distribution $P(m)$ for $\hat{J}_3$ is therefore binomial since

$$P(m) = (\cos\tfrac{1}{2}\theta)^{4j} C^{2j}_{j+m} (\tan\tfrac{1}{2}\theta)^{2(j+m)}. \tag{3.8.8}$$

As with the coherent states and the squeezed states, the angular momentum coherent states are not mutually orthogonal, the overlap between the states $|\theta, \phi\rangle$ and $|\theta', \phi'\rangle$ being

$$\langle\theta, \phi\,|\,\theta', \phi'\rangle = [\cos^2\tfrac{1}{2}\theta\cos^2\tfrac{1}{2}\theta'\{1 + \exp[\mathrm{i}(\phi - \phi')]\tan\tfrac{1}{2}\theta\tan\tfrac{1}{2}\theta'\}]^j. \tag{3.8.9}$$

Hence

$$|\langle\theta, \phi\,|\,\theta', \phi'\rangle|^2 = (\cos\tfrac{1}{2}\Theta)^{4j}, \tag{3.8.10}$$

where Θ is the angle between the directions defined by the angles (θ, ϕ) and (θ', ϕ'). Despite the lack of orthogonality of the angular momentum coherent states, they form an overcomplete set in that the identity on the space with total angular momentum j can be resolved as

$$\frac{(2j+1)}{4\pi}\int \mathrm{d}\Omega\,|\theta, \phi\rangle\langle\theta, \phi| = \mathbf{1}, \tag{3.8.11}$$

where the integration $\int \mathrm{d}\Omega = \int_0^{2\pi}\mathrm{d}\phi\int_0^{\pi}\sin\theta\,\mathrm{d}\theta$ is over the whole surface of a sphere. We prove this by first inserting the definition (3.8.7) into the left-hand side of (3.8.11) and carrying out the integration over ϕ to obtain

$$\frac{(2j+1)}{4\pi}\int \mathrm{d}\Omega\,|\theta, \phi\rangle\langle\theta, \phi| = (2j+1)\sum_{m=-j}^{j} C^{2j}_{j+m}\,|j, m\rangle\langle j, m|$$
$$\times\int_0^{\pi}\mathrm{d}\theta(\sin\tfrac{1}{2}\theta)^{2(j+m)+1}(\cos\tfrac{1}{2}\theta)^{2(j-m)+1}. \tag{3.8.12}$$

We show in Appendix 3 with the aid of the gamma and beta functions that the value of the integral in (3.8.12) is $[(2j+1)C^{2j}_{j+m}]^{-1}$ and therefore

$$\frac{(2j+1)}{4\pi}\int \mathrm{d}\Omega\,|\theta,\phi\rangle\langle\theta,\phi| = \sum_{m=-j}^{j}|j,m\rangle\langle j,m| = \mathbf{1}. \qquad (3.8.13)$$

It follows that the trace of an operator $\hat{A}$ can be expressed as

$$\mathrm{Tr}(\hat{A}) = \frac{(2j+1)}{4\pi}\int \mathrm{d}\Omega\,\langle\theta,\phi|\,\hat{A}\,|\theta,\phi\rangle. \qquad (3.8.14)$$

The state $|j,-j\rangle$ is an eigenstate of $\hat{J}_3$ with eigenvalue $-j$ and a right eigenstate of $\hat{J}_-$ with eigenvalue 0. It therefore follows that $|\theta,\phi\rangle$ is an eigenstate of $\hat{R}(\theta,\phi)\hat{J}_3\hat{R}(-\theta,\phi)$ with eigenvalue $-j$ and a right eigenstate of $\hat{R}(\theta,\phi)\hat{J}_-\hat{R}(-\theta,\phi)$ with eigenvalue 0 since

$$\begin{aligned}\hat{R}(\theta,\phi)\hat{J}_3\hat{R}(-\theta,\phi)|\theta,\phi\rangle &= \hat{R}(\theta,\phi)\hat{J}_3\hat{R}(-\theta,\phi)\hat{R}(\theta,\phi)|j,-j\rangle \\ &= -j\hat{R}(\theta,\phi)|j,-j\rangle = -j|\theta,\phi\rangle, \qquad (3.8.15)\end{aligned}$$

and

$$\hat{R}(\theta,\phi)\hat{J}_-\hat{R}(-\theta,\phi)|\theta,\phi\rangle = \hat{R}(\theta,\phi)\hat{J}_-\,|j,-j\rangle = 0. \qquad (3.8.16)$$

The transformed angular momentum operators in (3.8.15) and (3.8.16) have the explicit forms

$$\begin{aligned}\hat{R}(\theta,\phi)\hat{J}_3\hat{R}(-\theta,\phi) &= \hat{J}_3 - \tfrac{1}{2}\theta\left[\exp(-\mathrm{i}\phi)\hat{J}_+ + \exp(\mathrm{i}\phi)\hat{J}_-\right] - \tfrac{1}{2}\theta^2\hat{J}_3 + \cdots \\ &= \hat{J}_3\cos\theta - \tfrac{1}{2}\left[\exp(-\mathrm{i}\phi)\hat{J}_+ + \exp(\mathrm{i}\phi)\hat{J}_-\right]\sin\theta \qquad (3.8.17)\end{aligned}$$

and

$$\begin{aligned}\hat{R}(\theta,\phi)\hat{J}_-\hat{R}(-\theta,\phi) = \exp(-\mathrm{i}\phi)\Big[&\exp(\mathrm{i}\phi)\hat{J}_-\cos^2\tfrac{1}{2}\theta \\ &-\exp(-\mathrm{i}\phi)\hat{J}_+\sin^2\tfrac{1}{2}\theta + \hat{J}_3\sin\theta\Big], \qquad (3.8.18)\end{aligned}$$

where we have again used (3.3.9) together with the commutators (3.2.13) and (3.2.14). The state $|\theta,\phi\rangle$ is also a left eigenstate with zero eigenvalue of the transformed operator

$$\begin{aligned}\hat{R}(\theta,\phi)\hat{J}_+\hat{R}(-\theta,\phi) = \exp(\mathrm{i}\phi)\Big[&\exp(-\mathrm{i}\phi)\hat{J}_+\cos^2\tfrac{1}{2}\theta \\ &-\exp(\mathrm{i}\phi)\hat{J}_-\sin^2\tfrac{1}{2}\theta + \hat{J}_3\sin\theta\Big]. \qquad (3.8.19)\end{aligned}$$

The expectation values of the components of the angular momentum vector $\hat{\mathbf{J}} = (\hat{J}_1, \hat{J}_2, \hat{J}_3)$ are

$$\begin{aligned}\langle\theta,\phi|\,\hat{J}_3\,|\theta,\phi\rangle &= \langle j,-j|\,\hat{R}(-\theta,\phi)\hat{J}_3\hat{R}(\theta,\phi)\,|j,-j\rangle \\ &= \langle j,-j|\Big(\hat{J}_3\cos\theta + \tfrac{1}{2}\big[\exp(-\mathrm{i}\phi)\hat{J}_+ + \exp(\mathrm{i}\phi)\hat{J}_-\big]\sin\theta\Big)\,|j,-j\rangle \\ &= -j\cos\theta, \end{aligned} \qquad (3.8.20)$$

$$\langle\theta,\phi|\,\hat{J}_1\,|\theta,\phi\rangle = j\sin\theta\cos\phi \qquad (3.8.21)$$

and

$$\langle\theta,\phi|\,\hat{J}_2\,|\theta,\phi\rangle = j\sin\theta\sin\phi, \qquad (3.8.22)$$

where we have used the inverses of the transformations (3.8.17) to (3.8.19) obtained by changing the sign of θ. These expectation values illustrate the action of the operator $\hat{R}(\theta,\phi)$ in causing a rotation of $\hat{\mathbf{J}}$ since

$$\langle j,-j|\hat{\mathbf{J}}\,|j,-j\rangle = (0,0,-j) \qquad (3.8.23)$$

and

$$\langle\theta,\phi|\hat{\mathbf{J}}\,|\theta,\phi\rangle = (j\sin\theta\cos\phi, j\sin\theta\sin\phi, -j\cos\theta). \qquad (3.8.24)$$

Angular momentum coherent states arise naturally in the Rabi oscillations of a two-level atom. For such a two-state system, $j = \frac{1}{2}$ and we make the identifications of the ground and excited states $|1\rangle$ and $|2\rangle$ with the angular momentum states $|\frac{1}{2},-\frac{1}{2}\rangle$ and $|\frac{1}{2},\frac{1}{2}\rangle$, respectively. The angular momentum coherent state for the two-level atom is then

$$|\theta,\phi\rangle = \cos\tfrac{1}{2}\theta\,|1\rangle + \exp(-\mathrm{i}\phi)\sin\tfrac{1}{2}\theta\,|2\rangle. \qquad (3.8.25)$$

For exact resonance between the atomic transition and the driving field, the Hamiltonian in the interaction picture is (see (2.2.29) with $\delta = 0$)

$$\hat{H}_\mathrm{I} = -\frac{\hbar V}{2}[\hat{\sigma}_+\exp(-\mathrm{i}\varphi) + \hat{\sigma}_-\exp(\mathrm{i}\varphi)] \qquad (3.8.26)$$

and therefore, using (3.3.56), (3.2.4), and (3.2.5), the time evolution operator is

$$\exp\big(-\mathrm{i}\hat{H}_\mathrm{I}t/\hbar\big) = \cos(Vt/2) + \mathrm{i}\sin(Vt/2)[\hat{\sigma}_+\exp(-\mathrm{i}\varphi) + \hat{\sigma}_-\exp(\mathrm{i}\varphi)]. \qquad (3.8.27)$$

If the atom is initially in its ground state, then the time-evolved state is

$$\exp\big(-\mathrm{i}\hat{H}_\mathrm{I}t/\hbar\big)\,|1\rangle = \cos(Vt/2)\,|1\rangle + \mathrm{i}\exp(-\mathrm{i}\varphi)\sin(Vt/2)\,|2\rangle. \quad (3.8.28)$$

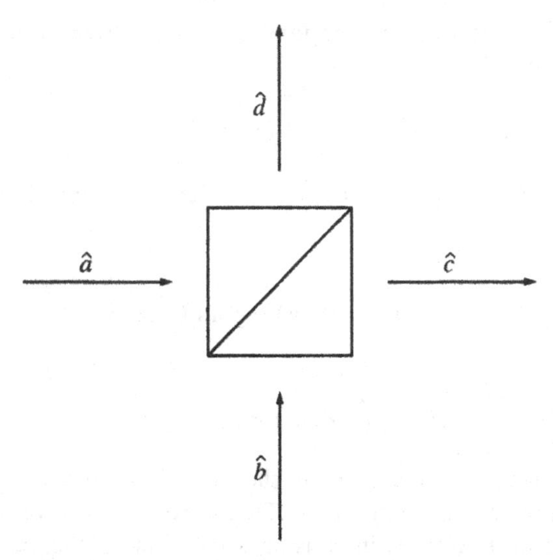

Fig. 3.1 The schematic of a beam-splitter showing the relationship between the incoming and outgoing modes.

Comparing this with (3.8.25), we see that this state is the angular momentum coherent state $|Vt, \varphi - \pi/2\rangle$. More generally, with non-zero detuning, the time-evolved state is an angular momentum coherent state multiplied by a time-dependent phase factor.

The unitary operator $\hat{R}(\theta, \phi)$ plays an important role in simple models of partially reflecting mirrors or beam-splitters. In Fig. 3.1 we illustrate the effect of a beam-splitter inducing coupling between input modes A and B with annihilation operators $\hat{a}$ and $\hat{b}$ to produce output modes C and D with annihilation operators $\hat{c}$ and $\hat{d}$. The boundary conditions at the surface of the beam-splitter lead to the relations

$$\hat{c} = t\hat{a} + r\hat{b}, \tag{3.8.29}$$

$$\hat{d} = r\hat{a} + t\hat{b}, \tag{3.8.30}$$

where t and r are complex transmission and reflection coefficients satisfying $|t|^2 + |r|^2 = 1$ and $rt^* + r^*t = 0$. Without loss of generality, we assume that t is real and that r is imaginary. These conditions imply that $[\hat{c}, \hat{c}^\dagger] = 1 = [\hat{d}, \hat{d}^\dagger]$ and $[\hat{c}, \hat{d}^\dagger] = 0 = [\hat{d}, \hat{c}^\dagger]$. The relations (3.8.29) and (3.8.30) have the form of a rotation and may be associated with the rotation operator $\hat{R}(\theta, \phi)$. In order to show this, we use a representation of the angular momentum operators in terms of the operators of the input modes, as in Appendix 5, given by

$$\hat{J}_3 = \tfrac{1}{2}(\hat{a}^\dagger\hat{a} - \hat{b}^\dagger\hat{b}), \tag{3.8.31}$$

$$\hat{J}_+ = \hat{a}^\dagger\hat{b}, \tag{3.8.32}$$

$$\hat{J}_- = \hat{b}^\dagger\hat{a}. \tag{3.8.33}$$

It is straightforward to verify that these operators do indeed satisfy the commutation relations (3.2.13) and (3.2.14). With the identifications (3.8.31) to (3.8.33), the operator $\hat{R}(\theta, -\pi/2)$ becomes

$$\hat{R}(\theta, -\pi/2) = \exp\left[\mathrm{i}\,\tfrac{1}{2}\theta(\hat{a}^\dagger\hat{b} + \hat{b}^\dagger\hat{a})\right] \tag{3.8.34}$$

and setting $t = \cos\frac{1}{2}\theta$ and $r = \mathrm{i}\sin\frac{1}{2}\theta$ allows the relations (3.8.29) and (3.8.30) to be written as the unitary transformations

$$\hat{c} = \hat{R}(-\theta, -\pi/2)\hat{a}\hat{R}(\theta, -\pi/2), \tag{3.8.35}$$

$$\hat{d} = \hat{R}(-\theta, -\pi/2)\hat{b}\hat{R}(\theta, -\pi/2). \tag{3.8.36}$$

If there are n photons in mode B and mode A is in its vacuum state then the state is an eigenstate of $\hat{J}_3$ in (3.8.31) with eigenvalue $-n/2$. We identify this state with the angular momentum state $|j, -j\rangle$ by setting $j = n/2$. The beam-splitter transforms this state into the angular momentum coherent state $|\theta, -\pi/2\rangle$ with the result that

$$\langle n/2, -n/2|\,\hat{R}(-\theta, -\pi/2)\hat{J}_3\hat{R}(\theta, -\pi/2)\,|n/2, -n/2\rangle = -\tfrac{1}{2}n\cos\theta. \tag{3.8.37}$$

It follows from (3.8.35) and (3.8.36) that this expectation value is one-half of the difference between the mean numbers of transmitted and reflected photons since $\cos\theta = |t|^2 - |r|^2$.

We can use the relations (3.8.29) and (3.8.30) to transform the state of the input modes into that of the output modes. This is done by expressing the input state in terms of the operators for modes A and B and the vacuum state $|0_A; 0_B\rangle$. The output state is found by applying the inverses of (3.8.29) and (3.9.30) to transform these operators into those for modes C and D and simply writing the vacuum state as $|0_C; 0_D\rangle$. As an example, consider the two-mode coherent state

$$|\alpha_A; \beta_B\rangle = \hat{D}_A(\alpha)\hat{D}_B(\beta)\,|0_A; 0_B\rangle \tag{3.8.38}$$

introduced in Section 3.6. Expressing the displacement operators in (3.8.38) in terms of the operators for modes C and D we obtain

$$\begin{aligned} |\alpha_A; \beta_B\rangle &= \hat{D}_C(t\alpha + r\beta)\hat{D}_D(r\alpha + t\beta)\,|0_C; 0_D\rangle \\ &= |(t\alpha + r\beta)_C; (r\alpha + t\beta)_D\rangle. \end{aligned} \tag{3.8.39}$$

This is simply the relationship between input and output fields found in classical optics. As a second example, consider the input state $|1_A; 1_B\rangle$. To determine the output state we write this state as $|1_A; 1_B\rangle = \hat{a}^\dagger\hat{b}^\dagger\,|0_A; 0_B\rangle$ and express $\hat{a}^\dagger$ and $\hat{b}^\dagger$ in terms of $\hat{c}^\dagger$ and $\hat{d}^\dagger$ to obtain

$$\begin{aligned} |1_A; 1_B\rangle &= \mathrm{i}\sqrt{2}\cos\tfrac{1}{2}\theta\sin\tfrac{1}{2}\theta(|2_c; 0_D\rangle + |0_C; 2_D\rangle) \\ &\quad + \left(\cos^2\tfrac{1}{2}\theta - \sin^2\tfrac{1}{2}\theta\right)|1_C; 1_D\rangle. \end{aligned} \tag{3.8.40}$$

It is interesting to note that if the probabilities for transmission and reflection are both 1/2 then the output state is $(\mathrm{i}/\sqrt{2})(|2_C;0_D\rangle + |0_C;2_D\rangle)$ and the probability for finding one photon in each of the output modes is zero.

Beam-splitters are often used in simple models of loss and finite detector efficiency. Suppose mode A is prepared in the state $|\psi_A\rangle$ while mode B is in the vacuum state $|0_B\rangle$. The expectation value of $\hat{c}^\dagger\hat{c}$, found by expressing $\hat{c}$ and $\hat{c}^\dagger$ in terms of the operators for modes A and B, is

$$\begin{aligned}\langle\hat{c}^\dagger\hat{c}\rangle &= \langle\psi_A;0_B|(t\hat{a}^\dagger + r^*\hat{b}^\dagger)(t\hat{a} + r\hat{b})|\psi_A;0_B\rangle \\ &= t^2\langle\hat{a}^\dagger\hat{a}\rangle;\end{aligned} \tag{3.8.41}$$

that is, the mean number in the input mode multiplied by the probability t^2 of transmission. In models of photodetection, this probability is the finite detector efficiency η. Similarly, $\langle\hat{c}^{\dagger l}\hat{c}^l\rangle = t^{2l}\langle\hat{a}^{\dagger l}\hat{a}^l\rangle$. The probability of finding m photons in mode C given that there were $l \geqslant m$ in mode A is given by the binomial factor $t^{2m}(1-t^2)^{l-m}C^l_m$. This follows from the form of the angular momentum coherent state (3.8.7).

The discrete mode analysis of the beam-splitter is readily extended to the continuum modes. The continuum mode input operators $\hat{a}(\omega)$ and $\hat{b}(\omega)$ are transformed into the output operators $\hat{c}(\omega)$ and $\hat{d}(\omega)$ by the action of the unitary operator

$$\hat{R}[\theta(\omega), -\pi/2] = \exp\left(\tfrac{1}{2}\mathrm{i}\int \mathrm{d}\omega\,\theta(\omega)[\hat{a}^\dagger(\omega)\hat{b}(\omega) + \hat{b}^\dagger(\omega)\hat{a}(\omega)]\right) \tag{3.8.42}$$

so that

$$\begin{aligned}\hat{c}(\omega') &= \hat{R}[-\theta(\omega), -\pi/2]\hat{a}(\omega')\hat{R}[\theta(\omega), -\pi/2] \\ &= \cos[\tfrac{1}{2}\theta(\omega')]\hat{a}(\omega') + \mathrm{i}\sin[\tfrac{1}{2}\theta(\omega')]\hat{b}(\omega')\end{aligned} \tag{3.8.43}$$

and

$$\begin{aligned}\hat{d}(\omega') &= \hat{R}[-\theta(\omega), -\pi/2]\hat{b}(\omega')\hat{R}[\theta(\omega), -\pi/2] \\ &= \cos[\tfrac{1}{2}\theta(\omega')]\hat{b}(\omega') + \mathrm{i}\sin[\tfrac{1}{2}\theta(\omega')]\hat{a}(\omega').\end{aligned} \tag{3.8.44}$$

As before, the properties of the output modes can be calculated by writing the input state in terms of the operators for the input modes and the vacuum state and then transforming the operators using the inverses of (3.8.43) and (3.8.44).

4

QUANTUM STATISTICS OF FIELDS

4.1 Introduction

The quantum description of the properties of the electromagnetic field is necessarily a probabilistic one in that we can only find the probability that a given property of the field has a particular value. One of the aims of modelling the electromagnetic field is to obtain the probability distribution of the possible values of a given property of the field. For most of this chapter, we will consider methods for describing the quantum statistics of a single mode of the electromagnetic field. Some of these methods may be extended to multimode fields and this is discussed briefly in the last section of this chapter.

In quantum theory, there is an infinity of different representations of the state of each mode of the electromagnetic field. All single-mode states can be expanded in terms of any complete set of basis states. For example, a normalized pure state can be written as a superposition of the number states so that

$$|\psi\rangle = \sum_{n=0}^{\infty} c_n |n\rangle, \tag{4.1.1}$$

where the complex expansion coefficients c_n satisfy the normalization condition $\sum_{n=0}^{\infty} |c_n|^2 = 1$. More generally, a mixed state with density matrix ρ can be written in the form

$$\rho = \sum_{n=0}^{\infty} \sum_{n'=0}^{\infty} \rho_{nn'} |n\rangle\langle n'|. \tag{4.1.2}$$

The orthonormality property $\langle n | n' \rangle = \delta_{nn'}$ of the number of states $|n\rangle$ together with the resolution of the identity operator, expressed as

$$\sum_{n=0}^{\infty} |n\rangle\langle n| = 1, \tag{4.1.3}$$

constitute a statement of the completeness of the discrete set $\{|n\rangle\}$. An important consequence of this completeness is that the trace of any operator $\hat{A}$ is simply

$$\mathrm{Tr}(\hat{A}) = \sum_{n=0}^{\infty} \langle n| \hat{A} |n\rangle. \tag{4.1.4}$$

An alternative complete set of basis states may be generated by the action of a unitary operator $\hat{U}$ on the number states. Since $\hat{U}\hat{U}^\dagger = \hat{U}^\dagger\hat{U} = 1$, the completeness of the set of states $\{\hat{U}|n\rangle\}$ follows from the completeness of the number states since $\langle n|\hat{U}^\dagger\hat{U}|n'\rangle = \langle n|n'\rangle = \delta_{nn'}$ and $\Sigma_{n=0}^{\infty}\hat{U}|n\rangle\langle n|\hat{U}^\dagger = \hat{U}\mathbf{1}\hat{U}^\dagger = \mathbf{1}$. An important example of a basis generated in this way is the set of displaced number states formed by the action of the Glauber displacement operator $\hat{D}(\alpha)$ discussed in Chapter 3 (see (3.6.1)).

A complete set of states can be continuous rather than discrete. The most important basis of this kind consists of the eigenstates $|x_\lambda\rangle$ of the quadrature operator $\hat{x}_\lambda = [\hat{a}\exp(-\mathrm{i}\lambda) + \hat{a}^\dagger\exp(\mathrm{i}\lambda)]/\sqrt{2}$ satisfying the eigenvalue equation $\hat{x}_\lambda|x_\lambda\rangle = x_\lambda|x_\lambda\rangle$ (see (3.2.20) and Appendix 4). These states are delta-function orthonormal so that $\langle x_\lambda|x'_\lambda\rangle = \delta(x_\lambda - x'_\lambda)$. This, together with the resolution

$$\int_{-\infty}^{\infty} \mathrm{d}x_\lambda\, |x_\lambda\rangle\langle x_\lambda| = \mathbf{1} \tag{4.1.5}$$

of the identity, constitute a statement of the completeness of the continuous set $\{|x_\lambda\rangle\}$. As with the number state basis, the completeness of this set allows the trace of an operator $\hat{A}$ to be written

$$\mathrm{Tr}(\hat{A}) = \int_{-\infty}^{\infty} \mathrm{d}x_\lambda\, \langle x_\lambda|\hat{A}|x_\lambda\rangle, \tag{4.1.6}$$

provided that the integral is defined. Any pure state $|\psi\rangle$ can be expanded as

$$|\psi\rangle = \int_{-\infty}^{\infty} \mathrm{d}x_\lambda\, \psi(x_\lambda)|x_\lambda\rangle, \tag{4.1.7}$$

where $\psi(x_\lambda) = \langle x_\lambda|\psi\rangle$ is analogous to the coordinate representation of the states of the harmonic oscillator.

The state of the field can also be expressed in terms of an overcomplete set of basis states. Such a set allows a resolution of the identity but two different states in the overcomplete set will not, in general, be orthogonal. The prime example of this is the set of coherent states

$$|\alpha\rangle = \hat{D}(\alpha)|0\rangle = \exp(-|\alpha|^2/2)\sum_{n=0}^{\infty}\frac{\alpha^n}{\sqrt{(n!)}}|n\rangle \tag{4.1.8}$$

discussed in Section 3.6. We have already seen that the coherent states are not orthogonal since $|\langle\alpha|\alpha'\rangle|^2 = \exp(-|\alpha-\alpha'|^2)$ (see (3.6.25)). However, the identity can be resolved in terms of the coherent states since

$$\frac{1}{\pi}\int_{-\infty}^{\infty} \mathrm{d}^2\alpha\, |\alpha\rangle\langle\alpha| = \mathbf{1}. \tag{4.1.9}$$

Furthermore, the trace of an operator $\hat{A}$ is given in terms of the coherent states by

$$\mathrm{Tr}(\hat{A}) = \frac{1}{\pi}\int_{-\infty}^{\infty} \mathrm{d}^2\alpha\, \langle\alpha|\hat{A}|\alpha\rangle = \sum_{n=0}^{\infty}\langle n|\hat{A}|n\rangle. \tag{4.1.10}$$

Any pure state $|\psi\rangle$ can be expressed in terms of the coherent states either as a double integral over the whole complex α-plane,

$$|\psi\rangle = \int_{-\infty}^{\infty} \mathrm{d}^2\alpha\, c(\alpha)|\alpha\rangle, \tag{4.1.11}$$

or as a line integral along a suitably chosen contour C,

$$|\psi\rangle = \int_{\mathrm{C}} \mathrm{d}\alpha\, b(\alpha) \exp(|\alpha|^2/2)|\alpha\rangle, \tag{4.1.12}$$

where $b(\alpha)$ is analytic apart from poles. The factor $\exp(|\alpha|^2/2)$ in (4.1.12) ensures that if $|\alpha\rangle$ is expanded in terms of the number of states as in (4.1.8), or any other complete basis, then the integrand is an analytic function of α, apart from poles. Neither of the above representations is unique in the sense that there are infinitely many functions $c(\alpha)$ and contours C corresponding to the same state $|\psi\rangle$. For example, the vacuum state $|0\rangle$ is obtained from (4.1.11) with any choice of $c(\alpha)$ which is a suitably well-behaved function of $|\alpha|$ only. Likewise $|0\rangle$ is obtained from (4.1.12) with $b(\alpha) = (2\pi \mathrm{i}\alpha)^{-1}$ if C is any contour enclosing the origin. The extension to mixed states is achieved by integrating $|\alpha\rangle\langle\beta|$ multiplied by a function of α and β either over the whole complex α- and β-planes or along suitably chosen contours in these planes. These are also simpler representations of any state in the form of quasi-probability distributions over the complex α-plane. We discuss this in Section 4.5.

4.2 Photon number statistics

The aim of this section is to develop methods for studying the photon number probability distribution $P(n)$ for a single field mode. A convenient quantity for ease of calculation is the moment generating function $M(\mu)$ defined for $0 \leqslant \mu \leqslant 2$ as

$$M(\mu) = \sum_{n=0}^{\infty} P(n)(1-\mu)^n. \tag{4.2.1}$$

A number of properties of $M(\mu)$ follow from this definition. We see immediately that $M(0) = 1$, while the derivatives of $M(\mu)$ evaluated at $\mu = 1$ are related to the probabilities $P(n)$ by

$$\frac{1}{n!}\left(-\frac{\mathrm{d}}{\mathrm{d}\mu}\right)^n M(\mu)\bigg|_{\mu=1} = P(n). \tag{4.2.2}$$

Further, the mth moment of the photon number operator is given by the derivative

$$\left((\mu-1)\frac{\mathrm{d}}{\mathrm{d}\mu}\right)^m M(\mu)\bigg|_{\mu=0} = \sum_{n=0}^{\infty} P(n) n^m = \langle \hat{n}^m \rangle. \tag{4.2.3}$$

We note that $M(2)$ is the difference between the probability that the photon number is even and the probability that it is odd.

The moment generating function $M(\mu)$ can also be expressed in terms of the mth factorial moment $n^{(m)}$ defined as

$$n^{(m)} = \sum_{n=0}^{\infty} P(n)n(n-1)\ldots(n-m+1) = \sum_{n=m}^{\infty} P(n)\frac{n!}{(n-m)!}, \quad (4.2.4)$$

where we set $n^{(0)} = 1$. In order to do this, we first invert (4.2.4) to express $P(n)$ in terms of the factorial moments and then substitute the result into (4.2.1). Consider the sum

$$\sum_{m=0}^{\infty} \frac{(-1)^m}{m!} n^{(n+m)} = \sum_{m=0}^{\infty} \frac{(-1)^m}{m!} \sum_{l=n+m}^{\infty} P(l)\frac{l!}{(l-n-m)!}, \quad (4.2.5)$$

using (4.2.4). Inverting the order of the summations, we have

$$\sum_{m=0}^{\infty} \frac{(-1)^m}{m!} n^{(n+m)} = \sum_{l=n}^{\infty} P(l)l! \sum_{m=0}^{l-n} \frac{(-1)^m}{m!(l-n-m)!}$$

$$= \sum_{l=n}^{\infty} P(l)\frac{l!}{(l-n)!}\delta_{ln} = P(n)n!, \quad (4.2.6)$$

the Kronecker delta arising from the binomial expansion of $(1-1)^{l-n}$. Hence the inversion of (4.2.4) is

$$P(n) = \frac{1}{n!}\sum_{m=0}^{\infty} \frac{(-1)^m}{m!} n^{(n+m)}, \quad (4.2.7)$$

and on substituting this into (4.2.1) we obtain

$$M(\mu) = \sum_{n=0}^{\infty} \frac{1}{n!}(1-\mu)^n \sum_{m=0}^{\infty} \frac{(-1)^m}{m!} n^{(n+m)}. \quad (4.2.8)$$

Writing $k = n + m$, inverting the order of the summations, and using the binomial expansion of $(1 - 1 + \mu)^k$, we obtain

$$M(\mu) = \sum_{k=0}^{\infty} n^{(k)} \frac{(-\mu)^k}{k!}. \quad (4.2.9)$$

From this expansion, we find that the factorial moment $n^{(m)}$ is given by the derivative

$$n^{(m)} = \left(-\frac{\mathrm{d}}{\mathrm{d}\mu}\right)^m M(\mu)\bigg|_{\mu=0}. \quad (4.2.10)$$

It is also straightforward, of course, to derive this result directly from (4.2.1). The properties above apply to any counting distribution $P(n)$. They can be used, for example, to study the photoelectron counting statistics arising in the photodetection of the full electromagnetic field.

For a single field mode, the photon number operator is expressed in terms of the creation operator $\hat{a}^\dagger$ and annihilation operator $\hat{a}$ for the mode as simply $\hat{n} = \hat{a}^\dagger\hat{a}$. It is convenient to introduce the operator

$$\hat{M}(\mu) = (1-\mu)^{\hat{a}^\dagger\hat{a}} = \exp\{[\ln(1-\mu)]\hat{a}^\dagger\hat{a}\}. \tag{4.2.11}$$

Clearly $\hat{M}(\mu)$ commutes with the number operator $\hat{n}$ and is therefore diagonal in the number state basis. Hence

$$\hat{M}(\mu) = \sum_{n=0}^{\infty} (1-\mu)^n \,|n\rangle\langle n|, \tag{4.2.12}$$

and consequently

$$\begin{aligned}\langle \hat{M}(\mu)\rangle &= \mathrm{Tr}(\rho\hat{M}(\mu)) \\ &= \sum_{n=0}^{\infty} (1-\mu)^n \langle n|\,\rho\,|n\rangle \\ &= M(\mu),\end{aligned} \tag{4.2.13}$$

showing that the expectation value of $\hat{M}(\mu)$ is the moment generating function $M(\mu)$. An alternative form of $M(\mu)$ may be obtained by using the normal ordering theorem (3.3.37) with $\theta = \ln(1-\mu)$ to give

$$\hat{M}(\mu) = \exp\{[\ln(1-\mu)]\hat{a}^\dagger\hat{a}\} = :\exp(-\mu\hat{a}^\dagger\hat{a}): \tag{4.2.14}$$

and hence

$$M(\mu) = \langle :\exp(-\mu\hat{a}^\dagger\hat{a}):\rangle. \tag{4.2.15}$$

Expanding the normal ordered form of $\hat{M}(\mu)$ as a power series in μ, we find

$$\hat{M}(\mu) = \sum_{m=0}^{\infty} \frac{(-\mu)^m}{m!} :(\hat{a}^\dagger\hat{a})^m: = \sum_{m=0}^{\infty} \frac{(-\mu)^m}{m!} \hat{a}^{\dagger m}\hat{a}^m, \tag{4.2.16}$$

where operators raised to the power of zero are defined to be the identity operator. Using (3.2.16) and (3.2.17) for the action of the creation and annihilation operators on the number states, it is straightforward to show that $\langle \hat{a}^{\dagger m}\hat{a}^m\rangle = n^{(m)}$ and hence that (4.2.15) is equivalent to (4.2.9).

We now give three explicit examples of $M(\mu)$ for important states of a single field mode: the coherent state, the thermal state, and the squeezed vacuum

⟩, the photon number probability distribution is

$$) = \exp(-|\alpha|^2)\frac{|\alpha|^{2n}}{n!} \qquad (4.2.17)$$

$= |\alpha|^2$ (see (3.6.7) and (3.6.8)). Inserting (4.2.17) nent generating function

$$ -|\alpha|^2)\frac{|\alpha|^{2n}}{n!}(1-\mu)^n = \exp(-\mu|\alpha|^2). \qquad (4.2.18)$$

normal ordered form of $\hat{M}(\mu)$ given by (4.2.14) he coherent states are right eigenstates of the .6.5)) to give

$$ \hat{a}^\dagger\hat{a}):|\alpha\rangle = \exp(-\mu|\alpha|^2) = \exp(-\mu\bar{n}). \qquad (4.2.19)$$

d, for example, that the mth factorial moment for

$$)^m \exp(-\mu|\alpha|^2)\Big|_{\mu=0} = |\alpha|^{2m}. \qquad (4.2.20)$$

ıermal field with mean photon number $\bar{n}$ has a probability distribution given by

$$ (n) = \frac{\bar{n}^n}{(\bar{n}+1)^{n+1}}, \qquad (4.2.21)$$

) into (4.2.1), we obtain the moment generating

$$ \sum_{=0}^{\infty} \frac{1}{(\bar{n}+1)}\left(\frac{\bar{n}}{\bar{n}+1}\right)^n (1-\mu)^n$$

$$ \frac{1}{\bar{n}+1)}\left(1-(1-\mu)\frac{\bar{n}}{\bar{n}+1}\right)^{-1}$$

$$ \frac{1}{+\mu\bar{n}}. \qquad (4.2.22)$$

ial moment is

$$ \left(\frac{d}{d\mu}\right)^m \frac{1}{1+\mu\bar{n}}\Big|_{\mu=0} = m!\,\bar{n}^m. \qquad (4.2.23)$$

The simple form of the factorial moments in the two previous examples is the result, in part, of the fact that $M(\mu)$ is a function only of the product $\mu\bar{n}$. This is not always the case as may be seen in the final example: the squeezed vacuum state $|\zeta\rangle$ (see (3.7.5)) with mean photon number $\bar{n} = \sinh^2 r$ given by

$$|\zeta\rangle = (\cosh r)^{-1/2} \sum_{k=0}^{\infty} [-\exp(\mathrm{i}\varphi)\tanh r]^k \frac{\sqrt{[(2k)!]}}{2^k k!} |2k\rangle, \quad (4.2.24)$$

for which the photon number probability distribution is

$$P(2n) = |\langle 2n | \zeta \rangle|^2 = \operatorname{sech} r (\tanh r)^{2n} \frac{(2n)!}{2^{2n}(n!)^2}, \quad (4.2.25)$$

$$P(2n+1) = 0, \quad (4.2.26)$$

for $n = 0, 1, 2, \ldots$. Inserting these into (4.2.1), we obtain the moment generating function

$$\begin{aligned} M(\mu) &= \operatorname{sech} r \sum_{n=0}^{\infty} (\tanh r)^{2n} \frac{(2n)!}{2^{2n}(n!)^2} (1-\mu)^{2n} \\ &= \operatorname{sech} r \left[1 - \tanh^2 r (1-\mu)^2\right]^{-1/2} \\ &= (1 + 2\mu \sinh^2 r - \mu^2 \sinh^2 r)^{-1/2} \\ &= (1 + 2\mu\bar{n} - \mu^2 \bar{n})^{-1/2}. \end{aligned} \quad (4.2.27)$$

We see immediately that $M(0) = 1$, as it must be for all photon statistics, and that $M(2) = 1$, reflecting the fact that only even photon numbers are possible. The factorial moments $n^{(m)}$ can be calculated directly from (4.2.27) by application of (4.2.10). However, a general expression for $n^{(m)}$ can be found by noticing that (4.2.27) may be cast into the form of a generating function for the Legendre polynomials $P_m(z)$ satisfying

$$(1 - 2tz + t^2)^{-1/2} = \sum_{m=0}^{\infty} t^m P_m(z). \quad (4.2.28)$$

Comparing the left-hand side of (4.2.28) with (4.2.27), we have

$$M(\mu) = \sum_{m=0}^{\infty} (\mathrm{i}\mu\bar{n}^{1/2})^m P_m(\mathrm{i}\bar{n}^{1/2}), \quad (4.2.29)$$

where $P_m(\mathrm{i}\bar{n}^{1/2})$ is evaluated by inserting $\mathrm{i}\bar{n}^{1/2}$ into the expression for the mth Legendre polynomial. The series (4.2.29) is strictly convergent only if μ is small enough. In general, therefore, we cannot use (4.2.29) to calculate the probabilities $P(n)$ since this involves derivatives of $M(\mu)$ evaluated at $\mu = 1$. However, moments of the number operator and the factorial moments are found from derivatives evaluated at $\mu = 0$ where (4.2.29) and its derivatives are convergent. From (4.2.29) and (4.2.9), we find that the mth factorial moment is

$$n^{(m)} = (-\mathrm{i}\bar{n}^{1/2})^m m!\, P_m(\mathrm{i}\bar{n}^{1/2}). \tag{4.2.30}$$

Using $P_1(z) = z$, $P_2(z) = (3z^2 - 1)/2$, and $P_3(z) = (5z^3 - 3z)/2$ we find that the first three factorial moments for the squeezed vacuum state are

$$n^{(1)} = \bar{n}, \quad n^{(2)} = \bar{n}(3\bar{n} + 1), \quad n^{(3)} = 3\bar{n}^2(5\bar{n} + 3). \tag{4.2.31}$$

In the same way, any desired order of factorial moment can be found using the expression for the appropriate Legendre polynomial.

As an illustration of the use of $M(\mu)$, we consider the phenomenon of Bernoulli sampling arising from the finite efficiency of photodetection as modelled, for example, by the action of a beam-splitter (see Section 3.8). If the probability of the detection of a single photon is η, then the probability of the successful detection of m photons given that the field contains $l \geqslant m$ photons is given by the binomial factor $\eta^m(1-\eta)^{l-m} l!/[m!(l-m)!]$. If the probability that the field contains l photons is $P(l)$ then the probability $P_{\mathrm{d}}(m)$ of detecting m photons is given by

$$P_{\mathrm{d}}(m) = \sum_{l=m}^{\infty} \eta^m (1-\eta)^{l-m} \frac{l!}{m!\,(l-m)!} P(l). \tag{4.2.32}$$

Consequently the moment generating function $M_{\mathrm{d}}(\mu)$ for the number of detected photons is, from (4.2.1),

$$M_{\mathrm{d}}(\mu) = \sum_{m=0}^{\infty} \sum_{l=m}^{\infty} \eta^m (1-\eta)^{l-m} \frac{l!}{m!\,(l-m)!} P(l)(1-\mu)^m. \tag{4.2.33}$$

Inverting the order of the summations, we find

$$\begin{aligned} M_{\mathrm{d}}(\mu) &= \sum_{l=0}^{\infty} P(l) \sum_{m=0}^{l} \frac{l!}{m!\,(l-m)!} [(1-\mu)\eta]^m (1-\eta)^{l-m} \\ &= \sum_{l=0}^{\infty} P(l)[(1-\mu)\eta + (1-\eta)]^l \\ &= \sum_{l=0}^{\infty} P(l)(1-\eta\mu)^l = M(\eta\mu), \end{aligned} \tag{4.2.34}$$

by comparison with (4.2.1). The effect of the finite detector efficiency is therefore simply to replace μ by $\eta\mu$. An immediate consequence of this is that, from (4.2.10), the mth factorial moment scales as η^m so that

$$n_{\mathrm{d}}^{(m)} = \eta^m n^{(m)}. \tag{4.2.35}$$

Applying the result (4.2.34) to each of the three examples considered above, we see that the effect of the finite detector efficiency on the measured statistics in the cases of coherent and thermal fields is to replace the mean photon number $\bar{n}$ by the mean detected photon number $\eta\bar{n}$ in (4.2.19) and (4.2.22). Bernoulli sampling of the Poisson and Bose–Einstein distributions therefore leaves the form of the distribution unchanged while reducing the mean. This simple behaviour follows from the dependence of their moment generating functions solely on $\mu\bar{n}$. For the squeezed vacuum, however, the moment generating function for the detected photon statistics is

$$M_{\mathrm{d}}(\mu) = (1 + 2\mu\eta\bar{n} - \mu^2\eta^2\bar{n})^{-1/2}, \tag{4.2.36}$$

which has a different functional form from that in (4.2.27) for the original squeezed vacuum. All the detected factorial moments and the photon number probabilities can be calculated from (4.2.36) using the methods described above. Here we illustrate the difference between the detected photon statistics and those of the original field by noting that, from (4.2.36),

$$M_{\mathrm{d}}(2) = [1 + 4\eta\bar{n}(1 - \eta)]^{-1/2}, \tag{4.2.37}$$

as compared with $M(2) = 1$ for the original squeezed vacuum. We recall that $M(2)$ is the difference between the probability that the number of photons is even and the probability that it is odd. While the original field contains an *even* number of photons, the value of $M_{\mathrm{d}}(2) < 1$ in (4.2.37) reflects the fact that there is a finite probability for registering an *odd* number of counts. For very high detector efficiencies for which $\eta \simeq 1$, we find from (4.2.37) that

$$M_{\mathrm{d}}(2) \simeq 1 - 2\bar{n}(1 - \eta), \tag{4.2.38}$$

showing that the probability for registering an odd number of counts is approximately $\bar{n}(1 - \eta)$. This is reasonable since $\bar{n}(1 - \eta)$ is the probability for failing to detect one of the photons if $\eta \simeq 1$.

It is also possible to use $M(\mu)$ to calculate the evolution of photon statistics in some cases, and we will discuss an example of this in Chapter 5.

4.3 Optical phase

As in classical optics, the phase of a field or a field mode is not a directly observable quantity; its value, relative to that of a reference field, must be inferred from an ensemble of measurements. Dirac envisaged phase as the quantity which is canonically conjugate to the photon number operator $\hat{n}$ associated with a given field mode. Unfortunately, such a phase operator cannot be represented within the conventional infinite-dimensional Hilbert space, nor in any extension of Hilbert space as used, for example, in representing the position and momentum operators, or indeed the quadrature operators $\hat{x}_\lambda$, and their eigenstates (see Section 3.2 and Appendix 4). In order to obtain the phase operator and its eigenstates, we require a larger state space than Hilbert space. This is constructed by considering the space Ψ_s spanned by the first $s+1$ photon number states $|0\rangle, |1\rangle, |2\rangle, \ldots, |s\rangle$, of the conventional harmonic oscillator state space. Within this space, we can construct all possible states and operators as functions of s. In order to complete the description of the state space, we require a prescription for taking the limit $s \to \infty$. This can be carried out in two ways: either we apply the limit $s \to \infty$ directly to the operators and states before calculating moments, or we can calculate moments as functions of s and only then let $s \to \infty$. For operators such as $\hat{n}$, $\hat{a}$, and $\hat{a}^\dagger$, these two limiting procedures give the same results and we can adopt either. The first produces the infinite Hilbert space of normalizable states and operators acting on it. For the operator representing phase, however, the two limiting procedures give quite different results. It is necessary in this case to use the second, more general procedure. The space Ψ so obtained contains all states in Hilbert space *and* those which are not normalizable in the limit $s \to \infty$ but which nevertheless yield finite moments for some operators. The space Ψ is precisely that required in order to represent the Hermitian phase operator conjugate to $\hat{n}$.

Two operators $\hat{A}$ and $\hat{B}$ acting only within the finite state space Ψ_s are said to be complementary or conjugate if, firstly, the eigenstates of $\hat{A}$ are equally weighted superpositions of the eigenstates of $\hat{B}$, and vice versa, and secondly, the operator $\hat{B}$ is the generator of shifts in the eigenvalue of any eigenstate of $\hat{A}$, and vice versa. The number operator $\hat{n}$ acting on Ψ_s has the photon number states $|n\rangle$ as its eigenstates with corresponding eigenvalues n (where $n = 0, 1, 2, \ldots, s$). An operator on Ψ_s conjugate to $\hat{n}$ will have normalized eigenstates $|\theta\rangle$ of the form

$$|\theta\rangle = (s+1)^{-1/2} \sum_{n=0}^{s} \exp(in\theta)|n\rangle, \tag{4.3.1}$$

which is an equally weighted superposition of the number states, and the number operator $\hat{n}$ generates shifts in the value θ according to

$$\exp(i\hat{n}\varphi)|\theta\rangle = |\theta + \varphi\rangle, \tag{4.3.2}$$

as required. The states $|\theta\rangle$ form an overcomplete set since they are not mutually orthogonal, their overlap $\langle\theta|\theta'\rangle$ being

$$\langle\theta|\theta'\rangle = (s+1)^{-1} \frac{1 - \exp[i(s+1)(\theta' - \theta)]}{1 - \exp[i(\theta' - \theta)]} \tag{4.3.3}$$

It is possible, however, to construct a complete orthonormal set of $s+1$ states $(m = 0, 1, 2, \ldots, s)$ spanning Ψ_s by taking

$$\theta_m = \theta_0 + \frac{2\pi m}{s+1}, \tag{4.3.4}$$

where θ_0 is arbitrary. A precise value of θ_0 determines which $s+1$ states $|\theta_m\rangle$ are chosen to span the space. We can, for example, express the number state $|n\rangle$ as a superposition of the states $|\theta_m\rangle$ as

$$|n\rangle = (s+1)^{-1/2} \sum_{m=0}^{s} \exp(-in\theta_m)|\theta_m\rangle. \tag{4.3.5}$$

The Hermitian optical phase operator $\hat{\phi}_\theta$ acting on Ψ_s is defined to be

$$\hat{\phi}_\theta = \sum_{m=0}^{s} \theta_m |\theta_m\rangle\langle\theta_m|. \tag{4.3.6}$$

It follows from the orthonormality of the states $|\theta_m\rangle$ that

$$\hat{\phi}_\theta|\theta_m\rangle = \theta_m|\theta_m\rangle, \tag{4.3.7}$$

so that the states $|\theta_m\rangle$ are eigenstates of the phase operator $\hat{\phi}_\theta$ with corresponding eigenvalues θ_m, and hence the states $|\theta_m\rangle$ are termed the phase states. The range of values of θ_m is restricted to the interval of length 2π determined by the choice of θ_0, which in the limit $s \to \infty$ will become the half-open interval $[\theta_0, \theta_0 + 2\pi)$. This is physically reasonable since states with phases differing by an integer multiple of 2π should be indistinguishable and must therefore have the same phase eigenvalue.

For practical purposes, it is often easier to work with the number state matrix elements of the phase operator obtained by using the definition (4.3.6) together with (4.3.1). We find that the diagonal and off-diagonal elements are

$$\langle n|\hat{\phi}_\theta|n\rangle = \theta_0 + \frac{\pi s}{s+1} \tag{4.3.8}$$

and

$$\langle n'|\hat{\phi}_\theta|n\rangle = \frac{2\pi}{s+1}\,\frac{\exp[i(n'-n)\theta_0]}{\exp[2\pi i(n'-n)/(s+1)]-1}, \tag{4.3.9}$$

respectively. From these a general expression for the expectation value of the phase operator can be obtained. A state in Ψ_s represented by the density matrix

$$\rho_s = \sum_{n=0}^{s}\sum_{n'=0}^{s} \rho_{nn'}|n\rangle\langle n'| \tag{4.3.10}$$

has the expectation value of $\hat{\phi}_\theta$ in Ψ_s given by

$$\langle\hat{\phi}_\theta\rangle_s = \mathrm{Tr}\{\rho_s\hat{\phi}_\theta\} = \sum_{n=0}^{s}\sum_{n'=0}^{s} \rho_{nn'}\langle n'|\hat{\phi}_\theta|n\rangle. \tag{4.3.11}$$

This value is the required function of s from which the expectation value of $\hat{\phi}_\theta$

in Ψ can be calculated by taking the limit as $s \to \infty$. Physical states have finite moments of the number operator, that is $\langle \hat{n}^q \rangle < \infty$ for all finite integers $q > 0$, and for such states the limit of (4.3.11) as $s \to \infty$ is well defined and is found by inserting (4.3.8) and (4.3.9) into (4.3.11) and taking the limit to give

$$\langle \hat{\phi}_\theta \rangle = \lim_{s \to \infty} \langle \hat{\phi}_\theta \rangle_s = \theta_0 + \pi + \sum_{n=0}^{\infty} \sum_{\substack{n'=0 \\ n' \neq n}}^{\infty} \frac{\mathrm{i}}{n - n'} \exp[\mathrm{i}(n' - n)\theta_0]\, \rho_{nn'}. \quad (4.3.12)$$

States which are diagonal in the number of state basis are states of random phase. This follows directly from the construction of $\hat{\phi}_\theta$ as complementary to $\hat{n}$, and for such states $\langle \hat{\phi}_\theta \rangle = \theta_0 + \pi$. This value lies at the mid-point of the eigenvalue range, θ_0 to $\theta_0 + 2\pi$, as might have been expected for states of random phase.

The commutator of $\hat{n}$ and $\hat{\phi}_\theta$ is well defined in Ψ_s and has diagonal elements in the number state and phase state bases given by

$$\langle n | \left[\hat{n}, \hat{\phi}_\theta\right] | n \rangle = 0 = \langle \theta_m | \left[\hat{n}, \hat{\phi}_\theta\right] | \theta_m \rangle, \quad (4.3.13)$$

which must be the case because $|n\rangle$ and $|\theta_m\rangle$ are eigenstates of $\hat{n}$ and $\hat{\phi}_\theta$, respectively. Summing (4.3.13) over either n or m gives zero for the trace of the commutator, as is the case for all commutators of operators which act on a finite-dimensional state space. The off-diagonal elements are non-zero and are given by

$$\langle n' | \left[\hat{n}, \hat{\phi}_\theta\right] | n \rangle = \frac{2\pi}{s+1} \frac{(n' - n)\exp[\mathrm{i}(n' - n)\theta_0]}{\exp[2\pi \mathrm{i}(n' - n)/(s+1)] - 1}. \quad (4.3.14)$$

and

$$\langle \theta_{m'} | \left[\hat{n}, \hat{\phi}_\theta\right] | \theta_m \rangle = \frac{2\pi}{s+1} \frac{(m - m')}{\exp[2\pi \mathrm{i}(m - m')/(s+1)] - 1}. \quad (4.3.15)$$

We calculate the expectation value of the commutator $[\hat{n}, \hat{\phi}_\theta]$ for a state in Ψ_s with density matrix (4.3.10) and then, as before, take the limit as $s \to \infty$. For physical states, the resulting expectation value is

$$\begin{aligned} \left\langle \left[\hat{n}, \hat{\phi}_\theta\right] \right\rangle &= \lim_{s \to \infty} \left\langle \left[\hat{n}, \hat{\phi}_\theta\right] \right\rangle_s \\ &= -\mathrm{i} \sum_{n=0}^{\infty} \sum_{n'=0}^{\infty} \rho_{nn'}(1 - \delta_{nn'})\exp[\mathrm{i}(n' - n)\theta_0]. \end{aligned} \quad (4.3.16)$$

This limiting value can also be obtained as the limit of a simpler expectation value in Ψ_s given by

$$\left\langle \left[\hat{n}, \hat{\phi}_\theta\right] \right\rangle = \lim_{s \to \infty} \mathrm{i}[1 - (s+1)\langle |\theta_0\rangle\langle \theta_0| \rangle]. \quad (4.3.17)$$

This equivalence only holds for the expectation value in physical states; in particular it does not hold for the expectation value in a phase state nor for general states in Ψ_s prior to the evaluation of the limit. In cases where conflicting results are obtained, the correct procedure is to use the expressions (4.3.13) to (4.3.15).

We can construct functions of the phase operator $\hat{\phi}_\theta$ as in Section 3.3, principal among these being the exponential functions $\exp(\pm i\hat{\phi}_\theta)$. These operators have the phase states as their eigenstates since

$$\exp\left(\pm i\hat{\phi}_\theta\right)|\theta_m\rangle = \exp(\pm i\theta_m)|\theta_m\rangle. \tag{4.3.18}$$

The properties of the operator $\exp(i\hat{\phi}_\theta)$ are found by considering its action on the photon number states $|n\rangle$. We have

$$\exp\left(i\hat{\phi}_\theta\right)|n\rangle = (s+1)^{-1/2}\sum_{m=0}^{s}\exp[-i(n-1)\theta_m]|\theta_m\rangle, \tag{4.3.19}$$

where we have used the expansion (4.3.5). For $n > 0$, the resulting state is simply the lowered number state $|n-1\rangle$. For $n = 0$, we obtain a state which it is tempting to label as $|-1\rangle$ but which must, in fact, be identified with the state $|s\rangle$ since

$$\begin{aligned}\exp\left(i\hat{\phi}_\theta\right)|0\rangle &= (s+1)^{-1/2}\sum_{m=0}^{s}\exp(i\theta_m)|\theta_m\rangle \\ &= (s+1)^{-1/2}\exp[i(s+1)\theta_0]\sum_{m=0}^{s}\exp(-is\theta_m)|\theta_m\rangle \\ &= \exp[i(s+1)\theta_0]|s\rangle.\end{aligned} \tag{4.3.20}$$

Therefore the number state representation of $\exp(i\hat{\phi}_\theta)$ is

$$\begin{aligned}\exp\left(i\hat{\phi}_\theta\right) &= |0\rangle\langle 1| + |1\rangle\langle 2| + \cdots + |s-1\rangle\langle s| \\ &\quad + \exp[i(s+1)\theta_0]|s\rangle\langle 0|,\end{aligned} \tag{4.3.21}$$

with $\exp(-i\hat{\phi}_\theta)$ given by the Hermitian conjugate of this expression. Hence the phase operator acts as the generator of shifts in the photon number. This confirms that $\hat{\phi}_\theta$ is the operator conjugate to $\hat{n}$. From the operators $\exp(\pm i\hat{\phi}_\theta)$, we can form usual trigonometric formulae. We then find that, for example,

$$\cos^2\left(\hat{\phi}_\theta\right) + \sin^2\left(\hat{\phi}_\theta\right) = 1 \tag{4.3.22}$$

and that all the other familiar trigonometrical identities apply to the operator functions of $\hat{\phi}_\theta$. Moreover, for all the number states $|n\rangle$,

$$\langle n|\cos^2\left(\hat{\phi}_\theta\right)|n\rangle = \tfrac{1}{2} = \langle n|\sin^2\left(\hat{\phi}_\theta\right)|n\rangle, \tag{4.3.23}$$

consistent with the number states being states of random phase.

We can define creation an annihilation operators acting on Ψ_s in terms of $\hat{\phi}_\theta$ and $\hat{n}$. The annihilation operator is

$$\hat{a} = \exp(\mathrm{i}\hat{\phi}_\theta)\hat{n}^{1/2}$$
$$= |0\rangle\langle 1| + \sqrt{2}\,|1\rangle\langle 2| + \cdots + \sqrt{s}\,|s-1\rangle\langle s|, \tag{4.3.24}$$

where

$$\hat{n}^{1/2} = \sum_{n=0}^{s} n^{1/2}|n\rangle\langle n|. \tag{4.3.25}$$

The creation operator $\hat{a}^\dagger$ is the Hermitian conjugate of $\hat{a}$. The action of $\hat{a}$ in (4.3.24) on the states in Ψ_s is the same as that of the conventional annihilation operator on the first $s+1$ number states of Hilbert space. Indeed, for physical states, the moments of products of the annihilation and creation operators, given by (4.3.24) and its Hermitian conjugate, become identical, in the limit $s \to \infty$, to those calculated using the conventional operators and the state in Hilbert space. For example, the commutator of $\hat{a}$ and $\hat{a}^\dagger$ obtained from (4.3.24) has the unfamiliar form

$$[\hat{a}, \hat{a}^\dagger] = \mathbf{1} - (s+1)|s\rangle\langle s|, \tag{4.3.26}$$

but all finite moments of this operator for physical states are nevertheless unity, as we now show. The qth power of the commutator (4.3.26) is found to be

$$[\hat{a}, \hat{a}^\dagger]^q = \mathbf{1} - [1 - (-s)^q]|s\rangle\langle s|. \tag{4.3.27}$$

The expectation value of this operator for any state in Ψ_s with density matrix (4.3.10) is

$$\langle[\hat{a}, \hat{a}^\dagger]^q\rangle_s = 1 - [1 - (-s)^q]\rho_{ss}. \tag{4.3.28}$$

In the limit $s \to \infty$, the second term in (4.3.28) tends to zero for all physical states. This follows directly from the requirement that all finite moments of the number operator are themselves finite. In particular, the moment $\langle\hat{n}^{q+1}\rangle > s^{q+1}\rho_{ss}$ but is finite as $s \to \infty$; this implies that $s^q\rho_{ss}$ tends to zero in this limit. We conclude that, for all physical states,

$$\langle[\hat{a}, \hat{a}^\dagger]^q\rangle = \lim_{s\to\infty}\langle[\hat{a}, \hat{a}^\dagger]^q\rangle_s = 1. \tag{4.3.29}$$

The moments of the phase operator may be found by calculating the expectation value of the appropriate power of $\hat{\phi}_\theta$ in Ψ_s and then taking the limit as $s \to \infty$, so that

$$\langle\hat{\phi}_\theta^l\rangle = \lim_{s\to\infty}\sum_{m=0}^{s}\theta_m^l\langle\theta_m|\rho|\theta_m\rangle. \tag{4.3.30}$$

However, for physical states, there is a more straightforward approach which involves applying the limit to the summation so as to convert it into a Riemann integral, as follows:

$$\langle \hat{\phi}_\theta^l \rangle = \int_{\theta_0}^{\theta_0+2\pi} \theta^l P(\theta)\, d\theta, \tag{4.3.31}$$

where $P(\theta)$ is the phase probability density function given by

$$\begin{aligned} P(\theta) &= \lim_{s\to\infty} \frac{(s+1)}{2\pi} \langle \theta | \rho | \theta \rangle \\ &= \frac{1}{2\pi} \sum_{n=0}^{\infty} \sum_{n'=0}^{\infty} \rho_{nn'} \exp[i(n'-n)\theta]. \end{aligned} \tag{4.3.32}$$

It must be emphasized that this procedure should only be applied to physical states, whereas (4.3.30) can be used to calculate the phase moments for any state.

We now examine the phase properties of particular states. As a first example, we consider the number states which, as mentioned previously, are states of random phase for which $\langle \hat{\phi}_\theta \rangle = \theta_0 + \pi$. The expectation value of $\hat{\phi}_\theta^2$ in Ψ_s is

$$\begin{aligned} \langle \hat{\phi}_\theta^2 \rangle_s &= \sum_{m=0}^{s} \theta_m^2 |\langle \theta_m | n \rangle|^2 \\ &= \sum_{m=0}^{s} \left(\theta_0 + \frac{2\pi m}{s+1} \right)^2 \frac{1}{s+1} \\ &= \theta_0^2 + \frac{2\pi\theta_0 s}{s+1} + \frac{4\pi^2}{3} \frac{s(s+\frac{1}{2})}{(s+1)^2}. \end{aligned} \tag{4.3.33}$$

This result becomes a true expectation value in the limit $s \to \infty$. In this limit, the variance of the phase is

$$\Delta \phi_\theta^2 = \lim_{s\to\infty} \left(\langle \hat{\phi}_\theta^2 \rangle_s - \langle \hat{\phi}_\theta \rangle_s^2 \right) = \pi^2/3. \tag{4.3.34}$$

Alternatively, since the number states are physical states, we can calculate the phase probability density function using (4.3.32) to obtain $P(\theta) = 1/2\pi$, obviously corresponding to a random distribution of phase. All the moments of $\hat{\phi}_\theta$ can now be calculated directly from (4.3.31) with the result that

$$\langle \hat{\phi}_\theta^l \rangle = \frac{1}{2\pi(l+1)} \left[(\theta_0 + 2\pi)^{l+1} - \theta_0^{l+1} \right], \tag{4.3.35}$$

which for $l = 2$ agrees with the limit as $s \to \infty$ of (4.3.33). The same phase properties are found for all states which are diagonal in the number state basis.

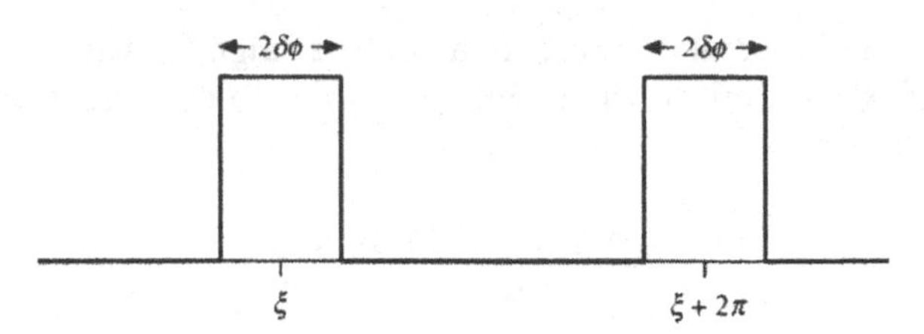

Fig. 4.1 A section of a 2π-periodic phase distribution.

The dependence of these moments on θ_0 may appear strange but is an inevitable consequence of the choice of the smallest eigenvalue θ_0 made in the definition of $\hat{\phi}_\theta$. As mentioned earlier, it is not meaningful to distinguish phases which differ by an integer multiple of 2π.

The significance of θ_0 in the interpretation of the phase properties of a state is more apparent when we consider states of partially determined phase. We illustrate this point with an elementary example. Consider a state that is an equally weighted superposition of all the phase states corresponding to phases within $\pm\delta\phi$ of some mean angle ξ. A section of the resulting 2π-periodic distribution of phases is shown in Fig. 4.1. The eigenvalues of $\hat{\phi}_\theta$ lie in the interval $[\theta_0, \theta_0 + 2\pi)$. Figures 4.2(a) and 4.2(b) show the effect of superimposing this range onto the phase distribution for different values of θ_0. For the choice $\xi - 2\pi + \delta\phi < \theta_0 < \xi - \delta\phi$ shown in Fig. 4.2(a), the calculated mean value of $\hat{\phi}_\theta$ is ξ while the variance is $\Delta\phi_\theta^2 = (\delta\phi)^2/3$. These results are as we might have inferred from inspection of Fig. 4.1. The expectation value of $\hat{\phi}_\theta$ does, in fact, depend on θ_0, for increasing θ_0 by 2π changes $\langle\hat{\phi}_\theta\rangle$ to $\xi + 2\pi$. A more serious situation is shown in Fig. 4.2(b) where we have chosen $\theta_0 = \xi$. The parts of the phase distribution associated with phases greater than or less than ξ appear at

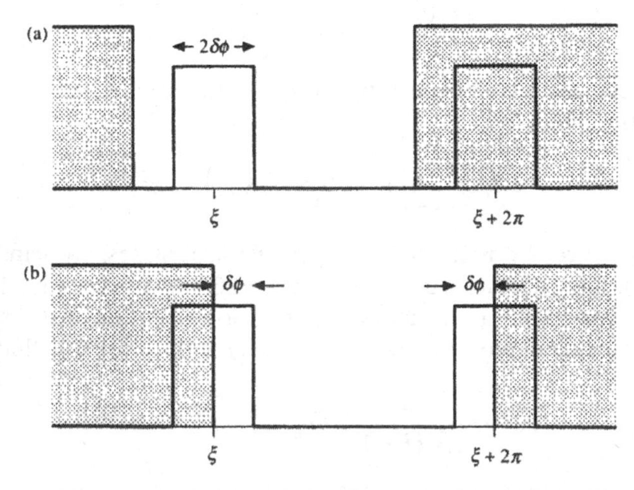

Fig. 4.2 Two possible ranges of phase eigenvalues superimposed on a 2π-periodic phase distribution.

opposite ends of the eigenvalue range of $\hat{\phi}_\xi$. The resulting expectation value and variance are

$$\langle \hat{\phi}_\xi \rangle = \xi + \pi \tag{4.3.36}$$

and

$$\Delta \phi_\xi^2 = \pi^2 - \pi \,\delta\phi + \tfrac{1}{3}(\delta\phi)^2. \tag{4.3.37}$$

If $\delta\phi$ is small then the variance (4.3.37) is approximately π^2 which is larger than the value $\pi^2/3$ associated with a state of random phase. This effect is not of quantum origin but arises even for classical phase distributions and is a consequence of the multivalued nature of phase. The expectation value and variance found in (4.3.36) and (4.3.37) are correct given the choice $\theta_0 = \xi$. The marked difference between the expressions found for the two choices of θ_0 highlights the need for caution when interpreting results obtained by employing the phase operator. It is usually the case that the most natural interpretation arises when θ_0 is chosen so as to minimize the calculated variance.

Finally, we consider the phase properties of the coherent states $|\alpha\rangle$ which in Hilbert space have the form

$$|\alpha\rangle = \exp(-|\alpha|^2/2) \sum_{n=0}^{\infty} \frac{|\alpha|^n}{\sqrt{(n!)}} \exp(in\xi)|n\rangle, \tag{4.3.38}$$

where $\alpha = |\alpha|\exp(i\xi)$. Since these states are physical states we do not need to construct states in Ψ_s for which the moments tend in the limit $s \to \infty$ to those of the coherent states. Instead we can apply the formula (4.3.32) for the phase probability density function for physical states to obtain

$$P(\theta) = \frac{1}{2\pi} \left| \sum_{n=0}^{\infty} \exp(-|\alpha|^2/2) \frac{|\alpha|^n}{\sqrt{(n!)}} \exp[in(\xi - \theta)] \right|^2. \tag{4.3.39}$$

This function is 2π periodic in θ with maxima at $\theta = \xi + 2k\pi$ for all integers k. For general values of the amplitude $|\alpha|$, $P(\theta)$ must be evaluated numerically. However, for large values of $|\alpha|$, we can obtain an approximate analytical expression for $P(\theta)$ near $\theta = \xi$ (or any of the maxima of $P(\theta)$). In this limit, the Poisson photon number distribution is well approximated by a Gaussian distribution

$$\exp(-|\alpha|^2)\frac{|\alpha|^{2n}}{n!} \simeq (2\pi|\alpha|^2)^{-1/2} \exp\left(-\frac{(|\alpha|^2 - n)^2}{2|\alpha|^2}\right). \tag{4.3.40}$$

Inserting the square root of (4.3.40) into (4.3.39) gives an approximate form for

$P(\theta)$. For values of θ near the maximum at ξ, the resulting summation over n in (4.3.39) can be approximated by the integral

$$(2\pi|\alpha|^2)^{-1/4}\sum_{n=0}^{\infty}\exp\left(-\frac{(|\alpha|^2-n)^2}{4|\alpha|^2}\right)\exp[\mathrm{i}n(\xi-\theta)]$$

$$\simeq(2\pi|\alpha|^2)^{-1/4}\int_{-\infty}^{\infty}\mathrm{d}x\exp\left(-\frac{(|\alpha|^2-x)^2}{4|\alpha|^2}\right)\exp[\mathrm{i}x(\xi-\theta)]$$

$$=(8\pi|\alpha|^2)^{1/4}\exp\left[-|\alpha|^2(\xi-\theta)^2\right]\exp\left[\mathrm{i}|\alpha|^2(\xi-\theta)\right]. \qquad (4.3.41)$$

Taking the modulus squared of this expression and inserting the result into (4.3.39) gives the form of $P(\theta)$ near $\theta=\xi$ as

$$P(\theta)\simeq\left(\frac{2|\alpha|^2}{\pi}\right)^{1/2}\exp\left[-2|\alpha|^2(\xi-\theta)^2\right]. \qquad (4.3.42)$$

The complete probability density function consists of a sequence of such Gaussians separated by 2π. From this we can calculate approximate forms for the phase moments. The comments made earlier in connection with states of partially determined phase still apply and these moments will, in general, depend on the choice of θ_0. However, if θ_0 is sufficiently different from ξ for the whole of the narrow Gaussian peak in $P(\theta)$ to lie in the eigenvalue range θ_0 to $\theta_0+2\pi$, then we find, within the above approximations, that

$$\langle\hat{\phi}_\theta\rangle=\xi \qquad (4.3.43)$$

and

$$\Delta\phi_\theta^2=(4|\alpha|^2)^{-1}. \qquad (4.3.44)$$

The bound on the product $\Delta n\cdot\Delta\phi_\theta$ is given by Heisenberg's uncertainty relation

$$\Delta n\cdot\Delta\phi_\theta\geqslant\tfrac{1}{2}\left|\left\langle\left[\hat{n},\hat{\phi}_\theta\right]\right\rangle\right|. \qquad (4.3.45)$$

For physical states, the right-hand side of this inequality can be found using (4.3.17). We obtain

$$\tfrac{1}{2}\left|\left\langle\left[\hat{n},\hat{\phi}_\theta\right]\right\rangle\right|=\tfrac{1}{2}\lim_{s\to\infty}\left[1-(s+1)|\langle\alpha|\theta_0\rangle|^2\right]$$

$$=\tfrac{1}{2}[1-2\pi P(\theta_0)], \qquad (4.3.46)$$

by (4.3.39). Within the limits of the approximations made in deriving (4.3.32) and for θ_0 sufficiently different from ξ, $P(\theta_0)\simeq 0$ so that the uncertainty relation for coherent states of large amplitude becomes

$$\Delta n\cdot\Delta\phi_\theta\geqslant\tfrac{1}{2}. \qquad (4.3.47)$$

For a coherent state, $\Delta n = |\alpha|$ (see (3.6.11)) so that, from (4.3.44), we see that large-amplitude coherent states are approximately number-phase minimum uncertainty states since $\Delta n \cdot \Delta\phi_\theta \simeq \frac{1}{2}$.

At the beginning of this section, we introduced the space Ψ_s and described two possible ways of taking the limit $s \to \infty$. We conclude this section by illustrating some of the inconsistencies that occur if we attempt to represent phase within Hilbert space by applying the limit directly to the states and operators as opposed to moments. We construct operators in Hilbert space from those in Ψ_s by taking the matrix elements of the operators in the number state representation in Ψ_s and letting s tend to infinity *before* calculating moments. The operators in Hilbert space are defined by the complete set of the matrix elements so obtained. For example, the Hilbert space analogue of $\exp(i\hat{\phi}_\theta)$ is

$$\widehat{\exp(i\phi_\theta)}_H = \lim_{N\to\infty}\left(\lim_{s\to\infty}\sum_{n=0}^{N}\sum_{n'=0}^{N}|n\rangle\langle n|\exp(i\hat{\phi}_\theta)|n'\rangle\langle n'|\right)$$

$$= \sum_{n=0}^{\infty}|n\rangle\langle n+1|, \tag{4.3.48}$$

where we have used (4.3.21). This operator is not unitary and does not commute with its Hermitian conjugate constructed in the same way from $\exp(-i\hat{\phi}_\theta)$. The fact that these operators do not commute means that they cannot be functions of a common operator and shows that an Hermitian phase operator cannot be constructed in Hilbert space. A similar procedure applied to the phase operator itself leads to

$$\hat{\phi}_{\theta H} = \lim_{N\to\infty}\left(\lim_{s\to\infty}\sum_{n=0}^{N}\sum_{n'=0}^{N}|n\rangle\langle n|\,\hat{\phi}_\theta\,|n'\rangle\langle n'|\right)$$

$$= \theta_0 + \pi + i\sum_{n=0}^{\infty}\sum_{\substack{n'=0\\ n\neq n'}}^{\infty}\frac{1}{n-n'}\exp[i(n'-n)\theta_0]\,|n'\rangle\langle n|, \tag{4.3.49}$$

where we have used (4.3.9). This is not the required Hermitian phase operator and, for example, $\exp(i\hat{\phi}_{\theta H})$ is *not* the operator $\widehat{\exp(i\phi_\theta)}_H$. From (4.3.49), we can see that $\hat{\phi}_{\theta H}$ cannot consistently represent phase, as follows. For all number states the expectation value of $\hat{\phi}_{\theta H}$ is $\theta_0 + \pi$ for any chosen value of θ_0. The only probability distribution consistent with this is the uniform distribution corresponding to a probability density $(2\pi)^{-1}$. Such a distribution must, as we have seen, give a phase variance of $\pi^2/3$. However, the variance of $\hat{\phi}_{\theta H}$ for the vacuum state has the value $\pi^2/6$. Thus taking the limit $s \to \infty$ before evaluating moments does not lead to consistent results for phase. Similar self-inconsistencies arise if we treat (4.3.48) and its Hermitian conjugate as phase operators.

In the remainder of this book, we restrict our attention to physical states of the field and represent them in Hilbert space. Although we cannot then explicitly represent the phase operator, we can calculate moments of the phase using (4.3.31) with the phase probability density function (4.3.32).

4.4 Characteristic functions

The preceding two sections have each dealt with the statistics of a single property of a field mode (the photon number and phase). However, there are methods for obtaining a more complete statistical description of the field. In this section and the next we discuss two such related methods based on characteristic functions and quasi-probability distributions, respectively. Both of these enable the expectation value of any operator function of $\hat{a}$ and $\hat{a}^\dagger$ to be calculated. A characteristic function or a quasi-probability distribution contains all the information necessary to reconstruct the density matrix for the state. In this sense, they are alternative complete descriptions of the state of the field. Both rely on the properties of the coherent states and the Glauber displacement operator $\hat{D}(\alpha)$ described in Chapter 3.

We define the p-ordered characteristic function to be

$$\begin{aligned}\chi(\xi,p) &= \mathrm{Tr}\left[\rho\hat{D}(\xi)\right]\exp\left(p|\xi|^2/2\right)\\ &= \mathrm{Tr}\left[\rho\exp(\xi\hat{a}^\dagger - \xi^*\hat{a})\right]\exp\left(p|\xi|^2/2\right),\end{aligned} \tag{4.4.1}$$

where we have used (3.3.30). Each value of the parameter p corresponds to a specific ordering of the creation and annihilation operators in this expression. In particular, the values $p = 1$, 0, and -1 correspond to normal, symmetric, and antinormal ordered characteristic functions, respectively:

$$\chi(\xi,1) = \mathrm{Tr}[\rho\exp(\xi\hat{a}^\dagger)\exp(-\xi^*\hat{a})], \tag{4.4.2}$$

$$\chi(\xi,0) = \mathrm{Tr}\left[\rho\exp(\xi\hat{a}^\dagger - \xi^*\hat{a})\right], \tag{4.4.3}$$

$$\chi(\xi,-1) = \mathrm{Tr}[\rho\exp(-\xi^*\hat{a})\exp(\xi\hat{a}^\dagger)], \tag{4.4.4}$$

where we have used the ordering theorems (3.3.31) and (3.3.32). The characteristic function $\chi(\xi,p)$ is always defined and is, in general, a complex-valued function. At $\xi = 0$, $\chi(0,p) = \mathrm{Tr}(\rho) = 1$ independently of the ordering parameter p. The characteristic function $\chi(\xi,0)$ is the expectation value of $\hat{D}(\xi)$ and, since $\hat{D}(\xi)$ is unitary, the magnitude of each of its eigenvalues is unity. Hence $|\chi(\xi,0)| \leqslant 1$ with $|\chi(\xi,0)|$ achieving its maximum value at $\xi = 0$. From this result and (4.4.1), we can bound the magnitude of $\chi(\xi,p)$ by

$$|\chi(\xi,p)| \leqslant \exp\left(p|\xi|^2/2\right). \tag{4.4.5}$$

Given the characteristic function $\chi(\xi,p)$, the expectation value of any p-ordered combination of $\hat{a}^\dagger$ and $\hat{a}$ can be calculated by differentiation with respect to ξ and ξ^*, treated as independent variables. The expectation value of the p-ordered product of $\hat{a}^{\dagger m}$ and $\hat{a}^n$ is

$$\langle \hat{a}^{\dagger m}\hat{a}^n \rangle_p = \left(\frac{\partial}{\partial \xi}\right)^m \left(-\frac{\partial}{\partial \xi^*}\right)^n \chi(\xi, p)\Big|_{\xi=0}. \tag{4.4.6}$$

This equation also serves as the *definition* of p-ordering. For $p=1$ and -1, we obtain $\langle \hat{a}^{\dagger m}\hat{a}^n\rangle$ and $\langle \hat{a}^n\hat{a}^{\dagger m}\rangle$, respectively, while for $p=0$, we obtain the expectation value of the symmetrized product $\langle S(\hat{a}^{\dagger m}\hat{a}^n)\rangle$ (see (3.3.17) and (3.3.18)). In calculating (4.4.6), it is more convenient to rewrite (4.4.1) in the equivalent form

$$\chi(\xi, p) = \mathrm{Tr}[\,\rho \exp(\xi \hat{a}^\dagger)\exp(-\xi^*\hat{a})]\exp\!\left[(p-1)|\xi|^2/2\right], \tag{4.4.7}$$

where we have again used the ordering theorem (3.3.31). As an example of the use of (4.4.6), we calculate the expectation value of the p-ordered product of $\hat{a}^\dagger$ and $\hat{a}$. We have, using (4.4.7),

$$\begin{aligned}\langle \hat{a}^\dagger \hat{a}\rangle_p = \Bigg\{ &\mathrm{Tr}[\,\rho \hat{a}^\dagger \exp(\xi\hat{a}^\dagger)\exp(-\xi^*\hat{a})\hat{a}] \\ &+\mathrm{Tr}[\,\rho \exp(\xi\hat{a}^\dagger)\exp(-\xi^*\hat{a})\hat{a}]\frac{(p-1)\xi^*}{2} \\ &-\mathrm{Tr}[\,\rho \hat{a}^\dagger \exp(\xi\hat{a}^\dagger)\exp(-\xi^*\hat{a})]\frac{(p-1)\xi}{2} \\ &-\mathrm{Tr}[\,\rho \exp(\xi\hat{a}^\dagger)\exp(-\xi^*\hat{a})]\frac{(p-1)}{2} \\ &\times\left(1+\frac{(p-1)|\xi|^2}{2}\right)\Bigg\}\exp\!\left[(p-1)|\xi|^2/2\right]\Bigg|_{\xi=0} \\ = \langle \hat{a}^\dagger\hat{a}\rangle + \tfrac{1}{2}(1-p).&\end{aligned} \tag{4.4.8}$$

For $p=1$, 0, and -1, we obtain $\langle \hat{a}^\dagger\hat{a}\rangle$, $\langle \frac{1}{2}(\hat{a}^\dagger\hat{a}+\hat{a}\hat{a}^\dagger)\rangle = \langle S(\hat{a}^\dagger\hat{a})\rangle$, and $\langle \hat{a}\hat{a}^\dagger\rangle$, respectively, as expected. Functions containing only $\hat{a}$ or only $\hat{a}^\dagger$ are, of course, independent of ordering and hence of p. For example,

$$\begin{aligned}\langle \hat{a}^{\dagger 2}\rangle_p &= \frac{\partial^2}{\partial \xi^2}\mathrm{Tr}[\,\rho\exp(\xi\hat{a}^\dagger)\exp(-\xi^*\hat{a})]\exp\!\left[(p-1)|\xi|^2/2\right]\Big|_{\xi=0} \\ &= \langle \hat{a}^{\dagger 2}\rangle.\end{aligned} \tag{4.4.9}$$

The characteristic function can also be used to generate p-ordered moments of the operator $\hat{x}_\lambda$ given by (3.2.20). Defining $\xi_\varphi = \mathrm{i}(\xi \exp(-\mathrm{i}\varphi) - \xi^* \exp(\mathrm{i}\varphi)]/\sqrt{2}$, we can rewrite the characteristic function (4.4.1) as

$$\chi(\xi, p) = \mathrm{Tr}\left\{\rho \exp\left[-\mathrm{i}\left(\xi_\lambda \hat{x}_\lambda + \xi_{\lambda+\pi/2}\hat{x}_{\lambda+\pi/2}\right)\right]\right\} \exp\left[\tfrac{1}{4}p\left(\xi_\lambda^2 + \xi_{\lambda+\pi/2}^2\right)\right]. \tag{4.4.10}$$

We obtain the p-ordered moment of $\hat{x}_\lambda^n$ by setting $\xi_{\lambda+\pi/2} = 0$ and differentiating with respect to ξ_λ, so that

$$\langle \hat{x}_\lambda^n \rangle_p = \left(\mathrm{i}\frac{\partial}{\partial \xi_\lambda}\right)^n \chi(\xi, p)\bigg|_{\xi=0}. \tag{4.4.11}$$

The characteristic function contains all the information necessary to reconstruct the density matrix, including the photon number statistics. This being the case, we expect to be able to construct the moment generating function $M(\mu)$, discussed in Section 4.2, from the characteristic function $\chi(\xi, p)$. We now show how this can be done. In Chapter 3, we introduced the operator

$$\hat{T}(p) = \frac{1}{\pi^2}\int_{-\infty}^{\infty} \mathrm{d}^2\xi \exp(\xi\hat{a}^\dagger - \xi^*\hat{a}) \exp(p|\xi|^2/2), \tag{4.4.12}$$

for $p < 1$, and showed that

$$\hat{T}(p) = \frac{2}{\pi(1-p)} :\exp\left(-\frac{2}{1-p}\hat{a}^\dagger\hat{a}\right): \tag{4.4.13}$$

(see (3.6.38)). Comparing this expression with that given by (4.2.14) for $\hat{M}(\mu)$, we see that

$$\hat{M}(\mu) = \frac{\pi}{\mu}\hat{T}\left(1 - \frac{2}{\mu}\right), \tag{4.4.14}$$

provided that the argument of $\hat{T}$ is less than unity, that is $\mu > 0$. Hence the moment generating function $M(\mu)$ can be expressed as

$$M(\mu) = \langle \hat{M}(\mu)\rangle = \frac{\pi}{\mu}\left\langle \hat{T}\left(1 - \frac{2}{\mu}\right)\right\rangle. \tag{4.4.15}$$

In order to relate this to the characteristic function $\chi(\xi, p)$, we note that the expectation value of $\hat{T}(p)$ is

$$\begin{aligned}\langle \hat{T}(p)\rangle &= \frac{1}{\pi^2}\int_{-\infty}^{\infty} \mathrm{d}^2\xi\, \mathrm{Tr}\left[\rho \exp(\xi\hat{a}^\dagger - \xi^*\hat{a})\right]\exp(p|\xi|^2/2)\\ &= \frac{1}{\pi^2}\int_{-\infty}^{\infty} \mathrm{d}^2\xi\, \chi(\xi, p),\end{aligned} \tag{4.4.16}$$

using the definition of the characteristic function (4.4.1). Hence, from (4.4.15),

$$M(\mu) = \frac{1}{\pi\mu}\int_{-\infty}^{\infty} \mathrm{d}^2\xi\, \chi(\xi, 1 - 2/\mu), \tag{4.4.17}$$

for $\mu > 0$. This form of the moment generating function is equivalent to its definition given in (4.2.1) and can be used to derive its properties. For example,

$$M(0) = \int_{-\infty}^{\infty} \mathrm{d}^2\xi \lim_{\mu\to 0} \frac{\chi(\xi, 1-2/\mu)}{\pi\mu}$$

$$= \int_{-\infty}^{\infty} \mathrm{d}^2\xi\, \chi(\xi,1) \lim_{\mu\to 0} \frac{\exp(-|\xi|^2/\mu)}{\pi\mu}. \tag{4.4.18}$$

The limit in the integrand of (4.4.18) is simply the two-dimensional delta function $\delta^{(2)}(\xi) \equiv \delta(\mathrm{Re}\,\xi)\delta(\mathrm{Im}\,\xi)$ (see Appendix 2), and hence $M(0) = \chi(0,1) = \mathrm{Tr}(\rho) = 1$, as previously demonstrated. Further, using the form (4.4.4) of $\chi(\xi,-1)$ and the cyclic property of the trace, we find

$$M(1) = \frac{1}{\pi}\int_{-\infty}^{\infty} \mathrm{d}^2\xi\, \mathrm{Tr}[\exp(\xi\hat{a}^\dagger)\rho\exp(-\xi^*\hat{a})]$$

$$= \frac{1}{\pi}\int_{-\infty}^{\infty} \mathrm{d}^2\xi \int_{-\infty}^{\infty} \mathrm{d}^2\alpha\, \frac{1}{\pi}\langle\alpha|\,\rho\,|\alpha\rangle \exp(\xi\alpha^* - \xi^*\alpha), \tag{4.4.19}$$

where the trace has been evaluated in the coherent state basis using (4.1.10) and we have used the property that the coherent states are right eigenstates of the annihilation operator (see (3.6.5)). Evaluating the integration over ξ first gives

$$M(1) = \int_{-\infty}^{\infty} \mathrm{d}^2\alpha\, \langle\alpha|\,\rho\,|\alpha\rangle\, \delta^{(2)}(\alpha) = \langle 0|\,\rho\,|0\rangle = P(0), \tag{4.4.20}$$

in agreement with (4.2.2).

As an example of a characteristic function, we find $\chi_n(\xi,p)$ for the number state $|n\rangle$. From (4.4.7) with $\rho = |n\rangle\langle n|$, we have

$$\chi_n(\xi,p) = \langle n|\exp(\xi\hat{a}^\dagger)\exp(-\xi^*\hat{a})|n\rangle \exp[(p-1)|\xi|^2/2]$$

$$= \sum_{l=0}^{\infty}\sum_{m=0}^{\infty} \frac{\xi^l(-\xi^*)^m}{l!\,m!} \langle n|\,\hat{a}^{\dagger l}\hat{a}^m\,|n\rangle \exp[(p-1)|\xi|^2/2]$$

$$= \sum_{l=0}^{n}\sum_{m=0}^{n} \frac{\xi^l(-\xi^*)^m}{l!\,m!}$$

$$\times \langle n-l|\left(\frac{n!}{(n-l)!}\right)^{1/2}\left(\frac{n!}{(n-m)!}\right)^{1/2} |n-m\rangle \exp[(p-1)|\xi|^2/2]$$

$$= \sum_{m=0}^{n} \frac{(-|\xi|^2)^m}{(m!)^2}\frac{n!}{(n-m)!} \exp[(p-1)|\xi|^2/2]$$

$$= L_n(|\xi|^2)\exp[(p-1)|\xi|^2/2], \tag{4.4.21}$$

where L_n is the Laguerre polynomial of order n. From this expression, we can obtain the characteristic function for any state, the density matrix of which is diagonal in the number state basis in the form of a weighted sum of the $\chi_n(\xi, p)$ in (4.4.21). For example, the characteristic function of a thermal state with mean photon number $\bar{n}$ is

$$\chi(\xi,p) = \sum_{n=0}^{\infty} P(n)\chi_n(\xi,p)$$

$$= \frac{1}{(1+\bar{n})} \sum_{n=0}^{\infty} \left(\frac{\bar{n}}{1+\bar{n}}\right)^n L_n(|\xi|^2) \exp\left[(p-1)|\xi|^2/2\right]. \quad (4.4.22)$$

This may be evaluated by using the expression for the generating function of the Laguerre polynomials given by $(1-z)^{-1}\exp[xz/(z-1)] = \sum_{n=0}^{\infty} z^n L_n(x)$ for $|z|<1$. We find

$$\chi(\xi,p) = \exp\left[-\tfrac{1}{2}(2\bar{n}+1-p)|\xi|^2\right]. \quad (4.4.23)$$

In general, the characteristic function will depend on both the modulus and argument of ξ. However, we see from the form of (4.4.21) that any density matrix diagonal in the number state basis will have a characteristic function which depends only on $|\xi|^2$. Consider such a function $\chi(\xi,p) = F(|\xi|^2)$ having the Taylor expansion

$$F(|\xi|^2) = \sum_{k=0}^{\infty} \frac{F^{(k)}(0)}{k!}|\xi|^{2k} = \sum_{k=0}^{\infty} \frac{F^{(k)}(0)}{k!}\xi^k\xi^{*k}. \quad (4.4.24)$$

We know that such a Taylor expansion must exist because the finiteness of all moments of the creation and annihilation operators implies, from (4.4.6), that $F(|\xi|^2)$ is regular at the origin. Differentiating (4.4.24), we obtain

$$\left(\frac{\partial}{\partial\xi}\right)^m \left(\frac{\partial}{\partial\xi^*}\right)^n F(|\xi|^2) = \sum_{k\geqslant m,n}^{\infty} \frac{F^{(k)}(0)}{k!} \frac{k!}{(k-m)!}\xi^{k-m} \frac{k!}{(k-n)!}\xi^{*k-n} \quad (4.4.25)$$

and hence, since this expression can be non-zero at $\xi=0$ only if $m=n$,

$$\left(\frac{\partial}{\partial\xi}\right)^m \left(\frac{\partial}{\partial\xi^*}\right)^n F(|\xi|^2)\bigg|_{\xi=0} = m!\,F^{(m)}(0)\,\delta_{mn}. \quad (4.4.26)$$

From (4.4.6) we therefore have that

$$\langle\hat{a}^{\dagger m}\hat{a}^n\rangle_p = (-1)^m m!\,F^{(m)}(0)\,\delta_{mn}, \quad (4.4.27)$$

showing that only moments of products of equal numbers of creation and annihilation operators are non-zero for density matrices which are diagonal in the number state basis. For the thermal state, we have from the characteristic function (4.4.23) that

$$\langle \hat{a}^{\dagger m}\hat{a}^m \rangle_p = m![\bar{n} + \tfrac{1}{2}(1-p)]^m. \tag{4.4.28}$$

For normal ordering, for which $p = 1$, we recover the mth factorial moment given by (4.2.23). In the case of the number state $|n\rangle$ with characteristic function given by (4.4.21), we find

$$\begin{aligned}\langle \hat{a}^{\dagger m}\hat{a}^m \rangle_p &= m!(-1)^m \frac{\mathrm{d}^m}{\mathrm{d}x^m}\{L_n(x)\exp[(p-1)x/2]\}\Big|_{x=0} \\ &= m!(-1)^m \sum_{k=0}^{m} L_n^{(k)}(0)\left(\frac{p-1}{2}\right)^{m-k} C_k^m,\end{aligned} \tag{4.4.29}$$

where $x = |\xi|^2$, $C_k^m = m!/k!(m-k)!$, and

$$L_n^{(k)}(x) = \frac{\mathrm{d}^k}{\mathrm{d}x^k} L_n(x) \tag{4.4.30}$$

is an associated Laguerre polynomial. At $x = 0$, $L_n^{(k)}(0) = (-1)^k C_k^n$ so that

$$\langle \hat{a}^{\dagger m}\hat{a}^m \rangle_p = m! \sum_{k=0}^{\min(n,m)} C_k^n C_k^m \left(\frac{1-p}{2}\right)^{m-k}, \tag{4.4.31}$$

where the upper limit of summation (the minimum of n and m) arises from the fact that $L_n^{(k)}(x)$ vanishes for $k > n$. It is possible to express (4.4.31) in closed form in terms of a Jacobi polynomial but it is much more convenient to obtain the moments directly from (4.4.31). If $p = 1$, a non-zero term can only arise when $k = m$, so that $\langle n|\,\hat{a}^{\dagger m}\hat{a}^m\,|n\rangle$ is zero if $m > n$ and equal to $m!\,C_m^n C_m^m = n!/(n-m)!$ if $m \leqslant n$. If $p = -1$, we obtain

$$\langle n|\,\hat{a}^m\hat{a}^{\dagger m}\,|n\rangle = m! \sum_{k=0}^{\min(n,m)} C_k^n C_k^m = m!\,C_n^{n+m} = \frac{(n+m)!}{n!}. \tag{4.4.32}$$

These results for $p = \pm 1$ agree with those found in Section 3.4. For $p = 0$, we illustrate the use of (4.4.31) to calculate the moment of $S(\hat{a}^{\dagger 2}\hat{a}^2)$. Inserting $p = 0$ and $m = 2$ into (4.4.31) we find

$$\langle n|\,S(\hat{a}^{\dagger 2}\hat{a}^2)\,|n\rangle = n^2 + n + \tfrac{1}{2}, \tag{4.4.33}$$

in agreement with (3.3.21). More generally, the p-ordered expectation value of $\hat{n}^2$ for the number state $|n\rangle$ is $\langle n|\,\hat{a}^{\dagger 2}\hat{a}^2\,|n\rangle_p = n^2 + n(1-2p) + \tfrac{1}{2}(p-1)^2$.

The action of a unitary transformation on a state can lead to a simple modification of the characteristic function so that we can easily find the characteristic function for the transformed state if we know that of the original state. Consider a state with density matrix $\tilde{\rho}$ generated from the density matrix ρ by the unitary transformation $\tilde{\rho} = \hat{U}\rho\hat{U}^\dagger$. The characteristic function for this state is

$$\begin{aligned}\tilde{\chi}(\xi, p) &= \mathrm{Tr}\left[\rho\hat{U}^\dagger \exp(\xi\hat{a}^\dagger - \xi^*\hat{a})\hat{U}\right]\exp(p|\xi|^2/2)\\ &= \mathrm{Tr}\left[\rho\exp\left(\xi\hat{U}^\dagger\hat{a}^\dagger\hat{U} - \xi^*\hat{U}^\dagger\hat{a}\hat{U}\right)\right]\exp(p|\xi|^2/2),\end{aligned} \tag{4.4.34}$$

where we have used the cyclic property of the trace operation and (3.3.14). This expression will be useful if $\tilde{\chi}(\xi, p)$ is simply related to the known $\chi(\xi, p)$. This occurs when $\hat{U}^\dagger\hat{a}\hat{U}$ is a linear combination of $\hat{a}$ and $\hat{a}^\dagger$, and there are three basic types of unitary transformation for which this condition is satisfied. The first of these is the transformation with $\hat{U} = \exp\{-\mathrm{i}\varphi\hat{a}^\dagger\hat{a}\}$ for which

$$\hat{U}^\dagger\hat{a}\hat{U} = \hat{a}\exp(-\mathrm{i}\varphi), \tag{4.4.35}$$

by (3.3.9). Inserting this and its Hermitian conjugate into (4.4.34), we obtain

$$\begin{aligned}\tilde{\chi}(\xi, p) &= \mathrm{Tr}\{\rho\exp[\xi\exp(\mathrm{i}\varphi)\hat{a}^\dagger - \xi^*\exp(-\mathrm{i}\varphi)\hat{a}]\}\exp(p|\xi|^2/2)\\ &= \chi(\xi\exp(\mathrm{i}\varphi), p).\end{aligned} \tag{4.4.36}$$

Setting $\varphi = \omega t$ in (4.4.35) gives the evolution of the annihilation operator for the free-field mode, and (4.4.36) then gives the corresponding evolution of the characteristic function. The second unitary transformation is produced by the action of the Glauber displacement operator $\hat{U} = \hat{D}(\alpha)$ for which $\hat{U}^\dagger\hat{a}\hat{U} = \hat{a} + \alpha$ (see (3.6.20)). Inserting this and its Hermitian conjugate into (4.4.34), we find

$$\tilde{\chi}(\xi, p) = \chi(\xi, p)\exp(\xi\alpha^* - \xi^*\alpha). \tag{4.4.37}$$

The argument of the exponential in (4.4.37) is purely imaginary so that transforming the density matrix using $\hat{D}(\alpha)$ corresponds to changing the phase of the characteristic function. Using (4.4.37) and (4.4.21) for $n = 0$, we find that the characteristic function for the coherent state $|\alpha\rangle = \hat{D}(\alpha)|0\rangle$ is

$$\chi_\alpha(\xi, p) = \exp\left[\tfrac{1}{2}(p-1)|\xi|^2\right]\exp(\xi\alpha^* - \xi^*\alpha), \tag{4.4.38}$$

from which any desired p-ordered moment can be found. For example, for normal ordering with $p = 1$ we have

$$\begin{aligned}\langle\alpha|\hat{a}^{\dagger m}\hat{a}^n|\alpha\rangle &= \left(\frac{\partial}{\partial\xi}\right)^m\left(-\frac{\partial}{\partial\xi^*}\right)^n \exp(\xi\alpha^* - \xi^*\alpha)\bigg|_{\xi=0}\\ &= \alpha^{*m}\alpha^n,\end{aligned} \tag{4.4.39}$$

in accordance with the fact that $\hat{a}|\alpha\rangle = \alpha|\alpha\rangle$. The third unitary transformation is produced by the action of the single-mode squeezing operator $\hat{U} = \hat{S}(\zeta)$ for which

$$\hat{U}^\dagger \hat{a} \hat{U} = \hat{a} \cosh r - \hat{a}^\dagger \exp(\mathrm{i}\varphi) \sinh r, \qquad (4.4.40)$$

where $\zeta = r\exp(\mathrm{i}\varphi)$ (see (3.7.10)). In this case, the relationship between the characteristic functions for the transformed state and the original state is only simple for $p = 0$. Inserting (4.4.40) into (4.4.34) with $p = 0$, we obtain

$$\begin{aligned}\tilde{\chi}(\xi, 0) &= \mathrm{Tr}\Big[\rho \exp\Big\{\xi\left[\hat{a}^\dagger \cosh r - \hat{a}\exp(-\mathrm{i}\varphi)\sinh r\right] \\ &\quad - \xi^*[\hat{a}\cosh r - \hat{a}^\dagger \exp(\mathrm{i}\varphi)\sinh r]\Big\}\Big] \\ &= \chi(\xi \cosh r + \xi^* \exp(\mathrm{i}\varphi)\sinh r, 0).\end{aligned} \qquad (4.4.41)$$

In particular, the squeezed vacuum state $|\zeta\rangle = \hat{S}(\zeta)|0\rangle$ has the characteristic function

$$\chi_\zeta(\xi, 0) = \exp\left[-\tfrac{1}{2}|\xi \cosh r + \xi^* \exp(\mathrm{i}\varphi)\sinh r|^2\right]. \qquad (4.4.42)$$

The p-ordered characteristic function for this state is found by multiplying (4.4.42) by $\exp(p|\xi|^2/2)$.

It is also possible to calculate the characteristic function if we know the number state matrix elements $\rho_{nn'}$ of the density matrix, since we can then write

$$\chi(\xi, p) = \sum_{n=0}^{\infty}\sum_{n'=0}^{\infty} \rho_{nn'} \langle n'|\hat{D}(\xi)|n\rangle \exp(p|\xi|^2/2). \qquad (4.4.43)$$

The diagonal number state matrix elements of $\hat{D}(\xi)$ are simply the characteristic functions for the corresponding number states with $p = 0$ and from (4.4.21) are therefore given by

$$\langle n|\hat{D}(\xi)|n\rangle = \chi_n(\xi, 0) = \exp(-|\xi|^2/2) L_n(|\xi|^2), \qquad (4.4.44)$$

where L_n is the Laguerre polynomial of order n. To calculate the off-diagonal elements we again use the normal ordered form (3.3.31) of $\hat{D}(\xi)$ to write

$$\begin{aligned}\langle n'|\hat{D}(\xi)|n'-l\rangle &= \exp(-|\xi|^2/2)\langle n'|\exp(\xi\hat{a}^\dagger) \\ &\quad \times \exp(-\xi^*\hat{a})\hat{a}^l|n'\rangle\left(\frac{(n'-l)!}{n'!}\right)^{1/2} \\ &= \left(\frac{(n'-l)!}{n'!}\right)^{1/2} \exp(-|\xi|^2/2)\left(-\frac{\partial}{\partial \xi^*}\right)^l \\ &\quad \times \langle n'|\exp(\xi\hat{a}^\dagger)\exp(-\xi^*\hat{a})|n'\rangle \\ &= \left(\frac{(n'-l)!}{n'!}\right)^{1/2} \exp(-|\xi|^2/2)(-\xi)^l\left(\frac{\partial}{\partial|\xi|^2}\right)^l L_{n'}(|\xi|^2),\end{aligned}$$

$$(4.4.45)$$

using (4.4.44). Hence if $n < n'$, the matrix element is

$$\langle n'|\hat{D}(\xi)|n\rangle = \left(\frac{n!}{n'!}\right)^{1/2} \exp(-|\xi|^2/2)(-\xi)^{n'-n} L_n^{(n'-n)}(|\xi|^2), \quad (4.4.46)$$

where $L_n^{(n'-n)}$ is an associated Laguerre polynomial. Similarly for $n > n'$, we find

$$\langle n'|\hat{D}(\xi)|n\rangle = \left(\frac{n'!}{n!}\right)^{1/2} \exp(-|\xi|^2/2)(\xi^*)^{n-n'} L_{n'}^{(n-n')}(|\xi|^2). \quad (4.4.47)$$

As an example of this approach, we consider the simple superposition state $c_0|0\rangle + c_1|1\rangle$ discussed at the end of Section 3.7. The non-zero density matrix elements of this state are $\rho_{00} = c_0^2$, $\rho_{11} = c_1^2$, and $\rho_{01} = \rho_{10} = c_0 c_1$, where as before we have chosen c_0 and c_1 to be real. From (4.4.43), (4.4.44), (4.4.46), and (4.4.47) we find that the characteristic function for this state is

$$\begin{aligned} \chi(\xi, p) &= \exp\left[(p-1)|\xi|^2/2\right] \\ &\quad \times \left[c_0^2 L_0(|\xi|^2) + c_1^2 L_1(|\xi|^2) + c_0 c_1 L_1^{(1)}(|\xi|^2)(\xi^* - \xi)\right] \\ &= \exp\left[(p-1)|\xi|^2/2\right]\left[1 - c_1^2|\xi|^2 - c_0 c_1(\xi^* - \xi)\right], \end{aligned} \quad (4.4.48)$$

where we have used $L_0(x) = 1$, $L_1(x) = 1 - x$, and hence $L_1^{(1)} = -1$.

4.5 Quasi-probability distributions

As an alternative to the characteristic function, it may be advantageous to employ a quasi-probability function obtained by taking the Fourier transform of the characteristic function. In contrast to the characteristic function, the quasi-probability distribution is real valued. It is similar to a true probability distribution for the field amplitude in that it is normalized and that moments of products of $\hat{a}$ and $\hat{a}^\dagger$ are calculated by evaluating an integral weighted by the quasi-probability distribution. In this sense, it is analogous to a classical phase-space probability distribution. However, the quasi-probability distribution is not always positive and consequently interpreting it as a probability distribution is not always possible.

We define the quasi-probability distribution $W(\alpha, p)$ corresponding to the p-ordered characteristic function $\chi(\xi, p)$ as

$$W(\alpha, p) = \frac{1}{\pi^2}\int_{-\infty}^{\infty} d^2\xi\, \chi(\xi, p) \exp(\alpha\xi^* - \alpha^*\xi). \quad (4.5.1)$$

This is a two-dimensional Fourier transform of $\chi(\xi, p)$ since $\alpha\xi^* - \alpha^*\xi = 2i(\alpha_i \xi_r - \alpha_r \xi_i)$, where $\alpha = \alpha_r + i\alpha_i$ and $\xi = \xi_r + i\xi_i$, the subscripts r and i denoting the real and imaginary parts, respectively. The integral (4.5.1) is not always well behaved. For some states and values of p, for example, $\chi(\xi, p)$

diverges as $|\xi|\to\infty$ and $W(\alpha,p)$ can then only be expressed in terms of generalized functions. Nevertheless, $W(\alpha,p)$ can still be used to calculate moments of products of $\hat{a}$ and $\hat{a}^\dagger$. The properties of $W(\alpha,p)$ stated above follow from the definition (4.5.1) and the properties of $\chi(\xi,p)$. To see that $W(\alpha,p)$ is always real valued consider its complex conjugate value given by

$$[W(\alpha,p)]^* = \frac{1}{\pi^2}\int_{-\infty}^{\infty} \mathrm{d}^2\xi [\chi(\xi,p)]^* \exp(\alpha^*\xi - \alpha\xi^*)$$
$$= \frac{1}{\pi^2}\int_{-\infty}^{\infty} \mathrm{d}^2\xi \,\mathrm{Tr}[\rho \exp(\xi^*\hat{a} - \xi\hat{a}^\dagger)]$$
$$\times \exp\left(p|\xi|^2/2\right)\exp(\alpha^*\xi - \alpha\xi^*). \tag{4.5.2}$$

Making the substitution $\xi = -\eta$, we obtain

$$[W(\alpha,p)]^* = \frac{1}{\pi^2}\int_{-\infty}^{\infty} \mathrm{d}^2\eta \,\mathrm{Tr}\left[\rho \exp(\eta\hat{a}^\dagger - \eta^*\hat{a})\right]$$
$$\times \exp\left(p|\eta|^2/2\right)\exp(\alpha\eta^* - \alpha^*\eta)$$
$$= W(\alpha,p), \tag{4.5.3}$$

from (4.5.1), showing that the value of $W(\alpha,p)$ is real for all complex α and real p. To see that $W(\alpha,p)$ is normalized, consider the integral

$$\int_{-\infty}^{\infty} \mathrm{d}^2\alpha\, W(\alpha,p) = \frac{1}{\pi^2}\int_{-\infty}^{\infty} \mathrm{d}^2\alpha \int_{-\infty}^{\infty} \mathrm{d}^2\xi\, \chi(\xi,p)\exp(\alpha\xi^* - \alpha^*\xi). \tag{4.5.4}$$

Carrying out the integration over α first gives

$$\int_{-\infty}^{\infty} \mathrm{d}^2\alpha\, W(\alpha,p) = \int_{-\infty}^{\infty} \mathrm{d}^2\xi\, \chi(\xi,p)\delta^{(2)}(\xi) = \chi(0,p) = 1. \tag{4.5.5}$$

Moments of any desired p-ordered product of $\hat{a}$ and $\hat{a}^\dagger$ can be obtained by evaluating the integral

$$\langle \hat{a}^{\dagger m}\hat{a}^n\rangle_p = \int_{-\infty}^{\infty} \mathrm{d}^2\alpha\, W(\alpha,p)\alpha^{*m}\alpha^n. \tag{4.5.6}$$

This result may be established by inserting the definition (4.5.1) of $W(\alpha,p)$ into the right-hand side of (4.5.6) giving

$$\frac{1}{\pi^2}\int_{-\infty}^{\infty} \mathrm{d}^2\alpha \int_{-\infty}^{\infty} \mathrm{d}^2\xi\, \chi(\xi,p)\alpha^{*m}\alpha^n \exp(\alpha\xi^* - \alpha^*\xi)$$
$$= \frac{1}{\pi^2}\int_{-\infty}^{\infty} \mathrm{d}^2\alpha \int_{-\infty}^{\infty} \mathrm{d}^2\xi\, \chi(\xi,p)\left(-\frac{\partial}{\partial\xi}\right)^m\left(\frac{\partial}{\partial\xi^*}\right)^n \exp(\alpha\xi^* - \alpha^*\xi). \tag{4.5.7}$$

Carrying out the integration over α first leads to a generalized function which is the derivative of $\delta^{(2)}(\xi)$ (see Appendix 2). The right-hand side of (4.5.7) then becomes

$$\int_{-\infty}^{\infty} d^2\xi\, \chi(\xi,p)\left(-\frac{\partial}{\partial\xi}\right)^m \left(\frac{\partial}{\partial\xi^*}\right)^n \delta^{(2)}(\xi) = \left(\frac{\partial}{\partial\xi}\right)^m \left(-\frac{\partial}{\partial\xi^*}\right)^n \chi(\xi,p)\bigg|_{\xi=0}$$
$$= \langle \hat{a}^{\dagger m}\hat{a}^n\rangle_p, \qquad (4.5.8)$$

where we have used (A2.20) and (4.4.6).

Given $W(\alpha,p)$, we can obtain $W(\alpha,p')$ for $p'<p$ as a convolution of $W(\alpha,p)$ with a Gaussian. From (4.4.1), the p-ordered and p'-ordered characteristic functions are related by

$$\chi(\xi,p') = \chi(\xi,p)\exp\left[-(p-p')|\xi|^2/2\right]. \qquad (4.5.9)$$

Hence from (4.5.1)

$$W(\alpha,p') = \frac{1}{\pi^2}\int_{-\infty}^{\infty} d^2\xi\, \chi(\xi,p)\exp\left[-(p-p')|\xi|^2/2\right]\exp(\alpha\xi^* - \alpha^*\xi), \qquad (4.5.10)$$

which is the Fourier transform of the product of $\chi(\xi,p)$ and the Gaussian $\exp[-(p-p')|\xi|^2/2]$. By the Fourier convolution theorem we find

$$W(\alpha,p') = \frac{2}{\pi(p-p')}\int_{-\infty}^{\infty} d^2\beta\, W(\beta,p)\exp\left(-\frac{2|\alpha-\beta|^2}{(p-p')}\right). \qquad (4.5.11)$$

As p' decreases, peaks in the quasi-probability distribution become broader.

Although we have defined $W(\alpha,p)$ in terms of $\chi(\xi,p)$, it is not necessary to calculate $\chi(\xi,p)$ first since there are other representations of $W(\alpha,p)$. The first of these, valid for all $p<1$, follows from the definitions (4.4.1) and (4.5.1) of $\chi(\xi,p)$ and $W(\alpha,p)$. Inserting (4.4.1) into (4.5.1) and rearranging gives

$$W(\alpha,p) = \frac{1}{\pi^2}\int_{-\infty}^{\infty} d^2\xi\, \mathrm{Tr}\{\rho\exp[\xi(\hat{a}^\dagger-\alpha^*) - \xi^*(\hat{a}-\alpha)]\}\exp(p|\xi|^2/2)$$
$$= \frac{1}{\pi^2}\int_{-\infty}^{\infty} d^2\xi\, \mathrm{Tr}\left[\rho\hat{D}(\alpha)\exp(\xi\hat{a}^\dagger - \xi^*\hat{a})\hat{D}^\dagger(\alpha)\right]\exp(p|\xi|^2/2), \qquad (4.5.12)$$

where we have used (3.6.20) and (3.6.22), the transformations induced by the displacement operator $\hat{D}(\alpha)$, with α replaced by $-\alpha$. We can express (4.5.12) in terms of the operator $\hat{T}(p)$ in (4.4.12) as

$$W(\alpha,p) = \mathrm{Tr}\left[\rho\hat{D}(\alpha)\hat{T}(p)\hat{D}^\dagger(\alpha)\right]$$
$$= \mathrm{Tr}\left[\hat{D}^\dagger(\alpha)\rho\hat{D}(\alpha)\hat{T}(p)\right], \qquad (4.5.13)$$

using the cyclic property of the trace operation. In Chapter 3, we showed that the operator $\hat{T}(p)$ is diagonal in the number state basis. Using (3.6.39) and evaluating the trace in (4.5.13) in the number state basis gives

$$W(\alpha, p) = \frac{2}{\pi(1-p)} \sum_{n=0}^{\infty} \left(\frac{p+1}{p-1}\right)^n \langle n| \hat{D}^{\dagger}(\alpha)\rho\hat{D}(\alpha)|n\rangle. \quad (4.5.14)$$

This is a representation of the quasi-probability distribution $W(\alpha, p)$, for $p < 1$, in terms of the diagonal matrix elements of the density matrix ρ in the displaced number state basis (see Section 3.6).

In practice it is often convenient to choose an appropriate value of p, usually $p = 1$, 0, or -1 corresponding to quasi-probability distributions giving normal symmetric, or antinormal ordered moments respectively. A special nomenclature has been adopted for quasi-probability distributions with these values of p. For $p = 1$, $W(\alpha, 1) = P(\alpha)$, the Glauber–Sudarshan P-representation. The density matrix ρ can be expressed directly in terms of the P-representation as

$$\rho = \int_{-\infty}^{\infty} \mathrm{d}^2\alpha\, P(\alpha)|\alpha\rangle\langle\alpha|, \quad (4.5.15)$$

where the $|\alpha\rangle$ are the coherent states. To justify this assertion, consider the expectation value of the normal ordered product

$$\begin{aligned}\langle \hat{a}^{\dagger m}\hat{a}^n\rangle &= \mathrm{Tr}(\rho\hat{a}^{\dagger m}\hat{a}^n) = \mathrm{Tr}(\hat{a}^n\rho\hat{a}^{\dagger m}) \\ &= \mathrm{Tr}\left(\int_{-\infty}^{\infty} \mathrm{d}^2\alpha\, P(\alpha)|\alpha\rangle\langle\alpha|\, \alpha^n\alpha^{*m}\right), \quad (4.5.16)\end{aligned}$$

where we have used $\hat{a}|\alpha\rangle = \alpha|\alpha\rangle$. The trace of the projector $|\alpha\rangle\langle\alpha|$ is unity and hence

$$\langle \hat{a}^{\dagger m}\hat{a}^n\rangle = \int_{-\infty}^{\infty} \mathrm{d}^2\alpha\, P(\alpha)\alpha^{*m}\alpha^n. \quad (4.5.17)$$

It is clear that $P(\alpha)$ in (4.5.17) has replaced $W(\alpha, 1)$ in (4.5.6). The equivalence of $W(\alpha, 1)$ and $P(\alpha)$ is established by substituting (4.5.15) into the characteristic function (4.4.2) and using the resulting expression for $\chi(\xi, 1)$ in (4.5.1). We find

$$\begin{aligned}W(\alpha, 1) &= \frac{1}{\pi^2}\int_{-\infty}^{\infty} \mathrm{d}^2\xi\, \mathrm{Tr}\left(\int_{-\infty}^{\infty} \mathrm{d}^2\beta\, P(\beta)|\beta\rangle\langle\beta|\exp(\xi\hat{a}^{\dagger})\exp(-\xi^*\hat{a})\right) \\ &\quad\times \exp(\alpha\xi^* - \alpha^*\xi) \\ &= \frac{1}{\pi^2}\int_{-\infty}^{\infty} \mathrm{d}^2\xi \int_{-\infty}^{\infty} \mathrm{d}^2\beta\, P(\beta)\exp[\xi(\beta^* - \alpha^*) - \xi^*(\beta - \alpha)] \\ &= P(\alpha), \quad (4.5.18)\end{aligned}$$

since carrying out the integration over ξ first again gives a two-dimensional delta function, in this case $\delta^{(2)}(\beta-\alpha)$. The P-representation is sometimes difficult to use because it can be highly singular. In particular, all pure states have P-representations which can only be expressed in terms of generalized functions. For example, the coherent state with density matrix $\rho=|\beta\rangle\langle\beta|$ has a P-representation $P(\alpha)=\delta^{(2)}(\alpha-\beta)$, as can be seen from (4.5.15), while the number states other than the vacuum state have P-representations which are sums of derivatives of delta functions.

For $p=0$, $W(\alpha,0)=W(\alpha)$, the Wigner function. We already have one representation of the Wigner function since putting $p=0$ into (4.5.14) gives

$$W(\alpha)=\frac{2}{\pi}\sum_{n=0}^{\infty}(-1)^n\langle n|\hat{D}^\dagger(\alpha)\rho\hat{D}(\alpha)|n\rangle$$
$$=\frac{2}{\pi}\mathrm{Tr}\left[\hat{D}^\dagger(\alpha)\rho\hat{D}(\alpha)(-1)^{\hat{a}^\dagger\hat{a}}\right], \tag{4.5.19}$$

where $(-1)^{\hat{a}^\dagger\hat{a}}$ is defined by (3.3.8). An alternative representation in terms of coherent state matrix elements of ρ can be obtained by evaluating the trace in (4.5.19) in the coherent state basis. We find

$$W(\alpha)=\frac{2}{\pi^2}\int_{-\infty}^{\infty}\mathrm{d}^2\beta\langle\beta|\hat{D}^\dagger(\alpha)\rho\hat{D}(\alpha)(-1)^{\hat{a}^\dagger\hat{a}}|\beta\rangle$$
$$=\frac{2}{\pi^2}\int_{-\infty}^{\infty}\mathrm{d}^2\beta\langle\beta|\hat{D}^\dagger(\alpha)\rho\hat{D}(\alpha)|-\beta\rangle, \tag{4.5.20}$$

using (3.6.34). As we showed in Chapter 3 (see (3.6.31)), the action of the displacement operator $\hat{D}(\alpha)$ on a coherent state $|\alpha'\rangle$ produces a displaced coherent state multiplied by a phase factor given by $\hat{D}(\alpha)|\alpha'\rangle=\exp[\frac{1}{2}(\alpha\alpha'^*-\alpha^*\alpha')]|\alpha+\alpha'\rangle$. Using this result in (4.5.20), we obtain

$$W(\alpha)=\frac{2}{\pi^2}\int_{-\infty}^{\infty}\mathrm{d}^2\beta\langle\alpha+\beta|\rho|\alpha-\beta\rangle\exp(\alpha^*\beta-\alpha\beta^*). \tag{4.5.21}$$

The Wigner function does not have the singular behaviour sometimes found for $P(\alpha)$, but it can have negative values. From (4.5.19), we can write

$$W(\alpha)=\frac{2}{\pi}\left(\sum_{n=0}^{\infty}\langle 2n|\hat{D}^\dagger(\alpha)\rho\hat{D}(\alpha)|2n\rangle\right.$$
$$\left.-\sum_{n=0}^{\infty}\langle 2n+1|\hat{D}^\dagger(\alpha)\rho\hat{D}(\alpha)|2n+1\rangle\right). \tag{4.5.22}$$

The first of these summations is the probability that the state with density matrix $\hat{D}^\dagger(\alpha)\rho\hat{D}(\alpha)$ has an even number of photons, and it therefore has a value between 0 and 1. Similarly, the second summation is the probability that this state has an odd number of photons. We infer that $W(\alpha)$ in (4.5.22) is a bounded function satisfying $|W(\alpha)| \leq 2/\pi$ for all α. If we view the Wigner function as a joint distribution for the two quadrature operators $\hat{x}_\lambda$ and $\hat{x}_{\lambda+\pi/2}$, then we find that $\frac{1}{2}\int_{-\infty}^{\infty} W(\alpha)\,\mathrm{d}x_{\lambda+\pi/2}$ is the true probability distribution for the quadrature component $\hat{x}_\lambda$. This means that, although the Wigner function can have negative values, the integral of the Wigner function over any quadrature variable x_λ, being a probability distribution, must be positive semi-definite. The property of producing a quadrature probability distribution by integrating over its in-quadrature variable is unique to the Wigner function.

For $p = -1$, $W(\alpha, -1) = Q(\alpha)$, the Husimi or Q-function. We obtain a simple representation for $Q(\alpha)$ in terms of the density matrix by inserting $p = -1$ into (4.5.14). Only the first term of the summation remains giving

$$Q(\alpha) = \frac{1}{\pi}\langle 0|\,\hat{D}^\dagger(\alpha)\rho\hat{D}(\alpha)\,|0\rangle = \frac{1}{\pi}\langle\alpha|\,\rho\,|\alpha\rangle. \tag{4.5.23}$$

An immediate consequence of (4.5.23) is that $Q(\alpha)$, being the expectation value of an Hermitian operator ρ with positive semi-definite eigenvalues, is itself positive semi-definite. The Q-function will have zeros for all pure states other than the coherent state (3.6.3) and the squeezed states (3.7.4) and (3.7.29) for which it is a Gaussian function. Unlike $P(\alpha)$ and $W(\alpha)$, $Q(\alpha)$ has an interpretation as a probability distribution and is associated with the probability for the results of joint measurements of the two in-quadrature components of the field, $\hat{x}_\lambda$ and $\hat{x}_{\lambda+\pi/2}$. It is straightforward to verify that antinormal ordered moments of $\hat{a}$ and $\hat{a}^\dagger$ are obtained from (4.5.6) with $p = -1$, as follows. Consider

$$\begin{aligned}\langle\hat{a}^m\hat{a}^{\dagger n}\rangle &= \mathrm{Tr}(\hat{a}^{\dagger n}\rho\hat{a}^m)\\ &= \frac{1}{\pi}\int_{-\infty}^{\infty}\mathrm{d}^2\alpha\,\langle\alpha|\,\hat{a}^{\dagger n}\rho\hat{a}^m\,|\alpha\rangle\\ &= \int_{-\infty}^{\infty}\mathrm{d}^2\alpha\,Q(\alpha)\alpha^{*n}\alpha^m,\end{aligned} \tag{4.5.24}$$

where we have evaluated the trace in the coherent state basis and used (4.5.23).

In the previous section, we expressed in (4.4.17) the moment generating function $M(\mu)$ in terms of the characteristic function $\chi(\xi, p)$. This result can be used to find a simple relationship between $M(\mu)$ and $W(\alpha, p)$. Comparing the definition of the quasi-probability distribution (4.5.1) with (4.4.17), we see that

$$M(\mu) = \frac{\pi}{\mu}W\left(0, 1 - \frac{2}{\mu}\right). \tag{4.5.25}$$

Hence the photon number probability distribution, the factorial moments, and other information contained in $M(\mu)$ can be obtained by evaluating derivatives of $W(0,p)$ with respect to the order parameter p. For example, $M(1) = \pi W(0,-1) = \pi Q(0)$ is the probability that the state has no photons, while $M(2) = \frac{1}{2}\pi W(0,0) = \frac{1}{2}\pi W(0)$ is the difference between the probability that there is an even number of photons and the probability that there is an odd number. There is also a simple relationship between $M(\mu)$ and the P-representation. This follows from (4.2.15) in which $M(\mu)$ is expressed as the expectation value of a normal ordered function of $\hat{a}^\dagger\hat{a}$. Hence

$$M(\mu) = \mathrm{Tr}\left[\rho :\exp(-\mu\hat{a}^\dagger\hat{a}):\right]$$

$$= \int_{-\infty}^{\infty} \mathrm{d}^2\alpha\, P(\alpha)\exp(-\mu|\alpha|^2), \qquad (4.5.26)$$

where we have used (4.2.19). This has an interesting interpretation if we write $\alpha = \sqrt{I}\exp(\mathrm{i}\varphi)$ and introduce an intensity quasi-probability distribution

$$P_I = \tfrac{1}{2}\int_0^{2\pi} P(\alpha)\,\mathrm{d}\varphi. \qquad (4.5.27)$$

With this definition, the moment generating function $M(\mu)$ in (4.5.26) becomes

$$M(\mu) = \int_0^{\infty} P_I \exp(-\mu I)\,\mathrm{d}I, \qquad (4.5.28)$$

the Laplace transform of P_I. Consequently, if $M(\mu)$ is known we can, in principle, invert the Laplace transform (4.5.28) to find P_I.

We conclude our discussion of quasi-probability distributions by calculating $W(\alpha, p)$ for some specific states. Consider first the photon number states $|n\rangle$. For $p < 1$, we obtain the quasi-probability distribution $W_n(\alpha, p)$ for the number state $|n\rangle$ by inserting the characteristic function (4.4.21) into the definition (4.5.1). We find

$$W_n(\alpha,p) = \frac{1}{\pi^2}\int_{-\infty}^{\infty} \mathrm{d}^2\xi\, L_n(|\xi|^2)\exp\left[(p-1)|\xi|^2/2\right]\exp(\alpha\xi^* - \alpha^*\xi). \qquad (4.5.29)$$

Writing $\xi = x\exp(\mathrm{i}\theta)$ and $\alpha = |\alpha|\exp(\mathrm{i}\varphi)$, the integration over θ becomes

$$\int_0^{2\pi} \exp[2\mathrm{i}|\alpha|x\sin(\theta-\varphi)]\,\mathrm{d}\theta = 2\pi J_0(2|\alpha|x), \qquad (4.5.30)$$

where J_0 is the Bessel function of order zero. Inserting this result into (4.5.29) gives (see Appendix 3)

$$W_n(\alpha,p) = \frac{2}{\pi}\int_0^{\infty} x\exp[-(1-p)x^2/2]L_n(x^2)J_0(2|\alpha|x)\,\mathrm{d}x$$

$$= \frac{2}{\pi(1-p)}(-1)^n\left(\frac{1+p}{1-p}\right)^n \exp\left(-\frac{2|\alpha|^2}{1-p}\right)L_n\left(\frac{4|\alpha|^2}{1-p^2}\right), \qquad (4.5.31)$$

for $p \neq -1$. For $p = -1$, $W_n(\alpha, -1) = Q_n(\alpha)$, the Q-function for the number state $|n\rangle$, is given by

$$Q_n(\alpha) = \frac{|\alpha|^{2n}}{n!\,\pi} \exp(-|\alpha|^2). \tag{4.5.32}$$

It can be verified by direct integration that (4.5.31) and (4.5.32) are normalized. For $p = 1$, we can only find a quasi-probability distribution $W_n(\alpha, 1) = P_n(\alpha)$ in terms of the delta function $\delta^{(2)}(\alpha)$ and its derivatives. Substituting (4.4.21) with $p = 1$ into (4.5.1) gives

$$\begin{aligned} P_n(\alpha) &= \frac{1}{\pi^2} \int_{-\infty}^{\infty} \mathrm{d}^2\xi \, L_n(|\xi|^2) \exp(\alpha\xi^* - \alpha^*\xi) \\ &= \frac{1}{\pi^2} \sum_{m=0}^{n} (-1)^m C_m^n \frac{1}{m!} \int_{-\infty}^{\infty} \mathrm{d}^2\xi \left(-\frac{\partial^2}{\partial\alpha\,\partial\alpha^*} \right)^m \exp(\alpha\xi^* - \alpha^*\xi) \\ &= \sum_{m=0}^{n} C_m^n \frac{1}{m!} \left(\frac{\partial^2}{\partial\alpha\,\partial\alpha^*} \right)^m \delta^{(2)}(\alpha), \end{aligned} \tag{4.5.33}$$

where we have used the series expansion of $L_n(|\xi|^2)$ as in (4.4.21). Although highly singular, $P_n(\alpha)$ can be used to calculate all the normal ordered moments of products of $\hat{a}$ and $\hat{a}^\dagger$ for the number state $|n\rangle$. Consider the moment $\langle n|\,\hat{a}^{\dagger k}\hat{a}^l|n\rangle$ which in Section 4.4 was shown to be $\delta_{kl} n!/(n-l)!$. Calculating this same moment using the P-representation (4.5.33) gives

$$\begin{aligned} \langle n|\,\hat{a}^{\dagger k}\hat{a}^l\,|n\rangle &= \sum_{m=0}^{n} C_m^n \frac{1}{m!} \int_{-\infty}^{\infty} \mathrm{d}^2\alpha\, \alpha^{*k}\alpha^l \left(\frac{\partial}{\partial\alpha\,\partial\alpha^*} \right)^m \delta^{(2)}(\alpha) \\ &= \sum_{m=0}^{n} C_m^n \frac{1}{m!} \left(\frac{\partial^2}{\partial\alpha\,\partial\alpha^*} \right)^m (\alpha^{*k}\alpha^l) \bigg|_{\alpha=0} \\ &= \sum_{m=0}^{n} C_m^n \frac{1}{m!} \delta_{kl}\delta_{ml}(m!)^2 \\ &= \delta_{kl} \frac{n!}{(n-l)!}, \end{aligned} \tag{4.5.34}$$

as before.

Next, consider the thermal state with mean photon number $\bar{n}$ and characteristic function given by (4.4.23). From (4.5.1), we have

$$W(\alpha, p) = \frac{1}{\pi^2} \int_{-\infty}^{\infty} \mathrm{d}^2\xi \exp\left[-\tfrac{1}{2}(2\bar{n} + 1 - p)|\xi|^2 \right] \exp(\alpha\xi^* - \alpha^*\xi). \tag{4.5.35}$$

For $p < 2\bar{n} + 1$, this expression is the Fourier transform of a Gaussian function and is therefore itself a Gaussian function given by

$$W(\alpha, p) = \frac{2}{\pi(2\bar{n} + 1 - p)} \exp\left(- \frac{2|\alpha|^2}{2\bar{n} + 1 - p} \right). \tag{4.5.36}$$

We recall from (4.5.25) that the moment generating function $M(\mu)$ can be retrieved from $W(0, p)$. From (4.5.36) we find

$$M(\mu) = \frac{\pi}{\mu} \frac{2}{\pi[2\bar{n} + 1 - (1 - 2/\mu)]} = \frac{1}{1 + \mu\bar{n}}, \tag{4.5.37}$$

in agreement with (4.2.22).

In our discussion of the characteristic function in Section 4.4, we noted that the action of a unitary transformation on a state could lead to a simple modification of $\chi(\xi, p)$. For some transformations, this feature leads to a corresponding simple modification of the quasi-probability distribution $W(\alpha, p)$. Consider a state with density matrix $\tilde{\rho}$ generated from the density matrix ρ by the unitary transformation $\tilde{\rho} = \hat{U}\rho\hat{U}^\dagger$. For $\hat{U} = \exp(-i\varphi\hat{a}^\dagger\hat{a})$, we found that the characteristic function became $\tilde{\chi}(\xi, p) = \chi(\xi \exp(i\varphi), p)$, as in (4.4.36). The corresponding quasi-probability distribution is

$$\begin{aligned}
\tilde{W}(\alpha, p) &= \frac{1}{\pi^2} \int_{-\infty}^{\infty} d^2\xi\, \chi(\xi \exp(i\varphi), p) \exp(\alpha\xi^* - \alpha^*\xi) \\
&= \frac{1}{\pi^2} \int_{-\infty}^{\infty} d^2\eta\, \chi(\eta, p) \exp[\alpha\eta^* \exp(i\varphi) - \alpha^*\eta \exp(-i\varphi)] \\
&= W(\alpha \exp(i\varphi), p),
\end{aligned} \tag{4.5.38}$$

where we have made the substitution $\eta = \xi \exp(i\varphi)$. The action of $\hat{U}$ is simply to rotate the quasi-probability distribution in the complex α-plane. For the state with density matrix $\tilde{\rho} = \hat{D}(\beta)\rho\hat{D}^\dagger(\beta)$, the modified characteristic function was found to be

$$\tilde{\chi}(\xi, p) = \chi(\xi, p) \exp(\xi\beta^* - \xi^*\beta). \tag{4.5.39}$$

The corresponding quasi-probability distribution is

$$\begin{aligned}
\tilde{W}(\alpha, p) &= \frac{1}{\pi^2} \int_{-\infty}^{\infty} d^2\xi\, \chi(\xi, p) \exp(\xi\beta^* - \xi^*\beta + \alpha\xi^* - \alpha^*\xi) \\
&= W(\alpha - \beta, p).
\end{aligned} \tag{4.5.40}$$

The effect of the displacement operator $\hat{D}(\beta)$ is to shift the quasi-probability distribution so that, for example, if $W(\alpha, p)$ is peaked at the origin, then $\tilde{W}(\alpha, p)$ is peaked at $\alpha = \beta$.

The third unitary transformation which we discussed in Section 4.4 was that produced by the action of the single-mode squeezing operator $\hat{S}(\zeta)$. For the transformed state with density matrix $\tilde{\rho} = \hat{S}(\zeta)\rho\hat{S}^{\dagger}(\zeta)$, the characteristic function was found to be

$$\tilde{\chi}(\xi,p) = \chi(\xi\cosh r + \xi^*\exp(\mathrm{i}\varphi)\sinh r, 0)\exp(p|\xi|^2/2), \quad (4.5.41)$$

where $\zeta = r\exp(\mathrm{i}\varphi)$. The relationship between $\tilde{\chi}(\xi,p)$ and $\chi(\xi,p)$ is only simple if $p = 0$, in which case (4.4.41) holds. The simplicity is retained for the Wigner function $W(\alpha)$ for which

$$\begin{aligned}\tilde{W}(\alpha) &= \frac{1}{\pi^2}\int_{-\infty}^{\infty} \mathrm{d}^2\xi\,\chi(\xi\cosh r + \xi^*\exp(\mathrm{i}\varphi)\sinh r, 0)\exp(\alpha\xi^* - \alpha^*\xi) \\ &= \frac{1}{\pi^2}\int_{-\infty}^{\infty} \mathrm{d}^2\eta\,\chi(\eta,0)\exp\{\alpha[\eta^*\cosh r - \eta\exp(-\mathrm{i}\varphi)\sinh r] \\ &\quad -\alpha^*[\eta\cosh r - \eta^*\exp(\mathrm{i}\varphi)\sinh r]\} \\ &= W(\alpha\cosh r + \alpha^*\exp(\mathrm{i}\varphi)\sinh r). \end{aligned} \quad (4.5.42)$$

This transformation is a shear in the complex α-plane. If, for example, $\varphi = 0$ then the transformed Wigner function has the form

$$\tilde{W}(\alpha) = W(\alpha_{\mathrm{r}}\exp(r) + \mathrm{i}\alpha_{\mathrm{i}}\exp(-r)), \quad (4.5.43)$$

corresponding to an area-preserving scaling of the real and imaginary axes in the complex α-plane.

Despite the lack of a simple scaling for quasi-probability distributions with $p \neq 0$, it is nevertheless possible to find $W(\alpha,p)$ for squeezed states, at least for some values of p. We conclude this section by finding $W(\alpha,p)$ for the squeezed vacuum state $|\zeta\rangle = \hat{S}(\zeta)|0\rangle$. At the end of Section 4.4, we showed that the characteristic function for this state is

$$\chi_{\zeta}(\xi,p) = \exp\left(-\tfrac{1}{2}|\xi\cosh r + \xi^*\sinh r|^2 + \tfrac{1}{2}p|\xi|^2\right), \quad (4.5.44)$$

where, for brevity, we have chosen $\varphi = 0$. Writing $\xi = \xi_{\mathrm{r}} + \mathrm{i}\xi_{\mathrm{i}}$, we find

$$\chi_{\zeta}(\xi,p) = \exp\{-\tfrac{1}{2}\xi_{\mathrm{r}}^2[\exp(2r) - p] - \tfrac{1}{2}\xi_{\mathrm{i}}^2[\exp(-2r) - p]\}, \quad (4.5.45)$$

and we see that $\chi_{\zeta}(\xi,p)$ is a two-dimensional Gaussian function if $p < \exp(-2r)$, the corresponding $W_{\zeta}(\alpha,p)$ also being a Gaussian function therefore. If $p > \exp(-2r)$, $W_{\zeta}(\alpha,p)$ can only be expressed in terms of derivatives to all orders of the delta function $\delta^{(2)}(\alpha)$. If we restrict ourselves to $p < \exp(-2r)$, then (4.5.1) gives

$$W_\zeta(\xi, p) = \frac{1}{\pi^2}\int_{-\infty}^{\infty} d\xi_r \int_{-\infty}^{\infty} d\xi_i \exp\{-\tfrac{1}{2}\xi_r^2[\exp(2r) - p]$$

$$-\tfrac{1}{2}\xi_i^2[\exp(-2r) - p]\}\exp[2i(\alpha_i\xi_r - \alpha_r\xi_i)]$$

$$= \frac{2}{\pi}\frac{1}{\{[\exp(2r) - p][\exp(-2r) - p]\}^{1/2}}$$

$$\times \exp\left[-2\left(\frac{\alpha_i^2}{[\exp(2r) - p]} + \frac{\alpha_r^2}{[\exp(-2r) - p]}\right)\right]. \quad (4.5.46)$$

For $p = -1$, this is the Q-function which, by (4.5.23), is $|\langle\alpha|\zeta\rangle|^2/\pi$. This Q-function agrees with the earlier expression (3.7.23) for $|\langle\alpha|\zeta\rangle|^2$. The Wigner function has the simple form

$$W_\zeta(\alpha) = \frac{2}{\pi}\exp\{-2[\exp(-2r)\alpha_i^2 + \exp(2r)\alpha_r^2]\}, \quad (4.5.47)$$

from which we will calculate the probability distribution $P_\zeta(x_\lambda)$ for the quadrature component x_λ by integrating over its in-quadrature variable $x_{\lambda+\pi/2}$. We find

$$P_\zeta(x_\lambda) = \tfrac{1}{2}\int_{-\infty}^{\infty} W_\zeta(\alpha)\, dx_{\lambda+\pi/2}, \quad (4.5.48)$$

where $x_\lambda = \sqrt{2}(\alpha_r\cos\lambda + \alpha_i\sin\lambda)$ and $x_{\lambda+\pi/2} = \sqrt{2}(\alpha_i\cos\lambda - \alpha_r\sin\lambda)$, so that

$$W_\zeta(\alpha) = \frac{2}{\pi}\exp\big[-\{x_\lambda^2[\exp(-2r)\sin^2\lambda + \exp(2r)\cos^2\lambda]$$

$$+x_{\lambda+\pi/2}^2[\exp(-2r)\cos^2\lambda + \exp(2r)\sin^2\lambda]$$

$$-2x_\lambda x_{\lambda+\pi/2}\sin\lambda\cos\lambda[\exp(2r) - \exp(-2r)]\}\big]. \quad (4.5.49)$$

Subsituting (4.5.49) into (4.5.48) and carrying out the integration over $x_{\lambda+\pi/2}$, we obtain

$$P_\zeta(x_\lambda) = \frac{1}{\sqrt{\{\pi[\exp(2r)\sin^2\lambda + \exp(-2r)\cos^2\lambda]\}}}$$

$$\times \exp\left(-\frac{x_\lambda^2}{[\exp(2r)\sin^2\lambda + \exp(-2r)\cos^2\lambda]}\right). \quad (4.5.50)$$

This is a Gaussian probability distribution with zero mean and variance equal to

$$\Delta x_\lambda^2 = \tfrac{1}{2}[\exp(2r)\sin^2\lambda + \exp(-2r)\cos^2\lambda], \tag{4.5.51}$$

and agrees with that found in (3.7.28) obtained using the quadrature representation of $|\zeta\rangle$. The value of the variance varies between $\frac{1}{2}\exp(2r)$ and $\frac{1}{2}\exp(-2r)$, the latter value being less than the value of $1/2$ associated with the vacuum and coherent states. This property is the distinctive feature of the squeezed states.

4.6 Statistics of multimode states

It is not always sufficient to consider only a single mode of the field. There are problems which require the treatment of either a small number of discrete modes or the full continuum of modes. Some of the methods described in this chapter can be extended to the calculation of the statistics of such multimode fields. The moment generating function can be generalized to provide the photoelectron counting statistics associated with states of the full electromagnetic field. The phase operator and its associated probability distribution have been applied to the problem of phase sum and difference statistics for two-mode states. In this section, we limit discussion to the characteristic functions and quasi-probability distributions for multimode states.

If the modes are uncorrelated then we can describe the statistics of each individual mode by constructing its characteristic function or quasi-probability distribution as described in the preceding sections. The resulting multimode characteristic function or quasi-probability distribution is simply the product of the individual single-mode functions. Of more interest are those states exhibiting correlations between modes. Consider two modes labelled A and B with annihilation operators $\hat{a}$ and $\hat{b}$, respectively. The two-mode state will exhibit correlations if the density matrix does not factorize into the product of two single-mode density matrices, that is if $\rho \neq \rho_A \otimes \rho_B$, where ρ is the total density matrix for the two-mode state and ρ_A and ρ_B are the density matrices for modes A and B obtained by taking the trace of ρ over the spaces B and A, respectively. For the two-mode state, the generalization of the characteristic function (4.4.1) is

$$\chi(\xi, \eta, p) = \mathrm{Tr}\left[\rho \exp(\xi\hat{a}^\dagger - \xi^*\hat{a}) \exp(\eta\hat{b}^\dagger - \eta^*\hat{b})\right] \exp\left[p(|\xi|^2 + |\eta|^2)/2\right]. \tag{4.6.1}$$

Clearly if the modes are uncorrelated so that $\rho = \rho_A \otimes \rho_B$, then (4.6.1) factorizes into the product of two single-mode characteristic functions. In general, however, no such factorization is possible. The quasi-probability distribution is defined to be the Fourier transform of $\chi(\xi,\eta,p)$ in (4.6.1) with respect to ξ and η in the form

$$W(\alpha,\beta,p)=\frac{1}{\pi^4}\int_{-\infty}^{\infty}\mathrm{d}^2\xi\int_{-\infty}^{\infty}\mathrm{d}^2\eta\,\chi(\xi,\eta,p)$$
$$\times\exp(\alpha\xi^*-\alpha^*\xi)\exp(\beta\eta^*-\beta^*\eta). \qquad (4.6.2)$$

The method of calculation of the p-ordered moments is a natural generalization of that described for a single-mode field, so that

$$\langle\hat{a}^{\dagger k}\hat{b}^{\dagger l}\hat{b}^m\hat{a}^n\rangle_p=\left(\frac{\partial}{\partial\xi}\right)^k\left(\frac{\partial}{\partial\eta}\right)^l\left(-\frac{\partial}{\partial\eta^*}\right)^m\left(-\frac{\partial}{\partial\xi^*}\right)^n\chi(\xi,\eta,p)\bigg|_{\xi=\eta=0}$$
$$=\int_{-\infty}^{\infty}\mathrm{d}^2\alpha\int_{-\infty}^{\infty}\mathrm{d}^2\beta\,W(\alpha,\beta,p)\alpha^{*k}\beta^{*l}\beta^m\alpha^n. \qquad (4.6.3)$$

The calculation of these functions is illustrated by two examples. First we consider the pure state

$$|\psi\rangle=c_0|0_A;1_B\rangle+c_1|1_A;0_B\rangle, \qquad (4.6.4)$$

where $|0_A;1_B\rangle$ is the state with no photons in mode A and one photon in mode B (see (3.4.4)). The characteristic function for this state is most readily calculated using the normal ordered form of the displacement operators so that

$$\chi(\xi,\eta,p)=\langle\psi|\exp(\xi\hat{a}^\dagger-\xi^*\hat{a})\exp(\eta\hat{b}^\dagger-\eta^*\hat{b})|\psi\rangle\exp\left[p(|\xi|^2+|\eta|^2)/2\right]$$
$$=\langle\psi|\exp(\xi\hat{a}^\dagger)\exp(-\xi^*\hat{a})\exp(\eta\hat{b}^\dagger)\exp(-\eta^*\hat{b})|\psi\rangle$$
$$\times\exp\left[(p-1)(|\xi|^2+|\eta|^2)/2\right]. \qquad (4.6.5)$$

This is evaluated by expanding the exponentials, inserting (4.6.4), and using the properties of the creation and annihilation operators on each mode. We obtain

$$\chi(\xi,\eta,p)=\left\{|c_0|^2(1-|\eta|^2)+|c_1|^2(1-|\xi|^2)\right.$$
$$\left.-(c_0c_1^*\xi\eta^*+c_0^*c_1\xi^*\eta)\right\}\exp\left[(p-1)(|\xi|^2+|\eta|^2)/2\right]. \qquad (4.6.6)$$

As expected, the entangled nature of the state $|\psi\rangle$ is reflected in the fact that the characteristic function (4.6.6) does not factorize into a product of a function of ξ and a function of η. We can now use (4.6.3) to calculate all the p-ordered moments of the creation and annihilation operators. As an example, we calculate the normal ordered moments by setting $p=1$ in (4.6.3). Using (4.6.6), we find that, apart from the identity operator for which $k = l = m = n = 0$, the only non-zero normal ordered moments are

$$\langle \hat{a}^\dagger \hat{a} \rangle = |c_1|^2, \tag{4.6.7}$$

$$\langle \hat{b}^\dagger \hat{b} \rangle = |c_0|^2, \tag{4.6.8}$$

$$\langle \hat{b}^\dagger \hat{a} \rangle = c_0^* c_1, \tag{4.6.9}$$

$$\langle \hat{a}^\dagger \hat{b} \rangle = c_0 c_1^*, \tag{4.6.10}$$

as may readily be verified directly from (4.6.4). The quasi-probability distribution $W(\alpha, \beta, p)$ is obtained using (4.6.6) and (4.6.2). For $p < 1$, we find

$$W(\alpha, \beta, p) = \left(\frac{2}{\pi(1-p)} \right)^2 \left[-\left(\frac{1+p}{1-p} \right) + \frac{4}{(1-p)^2} |c_0 \beta^* + c_1 \alpha^*|^2 \right] \times \exp\left(-\frac{2(|\alpha|^2 + |\beta|^2)}{(1-p)} \right). \tag{4.6.11}$$

We note that for $p > -1$, $W(\alpha, \beta, p)$ has negative values for sufficiently small $|c_0 \beta^* + c_1 \alpha^*|$, but for $p = -1$, corresponding to the positive semi-definite Q-function, $W(\alpha, \beta, -1) \geqslant 0$. For $p = 1$, the two-mode P-representation is expressible in terms of generalized functions as

$$P(\alpha, \beta) = \left[1 + \left(c_0 \frac{\partial}{\partial \beta} + c_1 \frac{\partial}{\partial \alpha} \right) \left(c_0^* \frac{\partial}{\partial \beta^*} + c_1^* \frac{\partial}{\partial \alpha^*} \right) \right] \delta^{(2)}(\alpha) \delta^{(2)} \beta. \tag{4.6.12}$$

As a second example, we consider the two-mode squeezed vacuum state

$$|\zeta_{AB}\rangle = \hat{S}_{AB}(\zeta) |0_A; 0_B\rangle = \exp\left(-\zeta \hat{a}^\dagger \hat{b}^\dagger + \zeta^* \hat{b} \hat{a} \right) |0_A; 0_B\rangle, \tag{4.6.13}$$

introduced in Section 3.7. For simplicity, we choose ζ to be real and calculate the characteristic function and quasi-probability distribution for this state. Setting $\zeta = r$, we find

$$\begin{aligned} \chi(\xi, \eta, p) &= \langle 0_A; 0_B | \hat{S}_{AB}^\dagger(r) \exp(\xi \hat{a}^\dagger - \xi^* \hat{a}) \exp(\eta \hat{b}^\dagger - \eta^* \hat{b}) \\ &\quad \times \hat{S}_{AB}(r) |0_A; 0_B\rangle \exp\left[p(|\xi|^2 + |\eta|^2)/2 \right] \\ &= \langle 0_A; 0_B | \exp\left[(\xi \cosh r + \eta^* \sinh r) \hat{a}^\dagger - (\xi^* \cosh r + \eta \sinh r) \hat{a} \right] \\ &\quad \times \exp\left[(\eta \cosh r + \xi^* \sinh r) \hat{b}^\dagger - (\eta^* \cosh r + \xi \sinh r) \hat{b} \right] |0_A; 0_B\rangle \\ &\quad \times \exp\left[p(|\xi|^2 + |\eta|^2)/2 \right], \end{aligned} \tag{4.6.14}$$

where we have used the unitary transformations (3.7.56) to (3.7.59). The expectation value in (4.6.14) may be evaluated either by normal ordering the exponential operators or by using the number state matrix elements of the Glauber displacement operator (4.4.44), (4.4.46), and (4.4.47). The resulting characteristic function is

$$\chi(\xi,\eta,p) = \exp\left\{-\tfrac{1}{2}\left[(|\xi|^2+|\eta|^2)(\cosh 2r - p) + (\eta\xi + \xi^*\eta^*)\sinh 2r\right]\right\}, \tag{4.6.15}$$

from which we obtain the quasi-probability distribution $W(\alpha,\beta,\mathrm{p})$ by Fourier transform. As with the single-mode squeezed states, this Fourier transform is only well behaved if $p < \exp(-2r)$, in which case

$$\begin{aligned} W(\alpha,\beta,p) = {} & \frac{4}{\pi^2[\exp(2r)-p][\exp(-2r)-p]} \\ & \times \exp\left(-\frac{2}{[\exp(2r)-p][\exp(-2r)-p]}\right. \\ & \left.\times\left[(\cosh 2r - p)(|\alpha|^2+|\beta|^2) + \sinh 2r(\alpha\beta+\beta^*\alpha^*)\right]\right). \end{aligned} \tag{4.6.16}$$

As expected neither $\chi(\xi,\eta,p)$ in (4.6.15) nor $W(\alpha,\beta,p)$ in (4.6.16) factorizes into a product of a function of α and a function of β corresponding to the properties of two independent single modes. This is again a consequence of the entangled nature of the state under consideration.

We can find the characteristic function and quasi-probability distribution for one of the modes from the two-mode functions. The characteristic function $\chi(\xi,p)$ for mode A is obtained from $\chi(\xi,\eta,p)$ by simply setting $\eta=0$, so that for the two-mode squeezed vacuum state we find from (4.6.15) with $\eta=0$ that

$$\chi(\xi,p) = \exp\left[-\tfrac{1}{2}(\cosh 2r - p)|\xi|^2\right]. \tag{4.6.17}$$

If we identify $\cosh 2r$ with $2\bar{n}+1$, we recognize (4.6.17) as the characteristic function for a single-mode thermal state given by (4.4.23). This confirms the single-mode thermal properties of the two-mode squeezed vacuum state discussed in Section 3.7. The quasi-probability distribution $W(\alpha,p)$ for mode A is obtained by integrating $W(\alpha,\beta,p)$ in (4.6.16) over the complex β-plane, so that

$$\begin{aligned} W(\alpha,p) &= \int_{-\infty}^{\infty} \mathrm{d}^2\beta\, W(\alpha,\beta,p) \\ &= \frac{2}{\pi(\cosh 2r - p)}\exp\left(-\frac{2|\alpha|^2}{(\cosh 2r - p)}\right). \end{aligned} \tag{4.6.18}$$

This is the quasi-probability distribution (4.5.36) obtained previously for a single-mode thermal state with $\cosh 2r = 2\bar{n} + 1$.

The procedure described above for obtaining and using the characteristic function and quasi-probability distribution is not restricted to one or two modes. The generalization to N discrete modes with annihilation operators $\hat{a}_j$ ($j = 1, 2, \ldots, N$) is achieved by introducing N complex variables ξ_j, one for each mode, so that the multimode characteristic function is

$$\chi(\{\xi_j\}, p) = \mathrm{Tr}\left[\rho \exp\left(\sum_{j=1}^{N} (\xi_j \hat{a}_j^\dagger - \xi_j^* \hat{a}_j)\right)\right] \exp\left(\frac{p}{2} \sum_{j=1}^{N} |\xi_j|^2\right). \quad (4.6.19)$$

The corresponding multimode quasi-probability distribution is obtained from $\chi(\{\xi_j\}, p)$ by calculating its $2N$-fold Fourier transform with transform variables α_j, so that

$$W(\{\alpha_j\}, p) = \frac{1}{\pi^{2N}} \int_{-\infty}^{\infty} \mathrm{d}^2\xi_1 \cdots \int_{-\infty}^{\infty} \mathrm{d}^2\xi_N \, \chi(\{\xi_j\}, p) \times \exp\left(\sum_{j=1}^{N} (\alpha_j \xi_j^* - \xi_j \alpha_j^*)\right). \quad (4.6.20)$$

Multimode p-ordered moments can be calculated either by differentiating the characteristic function or by evaluating integrals weighted by the quasi-probability distribution.

It is also possible to introduce characteristic functionals for states of the continuum field. These have a similar form to (4.6.19) with $\hat{b}(\omega)$ and $\xi(\omega)$ replacing $\hat{a}_j$ and ξ_j, respectively, and with frequency integrals in place of summations over j. A continuum quasi-probability functional can also be defined as a generalization of (4.6.20) with $\alpha(\omega)$ replacing α_j. Integrals over ξ_j then become functional integrals over $\xi(\omega)$.

5
DISSIPATIVE PROCESSES

5.1 Introduction

Until now we have assumed that the systems studied are completely isolated and evolve independently of their environments. In practice, however, dissipation is always present, originating from the coupling to the environment which comprises a much larger system or ensemble of states. This coupling is typically weak compared with couplings within the system of interest but may nevertheless profoundly affect the dynamics of that system. The effects of the dissipation do not, in general, depend on the precise form of the larger system or ensemble nor on the details of the coupling. The frequency dependence of the coupling to the larger system is usually all that is required to characterize the dissipation.

We will consider dissipation arising from either coupling between discrete states and continua of states, for example in photoionization, or coupling between simple quantum systems and their environment. We model this environment as a continuum of harmonic oscillators because these are relatively easy to deal with and, as we have said, the effects of the dissipation are largely independent of how the environment is modelled. We will find that many dissipative problems in quantum optics involve similar concepts and may be analysed by the same techniques. A number of idealized problems can be solved exactly and we will treat some of these here and in Chapter 6. Usually, however, it is necessary to develop approximation methods in order to obtain analytic solutions and these form the material of the present chapter.

5.2 Perturbation theory and Fermi's golden rule

We begin our discussion of dissipative processes by considering the short-time behaviour of a single discrete state $|1\rangle$, with energy $\hbar\omega_1$, weakly coupled to a continuum of states $\{|f\rangle\}$ with energies $\hbar\omega_f$. This combination of weak coupling and short times suggests that we adopt a perturbative approach. For definiteness, we study the problem of the photoionization of state $|1\rangle$ by a monochromatic classical field of frequency ω. The method may, however, be applied to other problems including spontaneous emission. The Hamiltonian in the rotating-wave approximation is

$$\hat{H} = \hbar\omega_1 |1\rangle\langle 1| + \hbar\int \omega_f |f\rangle\langle f| \mathrm{d}\omega_f$$
$$+ \hbar\int W_f\big[|1\rangle\langle f|\exp(\mathrm{i}\varphi_f)\exp(\mathrm{i}\omega t) + |f\rangle\langle 1|\exp(-\mathrm{i}\varphi_f)\exp(-\mathrm{i}\omega t)\big]\,\mathrm{d}\omega_f, \tag{5.2.1}$$

where $\langle 1 | 1 \rangle = 1$, $\langle 1 | f \rangle = 0$, $\langle f | f' \rangle = \delta(\omega_f - \omega_{f'})$, and $W_f \exp(\mathrm{i}\varphi_f)$ is the transition matrix element between the bound state $|1\rangle$ and the free-electron state $|f\rangle$. The wavefunction for the complete system is

$$|\psi(t)\rangle = b_1 \exp(-\mathrm{i}\,\omega_1 t)\,|1\rangle + \int b_f \exp(-\mathrm{i}\,\omega_f t)\,|f\rangle\, d\omega_f. \tag{5.2.2}$$

Inserting this wavefunction into the Schrödinger equation using the Hamiltonian (5.2.1) gives the coupled equations of motion

$$\dot{b}_1 = -\mathrm{i}\int W_f \exp(\mathrm{i}\varphi_f) b_f \exp(-\mathrm{i}\Delta_f t)\,\mathrm{d}\Delta_f, \tag{5.2.3}$$

$$\dot{b}_f = -\mathrm{i}W_f \exp(-\mathrm{i}\varphi_f) b_1 \exp(\mathrm{i}\Delta_f t), \tag{5.2.4}$$

where $\Delta_f = \omega_f - \omega_1 - \omega$ is the detuning between the state $|f\rangle$ and one-photon resonance with $|1\rangle$.

At short times, it is sufficient to adopt a perturbative approach to the solution of (5.2.3) and (5.2.4) with initial conditions $b_1(0) = 1$ and $b_f(0) = 0$. We obtain the first-order perturbative solution by substituting the initial condition $b_1(0) = 1$ into the right-hand side of (5.2.4) and integrating to give

$$b_f(t) = \frac{W_f}{\Delta_f} \exp(-\mathrm{i}\varphi_f)\left[1 - \exp(\mathrm{i}\Delta_f t)\right]. \tag{5.2.5}$$

Inserting this approximate solution into the right-hand side of (5.2.3), we find

$$\dot{b}_1 = -\mathrm{i}\int \frac{W_f^2}{\Delta_f}\left[\exp(-\mathrm{i}\Delta_f t) - 1\right]\mathrm{d}\Delta_f. \tag{5.2.6}$$

Integrating with respect to time and using $b_1(0) = 1$ gives

$$b_1(t) = 1 - \int \frac{W_f^2}{\Delta_f}\left(\frac{\left[1 - \exp(-\mathrm{i}\Delta_f t)\right]}{\Delta_f} - \mathrm{i}t\right)\mathrm{d}\Delta_f. \tag{5.2.7}$$

The probability for remaining in the bound state $|1\rangle$ is $|b_1(t)|^2$. In calculating this probability from $b_1(t)$ we must neglect terms of order W_f^4 because we have already neglected contributions of this order in obtaining (5.2.7). The first-order perturbative solution for the probability $|b_1(t)|^2$ is therefore

$$|b_1(t)|^2 = 1 - \int \frac{4W_f^2}{\Delta_f^2}\sin^2(\Delta_f t/2)\,\mathrm{d}\Delta_f. \tag{5.2.8}$$

This integral can, in principle, be evaluated once W_f^2 is specified. In practice, W_f^2 will often be a peaked function of Δ_f with a width $\Delta\omega$ which determines the effective range of integration. There are two important limiting cases which follow directly from (5.2.8). Firstly, for very short times for which $\Delta\omega \cdot t \ll 1$, the product $\Delta_f t$ in the integrand will always be small and we may then replace the sine by its argument. This gives

$$|b_1(t)|^2 = 1 - t^2 \int W_f^2 \, \mathrm{d}\Delta_f. \tag{5.2.9}$$

This t^2 dependence is characteristic of coherent interaction between the bound state and the continuum and may also be seen in the short-time evolution of the bound state coupled to a second discrete state. In dissipative processes, it is difficult to find direct evidence of the t^2 dependence because the coupling is usually weak and the continuum broad. This means that (5.2.9) only holds for extremely short times during which the effect is very small. In the limit of an unbounded flat continuum, this time interval shrinks to zero. Secondly, for times for which $\Delta\omega \cdot t \gg 1$ but t is still sufficiently small for the perturbative solution to hold, we can approximate W_f^2 by W_0^2, its value at $\Delta_f = 0$, and allow the range of integration to extend from $-\infty$ to $+\infty$. In this limit we find

$$|b_1(t)|^2 = 1 - 4W_0^2 \int_{-\infty}^{\infty} \frac{\sin^2(\Delta_f t/2)}{\Delta_f^2} \, \mathrm{d}\Delta_f = 1 - 2\pi W_0^2 t. \tag{5.2.10}$$

The rate at which population is ionized from state $|1\rangle$ is $2\pi W_0^2$, since

$$\frac{\mathrm{d}}{\mathrm{d}t} |b_1(t)|^2 = -2\pi W_0^2. \tag{5.2.11}$$

This is Fermi's golden rule. We emphasize that this rate is proportional to the modulus squared of the coupling between $|1\rangle$ and the continuum state with which it is resonant. We will see in the next section that when we go beyond a perturbative treatment of the decay of a bound state into a broad continuum, the bound state probability decays exponentially at the rate given by Fermi's golden rule. The perturbative solution (5.2.10) is simply the short-time expansion of this exponential behaviour.

5.3 Weisskopf–Wigner decay

In order to go beyond the short-time behaviour discussed in the previous section and to study the long-time behaviour and the final state spectrum, we require a non-perturbative solution of the amplitude equations (5.2.3) and (5.2.4). It is convenient to remove the explicit time dependence in these equations by making the substitution $c_f = b_f \exp(-\mathrm{i}\,\Delta_f t)$, and identifying c_1 with b_1. We obtain

$$\dot{c}_1 = -\mathrm{i} \int W_f \exp(\mathrm{i}\,\varphi_f) c_f \, \mathrm{d}\Delta_f, \tag{5.3.1}$$

$$\dot{c}_f = -\mathrm{i}\Delta_f c_f - \mathrm{i}W_f \exp(-\mathrm{i}\,\varphi_f) c_1, \tag{5.3.2}$$

with the initial conditions $c_1(0) = 1$ and $c_f(0) = 0$. This initial value problem is most easily solved by Laplace transforms, converting it into a purely algebraic one. Denoting the Laplace transforms of $c_1(t)$ and $c_f(t)$ by $\bar{c}_1(s)$ and $\bar{c}_f(s)$, respectively, we obtain from (5.3.1) and (5.3.2)

$$s\bar{c}_1(s) - 1 = -\mathrm{i}\int W_f \exp(\mathrm{i}\varphi_f)\bar{c}_f(s)\,\mathrm{d}\Delta_f, \tag{5.3.3}$$

$$(s + \mathrm{i}\Delta_f)\bar{c}_f(s) = -\mathrm{i}W_f \exp(-\mathrm{i}\varphi_f)\bar{c}_1(s). \tag{5.3.4}$$

Eliminating $\bar{c}_f(s)$ between these equations and solving for $\bar{c}_1(s)$ gives

$$\bar{c}_1(s) = [s + I(s)]^{-1} \tag{5.3.5}$$

where

$$I(s) = \int \frac{W_f^2}{s + \mathrm{i}\Delta_f}\,\mathrm{d}\Delta_f. \tag{5.3.6}$$

If we can invert (5.3.5), we will obtain $c_1(t)$ exactly. The simplest exactly soluble case is that of a flat unbounded continuum for which W_f^2 has the constant value W_0^2. Then $I(s)$ in (5.3.6) is given by

$$I(s) = W_0^2 \int_{-\infty}^{\infty} \frac{\mathrm{d}\Delta_f}{s + \mathrm{i}\Delta_f} = W_0^2\left(\int_0^{\infty} \frac{\mathrm{d}\Delta_f}{s + \mathrm{i}\Delta_f} + \int_0^{\infty} \frac{\mathrm{d}\Delta_f}{s - \mathrm{i}\Delta_f}\right)$$

$$= 2W_0^2 \int_0^{\infty} \frac{s}{s^2 + \Delta_f^2}\,\mathrm{d}\Delta_f = \pi W_0^2. \tag{5.3.7}$$

In performing this integral, we have assumed that the real part of s is positive. If $I(s)$ is required when s is purely imaginary, a limiting procedure will have to be carried out. We will see an example of this when obtaining the final state spectrum. Substituting (5.3.7) and (5.3.5) and inverting the Laplace transform we find

$$c_1(t) = \exp(-\pi W_0^2 t), \tag{5.3.8}$$

and hence the probability for remaining in the bound state $|1\rangle$ is

$$|c_1(t)|^2 = \exp(-2\pi W_0^2 t). \tag{5.3.9}$$

This is the Weisskopf–Wigner result for the decay of a discrete state coupled to a flat continuum. The decay is exponential at Fermi's rate $2\pi W_0^2$.

The form of the exact solution for a flat continuum suggests an approximation for dealing with broad continua with slowly varying couplings. We seek an exponentially decaying solution for $c_1(t)$ and consequently look for a single dominant pole at $s = -s_0$ in (5.3.5). The value of s_0 is a solution of the equation

$s_0 = I(s_0)$ with a positive real part corresponding to exponential *decay*. We assume that the magnitude of s_0 is small compared with the range of frequencies over which W_f^2 varies appreciably. This assumption leads us to replace $I(s)$ in (5.3.5) by its limiting value I_0 as $s \to 0^+$. A fuller account of this procedure together with a discussion of its validity can be found in Appendix 6. Replacing $I(s)$ by this limiting value is known as the pole approximation. Evaluating I_0, we have

$$I_0 = \lim_{s \to 0^+} I(s) = -\mathrm{i} \lim_{s \to 0^+} \int_{-\infty}^{\infty} \frac{W_f^2}{\Delta_f - \mathrm{i}s} \, \mathrm{d}\Delta_f$$

$$= \pi W_0^2 - \mathrm{i}\mathbb{P} \int_{-\infty}^{\infty} \frac{W_f^2}{\Delta_f} \, \mathrm{d}\Delta_f = \pi W_0^2 + \mathrm{i}\,\delta\omega, \tag{5.3.10}$$

where $\mathbb{P}$ denotes the principal part integral (see Appendix 7), and we have used (A7.7). Substituting this result into (5.3.5) and inverting, we see that the bound state amplitude is

$$c_1(t) = \exp\left[-(\pi W_0^2 + \mathrm{i}\,\delta\omega)t\right]. \tag{5.3.11}$$

The probability for remaining in the bound state decays exponentially as before but the amplitude oscillates at a frequency shifted by an amount $\delta\omega$. This shift, given by the principal part integral in (5.3.10), depends on the deviation of W_f^2 from a constant value, whereas the decay rate $2\pi W_0^2$ does not. In deriving the solution (5.3.11), we have assumed that the deviation from perfect flatness only results in a small correction to the Weisskopf–Wigner result for a flat continuum. The magnitude $\delta\omega$ of this correction must therefore be small compared with πW_0^2 in (5.3.10). If, once W_f^2 is specified, this is found not to be the case then the correction due to the deviation from a flat continuum has been treated in too simplistic a manner and the result for $\delta\omega$ is unreliable. An important example arises in the calculation of the spontaneous emission rate by this method which leads to the correct decay rate given by the Einstein A-coefficient but a divergent, and therefore incorrect, frequency shift $\delta\omega$.

The amplitude of state $|1\rangle$ in the time-evolved wavefunction (5.2.2) is $c_1 \exp(-\mathrm{i}\,\omega_1 t) = \exp(-\mathrm{i}\,\omega_1 t)\exp[-(\pi W_0^2 + \mathrm{i}\,\delta\omega)t]$. The effect on the amplitude of state $|1\rangle$ owing to the coupling to the continuum within the pole approximation may be accounted for by replacing the frequency ω_1 of $|1\rangle$ by the complex quantity $\Omega_1 = \omega_1 + \delta\omega - \mathrm{i}\pi W_0^2$. Alternatively, we can incorporate the small shift $\delta\omega$ into a redefinition of the bound state frequency ω_1 and account for the decay into the continuum in the equation of motion for c_1 by a term $-\pi W_0^2 c_1 = -\Gamma c_1$ so that

$$\dot{c}_1 = -\Gamma c_1 + \{\text{couplings to other states}\}. \tag{5.3.12}$$

The justification for this may be made clearer if the problem is treated entirely in the time domain, the corresponding approximation being known as the Markov approximation. Beginning with the equations of motion (5.2.3) and (5.2.4), we formally integrate (5.2.4) and substitute for $b_f(t)$ in (5.2.3) to obtain the following integro-differential equation for $b_1(t)$:

$$\dot{b}_1(t) = -\int_0^t b_1(t')K(t-t')\,\mathrm{d}t', \tag{5.3.13}$$

where the kernel $K(t-t')$ is given by

$$K(t-t') = \int W_f^2 \exp[-\mathrm{i}\Delta_f(t-t')]\,\mathrm{d}\Delta_f. \tag{5.3.14}$$

We note that the value of $\dot{b}_1(t)$ depends, in general, on the values of b_1 at all earlier times. However, if the kernel is sharply peaked at $t' = t$ then only values of t' close to t contribute significantly to the integral in (5.3.13). In particular if the continuum is flat with $W_f^2 = W_0^2$, then

$$K(t-t') = W_0^2 \int_{-\infty}^{\infty} \exp[-\mathrm{i}\Delta_f(t-t')]\,\mathrm{d}\Delta_f = 2\pi W_0^2\,\delta(t-t'), \tag{5.3.15}$$

and the delta function in $K(t-t')$ selects only the value $t' = t$ in the integrand. Using (5.3.15), (5.3.13) becomes

$$\dot{b}_1(t) = -\pi W_0^2 b_1(t), \tag{5.3.16}$$

where a factor of $1/2$ has arisen because the delta function is positioned at the upper limit of integration. The solution of (5.3.16) gives the Weisskopf–Wigner result (5.3.8) and (5.3.9).

When the continuum is broad with slowly varying coupling, the kernel, which is the Fourier transform of W_f^2, will be sharply peaked at $t' = t$. In the Markov approximation we replace the value $b_1(t')$ in the integrand of (5.3.13) by its value at $t' = t$. This converts (5.3.13) into the following differential equation for $b_1(t)$:

$$\dot{b}_1(t) = -b_1(t)\int_0^t K(t-t')\,\mathrm{d}t' = -b_1(t)\int_0^t K(\tau)\,\mathrm{d}\tau. \tag{5.3.17}$$

In making the Markov approximation, we have assumed that the kernel $K(\tau)$ is sharply peaked at $\tau = 0$ and consequently the integral in (5.3.17) should be independent of its upper limit $\tau = t$. In order to evaluate this integral, we allow the upper limit to tend to infinity. In doing so, we must ensure that the integral over τ converges and therefore we insert the convergence factor $\exp(-\epsilon\tau)$

letting $\epsilon \to 0^+$ after the integration has been performed. The integral in (5.3.17) becomes

$$\lim_{\epsilon \to 0^+} \int_0^\infty \left(\int W_f^2 \exp(-\mathrm{i}\Delta_f \tau - \epsilon \tau)\, \mathrm{d}\Delta_f \right) \mathrm{d}\tau$$

$$= \lim_{\epsilon \to 0^+} -\mathrm{i} \int \frac{W_f^2}{\Delta_f - \mathrm{i}\epsilon}\, \mathrm{d}\Delta_f$$

$$= \pi W_0^2 - \mathrm{i}\mathrm{P} \int \frac{W_f^2}{\Delta_f}\, \mathrm{d}\Delta_f = \pi W_0^2 + \mathrm{i}\, \delta\omega, \tag{5.3.18}$$

where we have used (A7.7) of Appendix 7. This result is precisely that obtained using the pole approximation earlier in this section (see (5.3.10)). The comments made there about the reliability of the expression for $\delta\omega$ continue to apply. We note that any additional terms in the equation for $b_1(t)$ representing couplings to other discrete states would be unaffected by the elimination of the continuum amplitudes within the Markov approximation. The coupling to the continuum gives rise to a term $-\Gamma b_1(t)$ and a shift in frequency of $\delta\omega$ in ω_1. Detunings between state $|1\rangle$ and other states will have to reflect this shift.

As an example of this approach, we consider the two-photon ionization of an atom by resonant excitation of an intermediate state $|1\rangle$ (see Fig. 5.1). The coupling between $|1\rangle$ and the continuum is due to a CW laser and leads to ionization from $|1\rangle$. The bound states $|0\rangle$ and $|1\rangle$ are resonantly coupled by a second CW laser which is turned on at $t = 0$. The equations of motion for the bound state amplitudes c_0 and c_1 may be written

$$\dot{c}_0 = -\mathrm{i}\frac{V}{2} \exp(\mathrm{i}\varphi) c_1, \tag{5.3.19}$$

$$\dot{c}_1 = -\Gamma c_1 - \mathrm{i}\frac{V}{2} \exp(-\mathrm{i}\varphi) c_0, \tag{5.3.20}$$

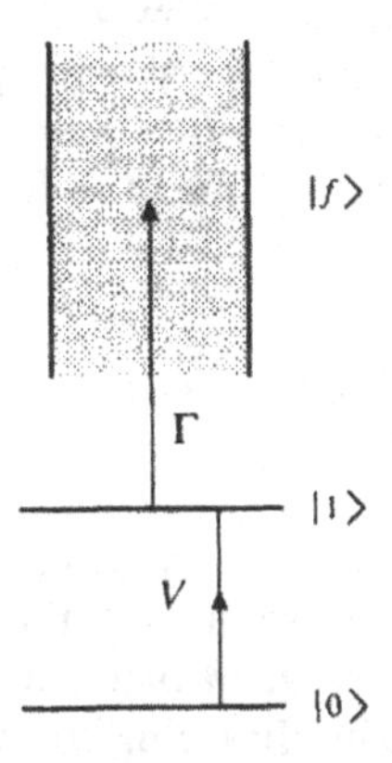

Fig. 5.1 Level scheme for resonant two-photon ionization.

where the effect of the ionizing laser is included following (5.3.12). Note that the shift $\delta\omega$ has been absorbed into a redefinition of the frequency ω_1 of state $|1\rangle$ and hence the second laser must be tuned to resonance between the frequency of state $|0\rangle$ and the *shifted* frequency of state $|1\rangle$. Eliminating c_1, we find the second-order equation

$$\ddot{c}_0 + \Gamma\dot{c}_0 + \frac{V^2}{4}c_0 = 0, \tag{5.3.21}$$

with the initial conditions $c_0(0) = 1$, $c_1(0) = 0 = \dot{c}_0(0)$. Writing

$$\Omega = \sqrt{(V^2 - \Gamma^2)}, \tag{5.3.22}$$

it is straightforward to show that the two independent solutions of (5.3.21) for $\Omega \neq 0$ are $\exp[(-\Gamma \pm i\Omega)t/2]$ and hence that

$$c_0(t) = \frac{(\Gamma + i\Omega)}{2i\Omega}\exp[-(\Gamma - i\Omega)t/2] - \frac{(\Gamma - i\Omega)}{2i\Omega}\exp[-(\Gamma + i\Omega)t/2]. \tag{5.3.23}$$

If $V > \Gamma$, Ω is real and the probability for remaining in the initial state $|0\rangle$ is

$$|c_0(t)|^2 = \frac{\exp(-\Gamma t)}{\Omega^2}[\Omega\cos(\Omega t/2) + \Gamma\sin(\Omega t/2)]^2. \tag{5.3.24}$$

This probability undergoes Rabi oscillations with frequency Ω and damping rate Γ. This is also true of the probability for being in the state $|1\rangle$ which is given by

$$|c_1(t)|^2 = \frac{V^2}{\Omega^2}\exp(-\Gamma t)\sin^2(\Omega t/2). \tag{5.3.25}$$

In Fig. 5.2, we plot $|c_0(t)|^2$ and $|c_1(t)|^2$ against Γt for $V = 7\Gamma$. As well as displaying the decaying Rabi oscillations of these functions, we also plot their sum to show that this, as expected, is a monotonically decreasing function of t. If $\Gamma > V$, so that the loss out of the two-state system is the dominant coupling,

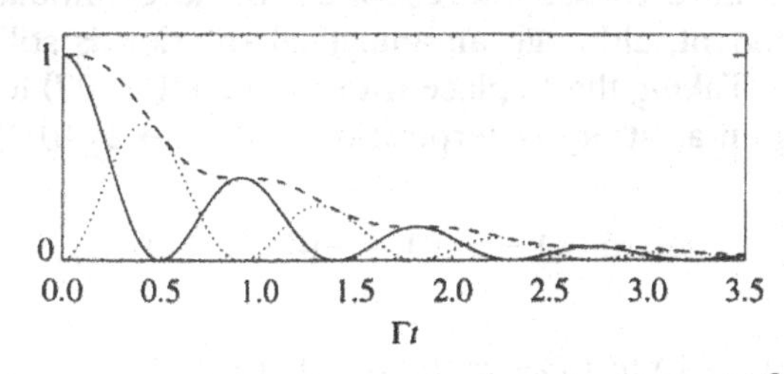

Fig. 5.2 The ground and intermediate state probabilities $|c_0(t)|^2$ (solid curve) and $|c_1(t)|^2$ (dotted curve), respectively, together with their sum (dashed curve) as a function of the scaled time.

then Ω is imaginary and the two independent solutions of (5.3.21) are both real, decaying exponentials. There are then no Rabi oscillations in either $|c_0(t)|^2$ or $|c_1(t)|^2$. The former decays monotonically from unity to zero, while the latter rises from zero to a single maximum as population is transferred from $|0\rangle$ to $|1\rangle$ before decaying to zero. Similar results hold for the special case $\Gamma = V$.

It may be readily shown that, in all cases, the total population $|c_0(t)|^2 + |c_1(t)|^2$ remaining in the two-state system satisfies

$$\frac{\mathrm{d}}{\mathrm{d}t}\left[|c_0(t)|^2 + |c_1(t)|^2\right] = -2\Gamma|c_1(t)|^2, \tag{5.3.26}$$

and hence, as remarked in connection with Fig. 5.2, it is a monotonically decreasing function of t. We would expect the rate of decay of this total population to be proportional to 2Γ and to $|c_1(t)|^2$, since the decay term in the *amplitude* equation (5.3.20) contains the rate Γ, and loss only occurs when $|1\rangle$ is populated; this is confirmed by (5.3.26).

The situation is not so straightforward, however, if we have more than one state coupled to a common continuum. In such cases, simply introducing a decay rate, as in (5.3.12), for *each* of the discrete states is not correct and will, in general, produce the wrong time evolution of the discrete state amplitudes. The reason for this is the omission of coherent coupling of the discrete states via the continuum, known as off-diagonal damping. As an example, we consider the case of two discrete states $|1\rangle$ and $|2\rangle$, state $|2\rangle$ having energy $\hbar\Delta$ greater than state $|1\rangle$, coupled to a common continuum of states $\{|f\rangle\}$. In an interaction picture in which the zero of energy is chosen to coincide with the energy of state $|1\rangle$, the equations of motion for the state amplitudes are

$$\dot{c}_1 = -\mathrm{i}W\int c_f\,\mathrm{d}\Delta_f, \tag{5.3.27}$$

$$\dot{c}_2 = -\mathrm{i}\Delta c_2 - \mathrm{i}V\int c_f\,\mathrm{d}\Delta_f, \tag{5.3.28}$$

$$\dot{c}_f = -\mathrm{i}\Delta_f c_f - \mathrm{i}Wc_1 - \mathrm{i}Vc_2. \tag{5.3.29}$$

For simplicity, we have chosen the coupling to the continuum to be real and frequency independent, although an analytical solution is still possible without this simplification. Taking the Laplace transforms of (5.3.27) to (5.3.29), with the initial state being an arbitrary superposition $c_1(0)|1\rangle + c_2(0)|2\rangle$, gives

$$s\bar{c}_1(s) - c_1(0) = -\mathrm{i}W\int \bar{c}_f(s)\,\mathrm{d}\Delta_f, \tag{5.3.30}$$

$$(s + \mathrm{i}\Delta)\bar{c}_2(s) - c_2(0) = -\mathrm{i}V\int \bar{c}_f(s)\,\mathrm{d}\Delta_f, \tag{5.3.31}$$

$$(s + \mathrm{i}\Delta_f)\bar{c}_f(s) = -\mathrm{i}W\bar{c}_1(s) - \mathrm{i}V\bar{c}_2(s). \tag{5.3.32}$$

Solving (5.3.32) for $\bar{c}_f(s)$ and substituting the result into (5.3.30) and (5.3.31) leads to the pair of coupled equations

$$(s+\Gamma_1)\bar{c}_1(s)+\Gamma_{12}\bar{c}_2(s)=c_1(0), \tag{5.3.33}$$

$$\Gamma_{12}\bar{c}_1(s)+(s+\mathrm{i}\Delta+\Gamma_2)\bar{c}_2(s)=c_2(0), \tag{5.3.34}$$

where $\Gamma_1=\pi W^2$, $\Gamma_2=\pi V^2$, and $\Gamma_{12}=\pi VW$. The terms containing Γ_{12} (the off-diagonal damping) induce transitions between the discrete states via the continuum so that, for example, if state $|2\rangle$ is initially unpopulated it does not remain so. If the off-diagonal damping were omitted, the population of each discrete state would undergo exponential decay and an initially unpopulated level would stay unpopulated. In order to simplify the analysis, we consider the case where $V=W$ so that $\Gamma_1=\Gamma_2=\Gamma_{12}=\Gamma$. The solution of (5.3.33) and (5.3.34) is then

$$\bar{c}_1(s)=\frac{(s+\mathrm{i}\Delta+\Gamma)c_1(0)-\Gamma c_2(0)}{s^2+s(\mathrm{i}\Delta+2\Gamma)+\mathrm{i}\Delta\Gamma}, \tag{5.3.35}$$

$$\bar{c}_2(s)=\frac{-\Gamma c_1(0)+(s+\Gamma)c_2(0)}{s^2+s(\mathrm{i}\Delta+2\Gamma)+\mathrm{i}\Delta\Gamma}. \tag{5.3.36}$$

In order to invert these transforms, we note that they have simple poles at s_+ and s_- given by

$$s_\pm=\tfrac{1}{2}[-(\mathrm{i}\Delta+2\Gamma)\pm\surd(4\Gamma^2-\Delta^2)], \tag{5.3.37}$$

so that

$$\begin{aligned}c_1(t)=\frac{1}{\surd(4\Gamma^2-\Delta^2)}\{&[(s_++\mathrm{i}\Delta+\Gamma)c_1(0)-\Gamma c_2(0)]\exp(s_+t)\\&+[\Gamma c_2(0)-(s_-+\mathrm{i}\Delta+\Gamma)c_1(0)]\exp(s_-t)\},\end{aligned} \tag{5.3.38}$$

$$\begin{aligned}c_2(t)=\frac{1}{\surd(4\Gamma^2-\Delta^2)}\{&[(s_++\Gamma)c_2(0)-\Gamma c_1(0)]\exp(s_+t)\\&+[\Gamma c_1(0)-(s_-+\Gamma)c_2(0)]\exp(s_-t)\}.\end{aligned} \tag{5.3.39}$$

The significance of off-diagonal damping will depend on the relative sizes of Δ and Γ. We illustrate this by considering two limiting cases. First we take $\Delta\gg\Gamma$ when $s_+\simeq-\Gamma$ and $s_-\simeq-\mathrm{i}\Delta-\Gamma$. In this limit, the amplitudes in (5.3.38) and (5.3.39) become

$$c_1(t)\simeq c_1(0)\exp(-\Gamma t), \tag{5.3.40}$$

$$c_2(t)\simeq c_2(0)\exp[-(\mathrm{i}\Delta+\Gamma)t]. \tag{5.3.41}$$

These are simply the solutions of (5.3.33) and (5.3.34) in the absence of off-diagonal damping ($\Gamma_{12} = 0$). If the difference between the energies of states $|1\rangle$ and $|2\rangle$ greatly exceeds their widths then the off-diagonal damping is unimportant and the states decay independently. The states are strongly coupled to widely separated parts of the continuum and transitions between them via the continuum are unlikely. At the other extreme, consider $\Delta \ll \Gamma$ when

$$s_+ \simeq -\frac{\mathrm{i}\Delta}{2} - \frac{\Delta^2}{8\Gamma}, \tag{5.3.42}$$

$$s_- \simeq -\frac{\mathrm{i}\Delta}{2} - 2\Gamma. \tag{5.3.43}$$

In this limit, the amplitudes (5.3.38) and (5.3.39) become

$$c_1(t) = \tfrac{1}{2}\exp(-\mathrm{i}\Delta t/2)$$
$$\times\{\exp(-2\Gamma t)[c_1(0)+c_2(0)] + \exp(-\Delta^2 t/8\Gamma)[c_1(0)-c_2(0)]\}, \tag{5.3.44}$$

$$c_2(t) = \tfrac{1}{2}\exp(-\mathrm{i}\Delta t/2)$$
$$\times\{\exp(-2\Gamma t)[c_1(0)+c_2(0)] - \exp(-\Delta^2 t/8\Gamma)[c_1(0)-c_2(0)]\}. \tag{5.3.45}$$

If off-diagonal damping were neglected, each discrete state amplitude would decay at rate Γ as found in the limit $\Delta \gg \Gamma$. In (5.3.44) and (5.3.45) the off-diagonal damping produces two different decay rates 2Γ and $\Delta^2/8\Gamma$, the first being twice the diagonal rate and the second being much smaller (since $\Delta \ll \Gamma$). If the initial state is $(|1\rangle + |2\rangle)/\sqrt{2}$ so that $c_1(0) = c_2(0)$, then the probability for remaining in either of the discrete states decays at the rate 4Γ, that is at twice the rate found in the absence of off-diagonal damping. If, on the other hand, the initial state is $(|1\rangle - |2\rangle)/\sqrt{2}$ so that $c_1(0) = -c_2(0)$, then the discrete state probabilities decay at the rate $\Delta^2/4\Gamma$ which is significantly less than that found if the off-diagonal damping is omitted. Off-diagonal damping will have a corresponding effect on the distribution at long times of occupied continuum states, or final state spectrum, as we shall see in the next section.

5.4 Final value theorem and spectra

The aim in attempting to solve non-perturbatively the amplitude equations of motion for a bound state coupled to a continuum was to obtain the form of the dynamics at long times. Using Laplace transforms reduces the problem to an algebraic one for the transformed amplitudes $\bar{c}_1(s)$ and $\bar{c}_f(s)$. The solution for $\bar{c}_1(s)$ is given exactly by (5.3.5) and (5.3.6) and, consequently, $\bar{c}_f(s)$ may be found from (5.3.4). Even when the inversion of the transform $\bar{c}_1(s)$ cannot be performed analytically, we can still obtain the final state spectrum $|c_f(t\to\infty)|^2$ analytically using the final value theorem. In Appendix 9, we show that if the transformed amplitude $\bar{c}_f(s)$ has only one simple pole on the imaginary axis in the complex s-plane at $s = \mathrm{i}\theta$, say, then the long-time behaviour of $c_f(t)$ is

$$c_f(t) = \lim_{s\to \mathrm{i}\theta} (s-\mathrm{i}\theta)\bar{c}_f(s)\exp(\mathrm{i}\theta t). \tag{5.4.1}$$

We note here that $\bar{c}_f(s)$ has no poles in the right-hand half of the complex s-plane since unphysical exponential growth is excluded, and that in (5.4.1) θ could be zero. The spectrum as $t \to \infty$ is therefore

$$\left|c_f(t \to \infty)\right|^2 = \left|\lim_{s \to \mathrm{i}\theta}(s - \mathrm{i}\theta)\bar{c}_f(s)\right|^2. \tag{5.4.2}$$

From (5.3.4) we have

$$\bar{c}_f(s) = -\frac{\mathrm{i}W_f \exp(-\mathrm{i}\varphi_f)\bar{c}_1(s)}{s + \mathrm{i}\Delta_f}, \tag{5.4.3}$$

which has a pole on the imaginary axis at $s = -\mathrm{i}\Delta_f$. Before applying the final value theorem, we must establish that $\bar{c}_1(s)$ has no poles on the imaginary axis, that is that $s + I(s)$ in (5.3.5) has no zeros there. In doing so, we use the form of $I(s)$ given in (5.3.6). Note that this expression for $I(s)$ is only valid for $\mathrm{Re}(s) > 0$. This is because as $t \to \infty$ the continuum amplitudes c_f tend to non-zero values and therefore their Laplace transforms $\bar{c}_f(s)$, used in the derivation of (5.3.6), only exist for $\mathrm{Re}(s) > 0$ (see Appendix 9). In order to evaluate $I(s)$ on the imaginary axis at $s = \mathrm{i}\theta$ using (5.3.6), we must write $s = \mathrm{i}\theta + \epsilon$ and take the limit $\epsilon \to 0^+$. Following this procedure we find

$$\begin{aligned} s + I(s) &= \lim_{\epsilon \to 0}\left(\epsilon + \mathrm{i}\theta - \mathrm{i}\int \frac{W_f^2}{\Delta_f + \theta - \mathrm{i}\epsilon}\,\mathrm{d}\Delta_f\right) \\ &= \mathrm{i}\theta - \mathrm{i}\mathbb{P}\int \frac{W_f^2}{\Delta_f + \theta}\,\mathrm{d}\Delta_f + \pi W_\theta^2, \end{aligned} \tag{5.4.4}$$

where we have again used (A7.7) of Appendix 7 and have written $W_\theta^2 = W_f^2(\Delta_f = -\theta)$. Note that if we were to employ the limit $\epsilon \to 0^-$, we would require the analytic continuation of $I(s)$ for $\mathrm{Re}(s) < 0$. Clearly $s + I(s)$ at $s = \mathrm{i}\theta$ evaluated in (5.4.4) using the above procedure will be non-zero, and hence $\bar{c}_1(s)$ will have no poles on the imaginary axis, if $W_\theta^2 \neq 0$. If W_f^2 has zeros then we must establish that the imaginary part of $s + I(s)$ is likewise non-zero at these zeros in order to employ the final value theorem. We will assume here that W_f^2 is non-zero everywhere. With this assumption, all the poles of $\bar{c}_1(s)$ have a negative real part and consequently $c_1(t)$ tends to zero in the long-time limit.

Applying the final value theorem (5.4.2) to (5.3.4), we find that the spectrum of continuum states is

$$\left|c_f(\infty)\right|^2 = W_f^2\left|\lim_{s + \mathrm{i}\Delta_f \to 0^+} \bar{c}_1(s)\right|^2. \tag{5.4.5}$$

The limit of $s + I(s)$ as $s + \mathrm{i}\Delta_f \to 0^+$ is given by

$$\lim_{s+\mathrm{i}\Delta_f \to 0^+} \left(s + \int \frac{W_g^2}{s + \mathrm{i}\Delta_g} \, \mathrm{d}\Delta_g \right) = -\mathrm{i}\Delta_f - \lim_{\epsilon \to 0^+} \mathrm{i} \int \frac{W_g^2}{\Delta_g - \Delta_f - \mathrm{i}\epsilon} \, \mathrm{d}\Delta_g. \quad (5.4.6)$$

Once again we can use (A7.7) to evaluate this limit and find

$$\lim_{s+\mathrm{i}\Delta_f \to 0^+} [s + I(s)] = -\mathrm{i}\Delta_f - \mathrm{i}\mathbb{P} \int \frac{W_g^2}{\Delta_g - \Delta_f} \, \mathrm{d}\Delta_g + \pi W_f^2, \quad (5.4.7)$$

Hence the spectrum is

$$|c_f(\infty)|^2 = \frac{W_f^2}{[\Delta_f - F(\Delta_f)]^2 + \pi^2 W_f^4}, \quad (5.4.8)$$

where

$$F(\Delta_f) = \mathbb{P} \int \frac{W_g^2}{\Delta_f - \Delta_g} \, \mathrm{d}\Delta_g. \quad (5.4.9)$$

We will rederive this spectrum in Section 6.5 of the next chapter and consider some specific examples of couplings there. Here we simply note that if the coupling is constant, so that $W_f^2 = W_0^2$ for all Δ_f, then the principal part integral in (5.4.9) is zero and the spectrum is a Lorentzian centred on the position $\Delta_f = 0$ in the continuum resonant with the discrete state $|1\rangle$ and with width $2\pi W_0^2$.

In the previous section we illustrated the importance of off-diagonal damping in the dynamics of two discrete states coupled to a common continuum. We can use the final value theorem to find the spectrum of continuum states in this problem. Applying (5.4.2) to (5.3.32) with $V = W$, we obtain

$$\begin{aligned} |c_f(\infty)|^2 &= W^2 |\bar{c}_1(-\mathrm{i}\Delta_f) + \bar{c}_2(-\mathrm{i}\Delta_f)|^2 \\ &= \frac{\Gamma}{\pi} \frac{|(\Delta_f - \Delta)c_1(0) + \Delta_f c_2(0)|^2}{\Delta_f^2(\Delta - \Delta_f)^2 + \Gamma^2(\Delta - 2\Delta_f)^2}, \end{aligned} \quad (5.4.10)$$

where we have used the expressions for $\bar{c}_1$ and $\bar{c}_2$ given by (5.3.35) and (5.3.36). If the state $|2\rangle$ is initially unoccupied so that $c_2(0) = 0$ and $c_1(0) = 1$ then the spectrum exhibits a zero at $\Delta_f = \Delta$, that is at the position in the spectrum corresponding to the energy of state $|2\rangle$. This is shown in Fig. 5.3 where we plot the spectrum (5.4.10) with $\Gamma = \Delta/2$. Similarly, if $c_1(0) = 0$ and $c_2(0) = 1$, the zero occurs at $\Delta_f = 0$ corresponding to the energy of state $|1\rangle$. More generally, the spectrum will have a zero at $\Delta_f = c_1(0)\Delta/[c_1(0) + c_2(0)]$ if this is real.

We consider again the two limits $\Delta \gg \Gamma$, for which off-diagonal damping is unimportant, and $\Delta \ll \Gamma$ for which it is crucial. In the first case, the amplitudes

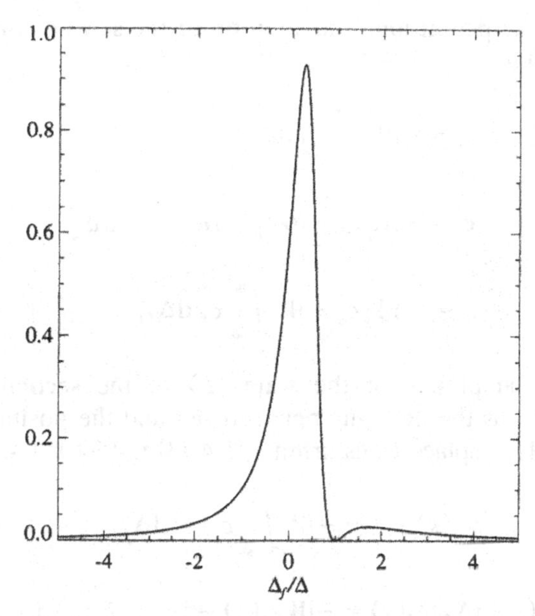

Fig. 5.3 Final state spectrum for resonant two-photon ionization.

of the discrete states decay independently at rate Γ and the final state spectrum becomes

$$|c_f(\infty)|^2 = \frac{\Gamma}{\pi}\left(\frac{|c_1(0)|^2}{\Delta_f^2+\Gamma^2} + \frac{|c_2(0)|^2}{(\Delta_f-\Delta)^2+\Gamma^2}\right), \tag{5.4.11}$$

consisting of two widely separated Lorentzians each of width 2Γ. In the second case ($\Delta \ll \Gamma$), the spectrum can be written

$$|c_f(\infty)|^2 = \frac{\Gamma}{\pi}\frac{|(\Delta_f-\Delta)c_1(0)+\Delta_f c_2(0)|^2}{\left[(\Delta_f-\Delta/2)^2+4\Gamma^2\right]\left[(\Delta_f-\Delta/2)^2+(\Delta^2/8\Gamma)^2\right]}, \tag{5.4.12}$$

which has a narrow Lorentzian on a broad background, and may have a zero, as before.

As a further example of the power of the final value theorem, we conclude this section with a brief discussion of a decay problem involving continuum–continuum interactions. We consider a discrete state $|1\rangle$ coupled to a flat continuum of states $\{|f\rangle\}$ with real, constant coupling W. The continuum states $\{|f\rangle\}$ are in turn coupled to a second continuum of states $\{|g\rangle\}$ with a real, constant coupling κ. This is the simplest model of above-threshold ionization in which an electron liberated by photoionization absorbs energy from the ionizing laser while near the parent ion. The addition of the second continuum modifies (5.3.1) and (5.3.2) to

$$\dot{c}_1 = -\mathrm{i}W\int_{-\infty}^{\infty} c_f\,\mathrm{d}\Delta_f, \tag{5.4.13}$$

$$\dot{c}_f = -\mathrm{i}\Delta_f c_f - \mathrm{i}Wc_1 - \mathrm{i}\kappa\int_{-\infty}^{\infty} c_g\,\mathrm{d}\Delta_g, \tag{5.4.14}$$

$$\dot{c}_g = -\mathrm{i}\Delta_g c_g - \mathrm{i}W\int_{-\infty}^{\infty} c_f\,\mathrm{d}\Delta_f, \tag{5.4.15}$$

where c_g is the amplitude of the state $|g\rangle$ of the second continuum and $\Delta_g = \omega_g - \omega_1 - 2\omega$ is the detuning between $|g\rangle$ and the position of two-photon resonance with $|1\rangle$. Laplace transforming (5.4.13) to (5.4.15), we obtain

$$s\bar{c}_1(s) - 1 = -\mathrm{i}W\int_{-\infty}^{\infty} \bar{c}_f(s)\,\mathrm{d}\Delta_f, \tag{5.4.16}$$

$$(s + \mathrm{i}\Delta_f)\bar{c}_f(s) = -\mathrm{i}W\bar{c}_1(s) - \mathrm{i}\kappa\int_{-\infty}^{\infty} \bar{c}_g(s)\,\mathrm{d}\Delta_g, \tag{5.4.17}$$

$$(s + \mathrm{i}\Delta_g)\bar{c}_g(s) = -\mathrm{i}\kappa\int_{-\infty}^{\infty} \bar{c}_f(s)\,\mathrm{d}\Delta_f. \tag{5.4.18}$$

Solving (5.4.18) for $\bar{c}_g(s)$, substituting the result into (5.4.17), and dividing by $(s + \mathrm{i}\Delta_f)$ gives

$$\begin{aligned}
\bar{c}_f(s) &= -\mathrm{i}\frac{W\bar{c}_1(s)}{s + \mathrm{i}\Delta_f} - \frac{\kappa^2}{s + \mathrm{i}\Delta_f}\int_{-\infty}^{\infty}\frac{\mathrm{d}\Delta_g}{s + \mathrm{i}\Delta_g}\int_{-\infty}^{\infty}\bar{c}_{f'}(s)\,\mathrm{d}\Delta_{f'} \\
&= -\mathrm{i}\frac{W\bar{c}_1(s)}{s + \mathrm{i}\Delta_f} - \frac{\pi\kappa^2}{s + \mathrm{i}\Delta_f}\int_{-\infty}^{\infty}\bar{c}_{f'}(s)\,\mathrm{d}\Delta_{f'},
\end{aligned} \tag{5.4.19}$$

as in (5.3.7). Integrating with respect to Δ_f, we find

$$\int_{-\infty}^{\infty}\bar{c}_f(s)\,\mathrm{d}\Delta_f = -\mathrm{i}\frac{\pi W\bar{c}_1(s)}{1 + \pi^2\kappa^2}. \tag{5.4.20}$$

Substituting this into (5.4.16) gives

$$\bar{c}_1(s) = \left(s + \frac{\pi W^2}{1 + \pi^2\kappa^2}\right)^{-1}, \tag{5.4.21}$$

which, on inversion, gives exponential decay of the discrete state amplitude with rate $\Gamma_\kappa = \pi W^2/(1 + \pi^2\kappa^2)$. Comparing this with the Weisskopf–Wigner result (5.3.8), we see that the decay proceeds *more slowly* in the presence of the coupling κ to the second continuum. Using (5.4.20) and (5.4.21) to express the right-hand side of (5.4.18) in terms of s, and using the final value theorem by allowing $s + \mathrm{i}\Delta_g$ to tend to zero as before, we find

$$|c_g(\infty)|^2 = \frac{\kappa^2 \Gamma_\kappa^2 / W^2}{\Delta_g^2 + \Gamma_\kappa^2}, \tag{5.4.22}$$

which is a Lorentzian spectrum of width $2\Gamma_\kappa$ centred at the position of two-photon resonance with the discrete state. Similarly, substituting (5.4.20) and (5.4.21) into (5.4.19) and using the final value theorem by allowing $s + \mathrm{i}\Delta_f$ to tend to zero gives

$$|c_f(\infty)|^2 = \frac{\Gamma_\kappa^2 / (\pi^2 W^2)}{\Delta_f^2 + \Gamma_\kappa^2} \tag{5.4.23}$$

which is also a Lorentzian spectrum of width $2\Gamma_\kappa$, this time centred at the one-photon resonance position. It is readily verified that all the population is transferred from the discrete state to the continua at long times, so that $\int_{-\infty}^{\infty} |c_g(\infty)|^2 \, \mathrm{d}\Delta_g + \int_{-\infty}^{\infty} |c_f(\infty)|^2 \, \mathrm{d}\Delta_f = 1$, and that the first and second continua share the population in the ratio $1 : \pi^2\kappa^2$.

5.5 Damping in the Heisenberg picture

So far in this chapter we have considered the problem of discrete states coupled to a continuum of states, leading to decay of the discrete state population. Damping is also important in the interaction between more complicated systems and their environments and may be treated by methods similar to those developed already in this chapter. In this section, we discuss two simple, commonly occurring examples: the damped harmonic oscillator, and the interaction between a two-state atom and the free-space electromagnetic field.

Many quantum optical experiments involve the use of cavities to enhance field strengths or to attempt to isolate a system from its environment. In practice, the cavity is imperfect and its modes couple to the environment around the cavity. This coupling is responsible for the leaking of energy from the cavity and also for the introduction of environmental fluctuations into the cavity which can have a profound effect on the quantum dynamics. A simple description of a damped cavity mode consists of a harmonic oscillator with annihilation and creation operators $\hat{a}(t)$ and $\hat{a}^\dagger(t)$ representing this mode coupled to a continuum of oscillators with annihilation and creation operators $\hat{b}(\Delta, t)$ and $\hat{b}^\dagger(\Delta, t)$ modelling the environment. In Chapter 3, we saw that operators of this type satisfy the equal-time commutation relations $[\hat{a}(t), \hat{a}^\dagger(t)] = 1$ and $[\hat{b}(\Delta, t), \hat{b}^\dagger(\Delta', t)] = \delta(\Delta - \Delta')$. The Hamiltonian for this model is

$$\begin{aligned} \hat{H}_\mathrm{I} &= \hbar \int \Delta \hat{b}^\dagger(\Delta, t)\hat{b}(\Delta, t)\, \mathrm{d}\Delta \\ &\quad + \hbar \int W(\Delta)\left\{\hat{a}^\dagger(t)\hat{b}(\Delta, t)\exp[-\mathrm{i}\varphi(\Delta)] + \hat{b}^\dagger(\Delta, t)\hat{a}(t)\exp[\mathrm{i}\varphi(\Delta)]\right\} \mathrm{d}\Delta, \end{aligned} \tag{5.5.1}$$

where $W(\Delta)\exp[-i\varphi(\Delta)]$ is the coupling and we have chosen an interaction picture in which the energy of the cavity mode is zero. The Heisenberg equations of motion for the annihilation operators $\hat{a}(t)$ and $\hat{b}(\Delta, t)$ are

$$\dot{\hat{a}}(t) = \frac{i}{\hbar}\left[\hat{H}_I, \hat{a}(t)\right] = -i\int W(\Delta)\exp[-i\varphi(\Delta)]\hat{b}(\Delta, t)\,d\Delta, \tag{5.5.2}$$

$$\dot{\hat{b}}(t) = \frac{i}{\hbar}\left[\hat{H}_I, \hat{b}(\Delta, t)\right] = -i\Delta\hat{b}(\Delta, t) - iW(\Delta)\exp[i\varphi(\Delta)]\hat{a}(t). \tag{5.5.3}$$

These equations are similar to (5.3.1) and (5.3.2) and we might therefore expect an exponentially decaying solution for $\hat{a}(t)$, with a form similar to (5.3.11) when the coupling is slowly varying. However, we can easily see that this cannot be the case since if both $\hat{a}(t)$ and $\hat{a}^\dagger(t)$ are exponentially decaying then the equal-time commutator $[\hat{a}(t), \hat{a}^\dagger(t)]$ must also be exponentially decaying, thus violating the commutation relation. The resolution of this problem is that the form of $\hat{a}(t)$ depends on the initial environment operators $\hat{b}(\Delta, 0)$ in such a way as to preserve the commutation relation. This feature did not arise in the treatment of Weisskopf–Wigner decay in Section 5.3 because in that case the initial amplitudes for the continuum states are zero. In the present problem, the environment forms a distinct quantum system the initial properties of which affect the dynamics. Consider first the case of constant coupling where $W(\Delta)\exp[-i\varphi(\Delta)]$ is a real constant W_0. We solve the Heisenberg equations (5.5.2) and (5.5.3) by Laplace transforms. The transformed equations of motion are

$$s\bar{\hat{a}}(s) - \hat{a}(0) = -iW_0\int \bar{\hat{b}}(\Delta, s)\,d\Delta, \tag{5.5.4}$$

$$(s + i\Delta)\bar{\hat{b}}(\Delta, s) - \hat{b}(\Delta, 0) = -iW_0\bar{\hat{a}}(s), \tag{5.5.5}$$

where $\bar{\hat{a}}(s)$ and $\bar{\hat{b}}(\Delta, s)$ denote the Laplace transforms of $\hat{a}(t)$ and $\hat{b}(\Delta, t)$, respectively. Eliminating $\bar{\hat{b}}(\Delta, s)$ between these equations and solving for $\bar{\hat{a}}(s)$ we obtain

$$\bar{\hat{a}}(s) = \frac{1}{s + W_0^2\int d\Delta/(s + i\Delta)}\left(\hat{a}(0) - iW_0\int\frac{\hat{b}(\Delta, 0)}{s + i\Delta}\,d\Delta\right). \tag{5.5.6}$$

The integral in the denominator may be evaluated as in (5.3.7) to give

$$\bar{\hat{a}}(s) = \frac{1}{s + \pi W_0^2}\left(\hat{a}(0) - iW_0\int\frac{\hat{b}(\Delta, 0)}{s + i\Delta}\,d\Delta\right). \tag{5.5.7}$$

This transform, and hence its inversion, depends on the initial properties of *both* the mode and the environment. The first term in (5.5.7) corresponds to the Weisskopf–Wigner solution for exponential decay arising from the single pole at

$s = -\pi W_0^2 = -\Gamma$. The second term is more complicated since in addition to this single pole, its integrand diverges at some point in the range of integration for any purely imaginary s. Unfortunately, this property precludes direct application of the final value theorem. We can, however, invert (5.5.7) to find $\hat{a}(t)$. In Appendix 9, we show that

$$\hat{a}(t) = \hat{a}(0)\exp(-\Gamma t) - \mathrm{i}W_0 \int \frac{\hat{b}(\Delta,0)}{(\Gamma - \mathrm{i}\Delta)}[\exp(-\mathrm{i}\Delta t) - \exp(-\Gamma t)]\,\mathrm{d}\Delta. \quad (5.5.8)$$

At long times, the first term on the right-hand side of (5.5.8) tends to zero so that $\hat{a}(t)$ depends only on the initial properties of the environment via the operators $\hat{b}(\Delta, 0)$. The integral term in (5.5.8) preserves the equal-time commutator $[\hat{a}(t), \hat{a}^\dagger(t)] = 1$, as we now show. The equal-time commutator can be written, using (5.5.8), as

$$[\hat{a}(t), \hat{a}^\dagger(t)] = [\hat{a}(0), \hat{a}^\dagger(0)]\exp(-2\Gamma t) + W_0^2 \int \mathrm{d}\Delta \int \mathrm{d}\Delta' \frac{[\hat{b}(\Delta,0), \hat{b}^\dagger(\Delta',0)]}{(\Gamma - \mathrm{i}\Delta)(\Gamma + \mathrm{i}\Delta')}$$

$$\times[\exp(-\mathrm{i}\Delta t) - \exp(-\Gamma t)][\exp(\mathrm{i}\Delta' t) - \exp(-\Gamma t)], \quad (5.5.9)$$

since $[\hat{a}(0), \hat{b}^\dagger(\Delta, 0)] = 0$. Using the commutators for $\hat{a}(0)$ and $\hat{b}(\Delta, 0)$ with their Hermitian conjugates, we find

$$[\hat{a}(t), \hat{a}^\dagger(t)] = \exp(-2\Gamma t) + [1 + \exp(-2\Gamma t)]\int_{-\infty}^{\infty} \frac{\Gamma/\pi}{(\Gamma^2 + \Delta^2)}\,\mathrm{d}\Delta$$

$$- \exp(-\Gamma t)\int_{-\infty}^{\infty} \frac{[\exp(\mathrm{i}\Delta t) + \exp(-\mathrm{i}\Delta t)]\Gamma/\pi}{(\Gamma^2 + \Delta^2)}\,\mathrm{d}\Delta. \quad (5.5.10)$$

The integrals in (5.5.10) may be evaluated by contour integration, as shown in Appendix 8, with the result that

$$[\hat{a}(t), \hat{a}^\dagger(t)] = \exp(-2\Gamma t) + [1 + \exp(-2\Gamma t)] - 2\exp(-\Gamma t)\cdot\exp(-\Gamma t) = 1. \quad (5.5.11)$$

When the coupling is weakly frequency dependent, we can use the pole approximation in Laplace space or the Markov approximation in the time domain as discussed in Section 5.3. In the following, we adopt the second of these approximations. We formally integrate (5.5.3) and obtain

$$\hat{b}(\Delta, t) = \hat{b}(\Delta, 0)\exp(-\mathrm{i}\Delta t) - \mathrm{i}W(\Delta)\exp[\mathrm{i}\varphi(\Delta)]\int_0^t \exp[-\mathrm{i}\Delta(t - t')]\hat{a}(t')\,\mathrm{d}t'. \quad (5.5.12)$$

Substituting this into (5.5.2), we find the integro-differential equation

$$\dot{\hat{a}}(t) = -\int_0^t \hat{a}(t')K(t-t')\,\mathrm{d}t' - \mathrm{i}\int W(\Delta)\exp[-\mathrm{i}\varphi(\Delta)]\exp(-\mathrm{i}\Delta t)\hat{b}(\Delta,0)\,\mathrm{d}\Delta, \tag{5.5.13}$$

where the kernel $K(t-t')$ is given by

$$K(t-t') = \int W^2(\Delta)\exp[-\mathrm{i}\Delta(t-t')]\,\mathrm{d}\Delta, \tag{5.5.14}$$

as in (5.3.14). The form of $\dot{\hat{a}}(t)$ depends on $\hat{a}(t)$ at all earlier times. In making the Markov approximation, we assume that $K(t-t')$ is sharply peaked at $t'=t$, so that $\hat{a}(t')$ in (5.5.13) may be approximated by $\hat{a}(t)$. The resulting integral is treated as in (5.3.18) giving

$$\dot{\hat{a}}(t) = -(\Gamma + \mathrm{i}\delta\omega)\hat{a}(t) + \hat{F}(t), \tag{5.5.15}$$

where $\Gamma = \pi W_0^2 = \pi W^2(\Delta = 0)$ and

$$\hat{F}(t) = -\mathrm{i}\int W(\Delta)\exp[-\mathrm{i}\varphi(\Delta)]\exp(-\mathrm{i}\Delta t)\hat{b}(\Delta,0)\,\mathrm{d}\Delta. \tag{5.5.16}$$

Equation (5.5.15) is of the Langevin type, developed for the study of Brownian motion, containing a damping term and a randomizing term. This equation can be solved by formal integration, but before doing this, consider the properties of the Langevin operator $\hat{F}(t)$. Since $\hat{F}(t)$ depends only on the initial environment operators $\hat{b}(\Delta,0)$, it commutes with the initial mode operators $\hat{a}(0)$ and $\hat{a}^\dagger(0)$. The unequal-time commutator $[\hat{F}(t),\hat{F}^\dagger(t')]$ is given by

$$\begin{aligned}[\hat{F}(t),\hat{F}^\dagger(t')] &= \int \mathrm{d}\Delta \int \mathrm{d}\Delta'\, W(\Delta)W(\Delta')\exp\{-\mathrm{i}[\varphi(\Delta)-\varphi(\Delta')]\} \\ &\quad \times [\hat{b}(\Delta,0),\hat{b}^\dagger(\Delta',0)]\exp(-\mathrm{i}\Delta t)\exp(\mathrm{i}\Delta' t') \\ &= \int W^2(\Delta)\exp[-\mathrm{i}\Delta(t-t')]\,\mathrm{d}\Delta = K(t-t'),\end{aligned} \tag{5.5.17}$$

which, within the Markov approximation, is a sharply peaked function of $t-t'$. Solving (5.5.15) for $\hat{a}(t)$, we obtain

$$\hat{a}(t) = \exp[-(\Gamma+\mathrm{i}\delta\omega)t]\left(\hat{a}(0) + \int_0^t \exp[(\Gamma+\mathrm{i}\delta\omega)t']\hat{F}(t')\,\mathrm{d}t'\right). \tag{5.5.18}$$

Once the initial states of the cavity and of the environment are specified, (5.5.18) and its Hermitian conjugate constitute a complete solution for the evolution of the mode from which moments of any mode observable can, in principle, be calculated. The solution (5.5.18) can be used to show that the

Markov approximation is consistent with the preservation of the commutator $[\hat{a}(t), \hat{a}^\dagger(t)] = 1$. Consider

$$[\hat{a}(t), \hat{a}^\dagger(t)] = \exp(-2\Gamma t)\left(1 + \int_0^t \mathrm{d}t' \int_0^t \mathrm{d}t'' \exp[(\Gamma + \mathrm{i}\delta\omega)t'] \times \exp[(\Gamma - \mathrm{i}\delta\omega)t''][\hat{F}(t'), \hat{F}^\dagger(t'')]\right)$$

$$= \exp(-2\Gamma t)\left(1 + \int_0^t \mathrm{d}t' \int_0^t \mathrm{d}t'' \exp[(\Gamma + \mathrm{i}\delta\omega)t'] \times \exp[(\Gamma - \mathrm{i}\delta\omega)t'']K(t' - t'')\right), \quad (5.5.19)$$

using (5.5.17). Within the Markov approximation, we use the fact that $K(t' - t'')$ is sharply peaked at $t' = t''$ to put $t'' = t'$ elsewhere in the integrand, giving

$$[\hat{a}(t), \hat{a}^\dagger(t)] = \exp(-2\Gamma t) \times \left(1 + \int_0^t \mathrm{d}t' \exp(2\Gamma t') \int_0^t \mathrm{d}t'' \int W(\Delta) \exp[-\mathrm{i}\Delta(t' - t'')] \,\mathrm{d}\Delta\right), \quad (5.5.20)$$

where we have used (5.5.14). The kernel, given by the innermost integral in (5.5.20), peaks at $t'' = t'$ and therefore, unless t' is zero or t, the integration over t'' will be essentially independent of its limits, the limiting cases making a negligible contribution to the integral. These features allow us to approximate the t'' integral by extending the limits to $\pm\infty$, giving the result $2\pi\delta(\Delta)$. Finally

$$[\hat{a}(t), \hat{a}^\dagger(t)] = \exp(-2\Gamma t)\left(1 + \int_0^t \exp(2\Gamma t') \cdot 2\Gamma \,\mathrm{d}t'\right) = 1. \quad (5.5.21)$$

The evolution of the statistical moments of the mode observables depends on the moments of $\hat{F}$ and $\hat{F}^\dagger$ and hence on the moments of the initial environment operators $\hat{b}(\Delta, 0)$ and $\hat{b}^\dagger(\Delta, 0)$. The most important example is when the environment is in thermodynamic equilibrium, in which case the lowest-order moments of the environment operators are (see Chapter 3)

$$\langle \hat{b}^\dagger(\Delta, 0)\hat{b}(\Delta', 0)\rangle = \bar{n}(\Delta)\delta(\Delta - \Delta'), \quad (5.5.22)$$

$$\langle \hat{b}(\Delta, 0)\hat{b}^\dagger(\Delta', 0)\rangle = (\bar{n}(\Delta) + 1)\delta(\Delta - \Delta'), \quad (5.5.23)$$

where $\bar{n}(\Delta)$ is the mean occupation number of a single mode at frequency Δ. Hence, from (5.5.16), the lowest-order non-zero moments of the Langevin operators are

$$\langle \hat{F}^\dagger(t')\hat{F}(t)\rangle = \int W^2(\Delta)\bar{n}(\Delta) \exp[-\mathrm{i}\Delta(t - t')] \,\mathrm{d}\Delta, \quad (5.5.24)$$

$$\langle \hat{F}(t)\hat{F}^\dagger(t')\rangle = \int W^2(\Delta)[\bar{n}(\Delta) + 1] \exp[-\mathrm{i}\Delta(t - t')] \,\mathrm{d}\Delta. \quad (5.5.25)$$

These moments allow us to calculate the time evolution of the mean photon number in the cavity mode from (5.5.18). We find

$$\langle \hat{a}^\dagger(t)\hat{a}(t)\rangle = \exp(-2\Gamma t)\bigg(\langle \hat{a}^\dagger(0)\hat{a}(0)\rangle + \int_0^t \mathrm{d}t' \int_0^t \mathrm{d}t'' \exp[(\Gamma + \mathrm{i}\delta\omega)t']$$

$$\times \exp[(\Gamma - \mathrm{i}\delta\omega)t''] \int W^2(\Delta)\bar{n}(\Delta)\exp[-\mathrm{i}\Delta(t'-t'')]\,\mathrm{d}\Delta\bigg), \quad (5.5.26)$$

where we have used the fact that the expectation value of $\hat{F}(t)$ is zero and that the mode and its environment are initially uncorrelated. The coupling $W^2(\Delta)$ has been assumed to be slowly varying; if $\bar{n}(\Delta)$ is also slowly varying, as it will be in most cases, then the integration over t'' in (5.5.26) may be approximated as described earlier in this section, with the result that

$$\langle \hat{a}^\dagger(t)\hat{a}(t)\rangle = \langle \hat{a}^\dagger(0)\hat{a}(0)\rangle \exp(-2\Gamma t) + \bar{n}(0)[1 - \exp(-2\Gamma t)], \quad (5.5.27)$$

where $\bar{n}(0) = \bar{n}(\Delta = 0)$ is the mean occupation number of the cavity mode in thermal equilibrium with the environment. It can be verified by calculating the higher-order moments that the cavity mode relaxes into thermal equilibrium with its environment. At low temperatures $\bar{n}(0) \simeq 0$ and the cavity mode relaxes to the vacuum state.

Modern experimental techniques have led to the possibility of preparing parts of the environment in states other than thermal equilibrium. For example, we could imagine illuminating the external surfaces of the mirrors with broad-band squeezed light. For this specially prepared environment, or 'rigged reservoir', the lowest-order moments of the environment operators are as in (5.2.22) and (5.2.23) with the addition of (see Section 3.7)

$$\langle \hat{b}(\Delta,0)\hat{b}(\Delta',0)\rangle = \bar{m}(\Delta)\delta(\Delta + \Delta' - 2\Delta_s), \quad (5.5.28)$$

$$\langle \hat{b}^\dagger(\Delta,0)\hat{b}^\dagger(\Delta',0)\rangle = \bar{m}^*(\Delta)\delta(\Delta + \Delta' - 2\Delta_s), \quad (5.5.29)$$

where Δ_s is the detuning between the cavity mode and the frequency at which the external field is squeezed. For a reservoir in the squeezed vacuum state (3.7.68), we find, on comparing (5.5.22), (5.5.23), (5.5.28), and (5.5.29) with (3.7.70) to (3.7.72), that $|\bar{m}|^2 = \bar{n}(\bar{n}+1)$. For more general states, the inequality $|\bar{m}|^2 \leqslant \bar{n}(\bar{n}+1)$ holds. In addition to (5.5.27), we now have

$$\langle \hat{a}^2(t)\rangle = \exp[-2(\Gamma + \mathrm{i}\delta\omega)t]\bigg(\langle \hat{a}^2(0)\rangle + \int_0^t \mathrm{d}t' \int_0^t \mathrm{d}t'' \exp[(\Gamma + \mathrm{i}\delta\omega)(t' + t'')]$$

$$\times \exp(-2\mathrm{i}\Delta_s t'') \int W(\Delta)W(-\Delta + 2\Delta_s)$$

$$\times \exp\{-\mathrm{i}[\varphi(\Delta) + \varphi(-\Delta + 2\Delta_s)]\}\bar{m}(\Delta)\exp[-\mathrm{i}\Delta(t'-t'')]\,\mathrm{d}\Delta\bigg).$$

(5.5.30)

If $\overline{m}(\Delta)$ and $W(\Delta)\exp[-\mathrm{i}\varphi(\Delta)]$ are slowly varying and Δ_s is small enough, the integration over t'' can be carried out as before and we obtain

$$\langle \hat{a}^2(t)\rangle = \exp[-2(\Gamma + \mathrm{i}\,\delta\omega)t]\Big(\langle \hat{a}^2(0)\rangle + \overline{m}(0)\int_0^t \exp(2\Gamma t') \times \exp[2\mathrm{i}(\delta\omega - \Delta_s)t']2\Gamma\exp[-2\mathrm{i}\varphi(0)]\,\mathrm{d}t'\Big). \quad (5.5.31)$$

This expression is particularly simple if we choose $\Delta_s = \delta\omega$, so that the frequency at which the external field is squeezed is equal to the shifted cavity mode frequency. In this case

$$\langle \hat{a}^2(t)\rangle = \exp(-2\mathrm{i}\,\delta\omega t) \times \{\exp(-2\Gamma t)\langle \hat{a}^2(0)\rangle + \overline{m}(0)\exp[-2\mathrm{i}\varphi(0)][1-\exp(-2\Gamma t)]\}. \quad (5.5.32)$$

The interaction of the cavity mode with the squeezed environment produces a long-time limit in which the uncertainties in the quadrature components $\hat{x}_0 = (\hat{a}+\hat{a}^\dagger)/\sqrt{2}$ and $\hat{x}_{\pi/2} = (\hat{a}-\hat{a}^\dagger)/\mathrm{i}\sqrt{2}$ of the cavity field are unequal. If, for simplicity, we set $\varphi(0)=0$ and assume that $\overline{m}(0)$ is real then the variances in these quadrature components tend to

$$\Delta x_0^2(t) = \langle \hat{x}_0^2(t)\rangle = \tfrac{1}{2}(1 + 2\overline{n}(0) + 2\overline{m}(0)\cos 2\,\delta\omega t) \quad (5.5.33)$$

and

$$\Delta x_{\pi/2}^2(t) = \langle \hat{x}_{\pi/2}^2(t)\rangle = \tfrac{1}{2}(1 + 2\overline{n}(0) - 2\overline{m}(0)\cos 2\,\delta\omega t). \quad (5.5.34)$$

If $\overline{m}(0) > \overline{n}(0)$, this limit exhibits squeezing with a minimum variance less than the vacuum value of $1/2$. The uncertainty product

$$\Delta x_0(t)\cdot\Delta x_{\pi/2}(t) = \tfrac{1}{2}\{[1+2\overline{n}(0)]^2 - 4\overline{m}^2(0)\cos^2 2\,\delta\omega t\}^{1/2} \quad (5.5.35)$$

has a minimum value of $\frac{1}{2}\{[1+2\overline{n}(0)]^2 - 4\overline{m}^2(0)\}^{1/2}$ which attains its minimum possible value of $1/2$ consistent with the Heisenberg uncertainty relation if $\overline{m}^2(0) = \overline{n}(0)[\overline{n}(0)+1]$, corresponding to the continuum squeezed state described in Section 3.7.

The equation of motion (5.5.15) for $\hat{a}(t)$ and its solution (5.5.18), together with a further application of the Markov approximation, enable us to establish a useful result relating the evolution of two-time averages to that of averages evaluated at a single time, known as the quantum regression theorem. Consider the evolution of the two-time average $\langle \hat{a}^\dagger(t)\hat{a}(t')\rangle$, where $t > t'$. We have, using (5.5.15),

$$\frac{\mathrm{d}}{\mathrm{d}t}\langle \hat{a}^\dagger(t)\hat{a}(t')\rangle = -(\Gamma - \mathrm{i}\,\delta\omega)\langle \hat{a}^\dagger(t)\hat{a}(t')\rangle + \langle \hat{F}^\dagger(t)\hat{a}(t')\rangle. \quad (5.5.36)$$

For t sufficiently greater than t', the last term in (5.5.36) is negligible, as we now show. Taking the solution for $\hat{a}(t')$ from (5.5.18), multiplying by $\hat{F}^\dagger(t)$ and evaluating the expectation value gives

$$\langle \hat{F}^\dagger(t)\hat{a}(t')\rangle = \int_0^{t'} \exp[(\Gamma + \mathrm{i}\delta\omega)(t'' - t')]\langle \hat{F}^\dagger(t)\hat{F}(t'')\rangle \, \mathrm{d}t'', \quad (5.5.37)$$

where we have again used the fact that $\langle \hat{F}^\dagger(t)\hat{a}(0)\rangle = \langle \hat{F}^\dagger(t)\rangle\langle \hat{a}(0)\rangle = 0$ because $\langle \hat{F}^\dagger(t)\rangle = 0$. From (5.5.24), assuming $\bar{n}(\Delta)$ is slowly varying, the correlation function $\langle \hat{F}^\dagger(t)\hat{F}(t'')\rangle$ will be sharply peaked at $t = t''$. However, since $t > t'$, this peak lies outside the range of integration over t''. Provided t exceeds t' by an amount greater than the width of the correlation function $\langle \hat{F}^\dagger(t)\hat{F}(t'')\rangle$, then the integral in (5.5.37), and hence $\langle \hat{F}^\dagger(t)\hat{a}(t')\rangle$ in (5.5.36), may be neglected. Within the Markov approximation, the width of $\langle \hat{F}^\dagger(t)\hat{F}(t'')\rangle$ is assumed to be much smaller than any other time scale and consequently $\langle \hat{F}^\dagger(t)\hat{a}(t')\rangle$ in (5.5.36) may be neglected for all $t > t'$. Hence the two-time average $\langle \hat{a}^\dagger(t)\hat{a}(t')\rangle$ for $t > t'$ obeys the equation of motion

$$\frac{\mathrm{d}}{\mathrm{d}t}\langle \hat{a}^\dagger(t)\hat{a}(t')\rangle = -(\Gamma - \mathrm{i}\delta\omega)\langle \hat{a}^\dagger(t)\hat{a}(t')\rangle \quad (5.5.38)$$

from which we deduce that $\langle \hat{a}^\dagger(t)\hat{a}(t')\rangle = \exp[-(\Gamma - \mathrm{i}\delta\omega)(t - t')]\langle \hat{a}^\dagger(t')\hat{a}(t')\rangle$. The equation of motion (5.5.38) for $\langle \hat{a}^\dagger(t)\hat{a}(t')\rangle$ is the same as that for $\langle \hat{a}^\dagger(t)\rangle$ since, from (5.5.15),

$$\frac{\mathrm{d}}{\mathrm{d}t}\langle \hat{a}^\dagger(t)\rangle = -(\Gamma - \mathrm{i}\delta\omega)\langle \hat{a}^\dagger(t)\rangle. \quad (5.5.39)$$

This fact is a statement of the quantum regression theorem for a cavity mode treated as a harmonic oscillator. Although we have established this theorem in the Heisenberg picture, we may, of course, use it in the Schrödinger picture also.

The relative simplicity of the above analysis of the damped harmonic oscillator is due to the equations of motion being linear in the annihilation operator $\hat{a}(t)$. In more complicated examples, the equations of motion may be non-linear but nevertheless amenable to study by similar methods. The most important case is when a two-state atom is coupled to the surrounding broad-band radiation field. Consider a two-state atom with ground state $|1\rangle$ and excited state $|2\rangle$. In Chapter 1, we introduced the convenient representation of the atom as a spin-$\frac{1}{2}$ system expressed in terms of the Pauli matrices

$$\hat{\sigma}_3 = |2\rangle\langle 2| - |1\rangle\langle 1|, \quad (5.5.40)$$

$$\hat{\sigma}_+ = |2\rangle\langle 1|, \quad (5.5.41)$$

$$\hat{\sigma}_- = |1\rangle\langle 2|. \quad (5.5.42)$$

The operator $\hat{\sigma}_3$ is the atomic inversion while $\hat{\sigma}_+$ and $\hat{\sigma}_-$ are proportional to the negative and positive frequency components of the atomic dipole. These operators obey the commutation relations $[\hat{\sigma}_+, \hat{\sigma}_-] = \hat{\sigma}_3$ and $[\hat{\sigma}_\pm, \hat{\sigma}_3] = \mp 2\hat{\sigma}_\pm$. The Hamiltonian describing the interaction between the atom and the field is

$$\hat{H}_\mathrm{I} = \hbar \int \Delta \hat{b}^\dagger(\Delta, t)\hat{b}(\Delta, t)\,\mathrm{d}\Delta + \hbar \int W(\Delta)$$
$$\times \left\{\hat{\sigma}_+(t)\hat{b}(\Delta, t)\exp[-\mathrm{i}\varphi(\Delta)] + \hat{b}^\dagger(\Delta, t)\hat{\sigma}_-(t)\exp[\mathrm{i}\varphi(\Delta)]\right\}\mathrm{d}\Delta, \quad (5.5.43)$$

where $W(\Delta)\exp[-\mathrm{i}\varphi(\Delta)]$ is the atom–field coupling and we have chosen an interaction picture in which the zero of energy corresponds to the transition frequency between $|1\rangle$ and $|2\rangle$. The Heisenberg equations of motion for the atomic operators and for the field-mode operators are

$$\dot{\hat{\sigma}}_-(t) = \mathrm{i} \int W(\Delta)\exp[-\mathrm{i}\varphi(\Delta)]\hat{\sigma}_3(t)\hat{b}(\Delta, t)\,\mathrm{d}\Delta, \quad (5.5.44)$$

$$\dot{\hat{\sigma}}_+(t) = -\mathrm{i} \int W(\Delta)\exp[\mathrm{i}\varphi(\Delta)]\hat{b}^\dagger(\Delta, t)\hat{\sigma}_3(t)\,\mathrm{d}\Delta, \quad (5.5.45)$$

$$\dot{\hat{\sigma}}_3(t) = -2\mathrm{i} \int W(\Delta)$$
$$\times \left\{\exp[-\mathrm{i}\varphi(\Delta)]\hat{\sigma}_+(t)\hat{b}(\Delta, t) - \exp[\mathrm{i}\varphi(\Delta)]\hat{b}^\dagger(\Delta, t)\hat{\sigma}_-(t)\right\}\mathrm{d}\Delta, \quad (5.5.46)$$

$$\dot{\hat{b}}(\Delta, t) = -\mathrm{i}\Delta\hat{b}(\Delta, t) - \mathrm{i}W(\Delta)\exp[\mathrm{i}\varphi(\Delta)]\hat{\sigma}_-(t), \quad (5.5.47)$$

$$\dot{\hat{b}}^\dagger(\Delta, t) = \mathrm{i}\Delta\hat{b}^\dagger(\Delta, t) + \mathrm{i}W(\Delta)\exp[-\mathrm{i}\varphi(\Delta)]\hat{\sigma}_+(t). \quad (5.5.48)$$

We proceed as before by formally integrating (5.5.47) and (5.5.48) to give

$$\hat{b}(\Delta, t) = \hat{b}(\Delta, 0)\exp(-\mathrm{i}\Delta t) - \mathrm{i}W(\Delta)\exp[\mathrm{i}\varphi(\Delta)]$$
$$\times \int_0^t \exp[-\mathrm{i}\Delta(t - t')]\hat{\sigma}_-(t')\,\mathrm{d}t', \quad (5.5.49)$$

$$\hat{b}^\dagger(\Delta, t) = \hat{b}^\dagger(\Delta, 0)\exp(\mathrm{i}\Delta t) + \mathrm{i}W(\Delta)\exp[-\mathrm{i}\varphi(\Delta)]$$
$$\times \int_0^t \exp[\mathrm{i}\Delta(t - t')]\hat{\sigma}_+(t')\,\mathrm{d}t'. \quad (5.5.50)$$

These are then substituted into (5.5.44) to (5.5.46), taking care to preserve the ordering of operator products. This is important because we do not yet have any information about the commutation relations between operators at different times. We obtain

$$\dot{\hat{\sigma}}_-(t) = \mathrm{i}\int W(\Delta)\exp[-\mathrm{i}\varphi(\Delta)]\hat{\sigma}_3(t)\hat{b}(\Delta,0)\exp(-\mathrm{i}\Delta t)\,\mathrm{d}\Delta$$
$$+\int_0^t \mathrm{d}t'\int \mathrm{d}\Delta\, W^2(\Delta)\exp[-\mathrm{i}\Delta(t-t')]\hat{\sigma}_3(t)\hat{\sigma}_-(t'), \tag{5.5.51}$$

$$\dot{\hat{\sigma}}_+(t) = -\mathrm{i}\int W(\Delta)\exp[\mathrm{i}\varphi(\Delta)]\hat{b}^\dagger(\Delta,0)\hat{\sigma}_3(t)\exp(\mathrm{i}\Delta t)\,\mathrm{d}\Delta$$
$$+\int_0^t \mathrm{d}t'\int \mathrm{d}\Delta\, W^2(\Delta)\exp[\mathrm{i}\Delta(t-t')]\hat{\sigma}_+(t')\hat{\sigma}_3(t), \tag{5.5.52}$$

$$\dot{\hat{\sigma}}_3(t) = -2\mathrm{i}\int W(\Delta)\{\exp[-\mathrm{i}\varphi(\Delta)]\hat{\sigma}_+(t)\hat{b}(\Delta,0)\exp(-\mathrm{i}\Delta t)$$
$$-\exp[\mathrm{i}\varphi(\Delta)]\hat{b}^\dagger(\Delta,0)\hat{\sigma}_-(t)\exp(\mathrm{i}\Delta t)\}\,\mathrm{d}\Delta - 2\int_0^t \mathrm{d}t'\int \mathrm{d}\Delta\, W^2(\Delta)$$
$$\times\{\exp[-\mathrm{i}\Delta(t-t')]\hat{\sigma}_+(t)\hat{\sigma}_-(t') + \exp[\mathrm{i}\Delta(t-t')]\hat{\sigma}_+(t')\hat{\sigma}_-(t)\}. \tag{5.5.53}$$

In each of the double integrals in (5.5.51) to (5.5.53) we note the appearance of the kernel $K(t-t')$, given by (5.5.14), or its complex conjugate. These integrals are treated by making the Markov approximation as before. For simplicity, we set the frequency shift $\delta\omega$ to zero; in practice, this can be achieved by moving to a new interaction picture in which the zero of energy corresponds to the *shifted* atomic transition frequency. The single integral terms in (5.5.51) to (5.5.53) can be expressed in terms of the Langevin operator $\hat{F}(t)$ in (5.5.36) and its Hermitian conjugate. We find

$$\dot{\hat{\sigma}}_-(t) = -\Gamma\hat{\sigma}_-(t) - \hat{\sigma}_3(t)\hat{F}(t), \tag{5.5.54}$$

$$\dot{\hat{\sigma}}_+(t) = -\Gamma\hat{\sigma}_+(t) - \hat{F}^\dagger(t)\hat{\sigma}_3(t), \tag{5.5.55}$$

$$\dot{\hat{\sigma}}_3(t) = -2\Gamma[\hat{\sigma}_3(t)+1] + 2\hat{\sigma}_+(t)\hat{F}(t) + 2\hat{F}^\dagger(t)\hat{\sigma}_-(t), \tag{5.5.56}$$

where, as before, $\Gamma = \pi W^2(0)$. Each of these equations has a similar structure to that obtained for the damped cavity mode annihilator operator $\hat{a}(t)$ in (5.5.15) consisting of a damping term and a Langevin term. However, in (5.5.54) to (5.5.56) the Langevin operators appear multiplied by atomic operators, a fact which complicates the problem because these atomic operators themselves depend on the Langevin operators $\hat{F}(t)$ and $\hat{F}^\dagger(t)$. Fortunately, the properties of the atom at any given time are described by the expectation values of the atomic operators. It is sufficient, therefore, to obtain equations of motion for

$\langle\hat{\sigma}_-(t)\rangle$, $\langle\hat{\sigma}_+(t)\rangle$, and $\langle\hat{\sigma}_3(t)\rangle$, but we cannot solve the equations obtained by taking the expectation values of (5.5.54) to (5.5.56) because we do not yet know how to deal with the expectation value of products of Langevin and atomic operators. It is tempting, perhaps, simply to decorrelate the atomic and Langevin operators at this stage by writing, for example, $\langle\hat{\sigma}_3(t)\hat{F}(t)\rangle \simeq \langle\hat{\sigma}_3(t)\rangle\langle\hat{F}(t)\rangle$. However, this is not possible because to do so would be to neglect part of the influence of the field on the atom so that, for example, the resulting equations are incorrect if the field is at finite temperature. More seriously, the form of (5.5.54) to (5.5.56) is dependent on the ordering of the operators in the Hamiltonian. Had we used symmetric ordering in the Hamiltonian then the equation of motion for $\hat{\sigma}_-(t)$ would contain only products of $\hat{\sigma}_3(t)$ and $\hat{F}(t)$, and decorrelating at this state would give $\langle\dot{\hat{\sigma}}_-(t)\rangle = 0$. We consider the problem of operator ordering in Appendix 10. In order to overcome the ordering problem, we first formally integrate (5.5.54) to (5.5.56) to give

$$\hat{\sigma}_-(t) = \exp(-\Gamma t)\left(\hat{\sigma}_-(0) - \int_0^t \exp(\Gamma t')\hat{\sigma}_3(t')\hat{F}(t')\,\mathrm{d}t'\right), \tag{5.5.57}$$

$$\hat{\sigma}_+(t) = \exp(-\Gamma t)\left(\hat{\sigma}_+(0) - \int_0^t \exp(\Gamma t')\hat{F}^\dagger(t')\hat{\sigma}_3(t')\,\mathrm{d}t'\right), \tag{5.5.58}$$

$$\hat{\sigma}_3(t) + 1 = \exp(-2\Gamma t)$$
$$\times\left(\hat{\sigma}_3(0) + 1 + 2\int_0^t \exp(2\Gamma t')\left[\hat{\sigma}_+(t')\hat{F}(t') + \hat{F}^\dagger(t')\hat{\sigma}_-(t')\right]\mathrm{d}t'\right), \tag{5.5.59}$$

and proceed to the next order of iteration. Substituting these into (5.5.54) to (5.5.56) and then taking the expectation value of each equation gives

$$\left\langle\dot{\hat{\sigma}}_-(t)\right\rangle = -\Gamma\langle\hat{\sigma}_-(t)\rangle - 2\int_0^t \exp[-2\Gamma(t-t')]$$
$$\times\left[\left\langle\hat{\sigma}_+(t')\hat{F}(t')\hat{F}(t)\right\rangle + \left\langle\hat{F}^\dagger(t')\hat{F}(t)\hat{\sigma}_-(t')\right\rangle\right]\mathrm{d}t', \tag{5.5.60}$$

$$\left\langle\dot{\hat{\sigma}}_+(t)\right\rangle = -\Gamma\langle\hat{\sigma}_+(t)\rangle - 2\int_0^t \exp[-2\Gamma(t-t')]$$
$$\times\left[\left\langle\hat{F}^\dagger(t)\hat{F}^\dagger(t')\hat{\sigma}_-(t')\right\rangle + \left\langle\hat{\sigma}_+(t')\hat{F}^\dagger(t)\hat{F}(t')\right\rangle\right]\mathrm{d}t', \tag{5.5.61}$$

$$\left\langle\dot{\hat{\sigma}}_3(t)\right\rangle = -2\Gamma[\langle\hat{\sigma}_3(t)\rangle + 1] - 2\int_0^t \exp[-\Gamma(t-t')]$$
$$\times\left[\left\langle\hat{F}^\dagger(t')\hat{\sigma}_3(t')\hat{F}(t)\right\rangle + \left\langle\hat{F}^\dagger(t)\hat{\sigma}_3(t')\hat{F}(t')\right\rangle\right]\mathrm{d}t'. \tag{5.5.62}$$

This process of formal integration and back-substitution could be continued, leading to equations of motion containing higher-order correlation functions of the Langevin and atomic operators representing the influence of the field, as modified by the atom, in turn modifying the atomic dynamics. However, the influence of the atom on the field is small and so we neglect higher-order corrections by decorrelating atomic and Langevin operators at this stage in (5.5.60) to (5.5.62). For example, $\langle\hat{\sigma}_+(t')\hat{F}^\dagger(t)\hat{F}(t')\rangle$ and $\langle\hat{F}^\dagger(t')\hat{\sigma}_3(t')\hat{F}(t)\rangle$ are approximated by $\langle\hat{\sigma}_+(t')\rangle\langle\hat{F}^\dagger(t)\hat{F}(t')\rangle$ and $\langle\hat{F}^\dagger(t')\hat{F}(t)\rangle\langle\hat{\sigma}_3(t')\rangle$, respectively, with similar expressions for the other correlation functions. The equations of motion, in this approximation, become

$$\langle\dot{\hat{\sigma}}_-(t)\rangle = -\Gamma\langle\hat{\sigma}_-(t)\rangle - 2\int_0^t \exp[-2\Gamma(t-t')] \times\left[\langle\hat{\sigma}_+(t')\rangle\langle\hat{F}(t')\hat{F}(t)\rangle + \langle\hat{F}^\dagger(t')\hat{F}(t)\rangle\langle\hat{\sigma}_-(t')\rangle\right]\mathrm{d}t', \quad (5.5.63)$$

$$\langle\dot{\hat{\sigma}}_+(t)\rangle = -\Gamma\langle\hat{\sigma}_+(t)\rangle - 2\int_0^t \exp[-2\Gamma(t-t')] \times\left[\langle\hat{F}^\dagger(t)\hat{F}^\dagger(t')\rangle\langle\hat{\sigma}_-(t')\rangle + \langle\hat{\sigma}_+(t')\rangle\langle\hat{F}^\dagger(t)\hat{F}(t')\rangle\right]\mathrm{d}t', \quad (5.5.64)$$

$$\langle\dot{\hat{\sigma}}_3(t)\rangle = -2\Gamma[\langle\hat{\sigma}_3(t)\rangle + 1] - 2\int_0^t \exp[-\Gamma(t-t')] \times[\langle\hat{F}^\dagger(t')\hat{F}(t)\rangle + \langle\hat{F}^\dagger(t)\hat{F}(t')\rangle]\langle\hat{\sigma}_3(t')\rangle\,\mathrm{d}t'. \quad (5.5.65)$$

Having made this decorrelation, the influence of the field on the atom is expressed only in terms of the correlation functions for the Langevin operators determined by the initial state of the field.

For a thermal field, the non-zero correlation functions are given by (5.5.24) and (5.5.25). Substituting (5.5.24) into (5.5.63) to (5.5.65) and once again making the Markov approximation, we obtain

$$\langle\dot{\hat{\sigma}}_-(t)\rangle = -\Gamma[2\bar{n}(0)+1]\langle\hat{\sigma}_-(t)\rangle, \quad (5.5.66)$$

$$\langle\dot{\hat{\sigma}}_+(t)\rangle = -\Gamma[2\bar{n}(0)+1]\langle\hat{\sigma}_+(t)\rangle, \quad (5.5.67)$$

$$\langle\dot{\hat{\sigma}}_3(t)\rangle = -2\Gamma\{[2\bar{n}(0)+1]\langle\hat{\sigma}_3(t)\rangle + 1\}, \quad (5.5.68)$$

where $\bar{n}(0)$ is the mean occupation number of a field mode the frequency of which coincides with the (shifted) atomic transition frequency. From these equations, we see that the rate of decay of the atomic dipole is enhanced by a factor $2\bar{n}(0)+1$, and that the steady-state solution has a zero dipole moment ($\langle\hat{\sigma}_-(\infty)\rangle = \langle\hat{\sigma}_+(\infty)\rangle = 0$) and inversion $\langle\hat{\sigma}_3(\infty)\rangle = -[2\bar{n}(0)+1]^{-1}$. This inversion corresponds to an excited state probability of $\bar{n}(0)/[2\bar{n}(0)+1]$. If we substitute for $\bar{n}(0)$ using the Bose–Einstein distribution (3.5.9), then we find that this steady-state probability coincides with that expected for a two-state system in thermal equilibrium with its environment (see (3.5.3)).

For a broad-band squeezed field, there are additional non-zero correlation functions, $\langle \hat{F}(t')\hat{F}(t)\rangle$ and its complex conjugate, given by

$$\langle \hat{F}(t')\hat{F}(t)\rangle = -\int d\Delta W(\Delta)\exp[-i\varphi(\Delta)]\exp(-i\Delta t')\int d\Delta' W(\Delta')$$

$$\times \exp[-i\varphi(\Delta')]\exp(-i\Delta' t)\langle \hat{b}(\Delta,0)\hat{b}(\Delta',0)\rangle$$

$$= -\int W(\Delta)W(-\Delta)\exp\{-i[\varphi(\Delta)+\varphi(-\Delta)]\}\overline{m}(\Delta)$$

$$\times \exp[-i\Delta(t'-t)]\,d\Delta, \tag{5.5.69}$$

using (5.5.28) with $\Delta_s = 0$, corresponding to resonance with the shifted atomic transition frequency. Assuming $\overline{m}(\Delta)$ is slowly varying and real and that $\varphi(\Delta)$ is zero in the region of $\Delta = 0$, we find within the Markov approximation that (5.5.63) and (5.5.65) become

$$\langle \dot{\hat{\sigma}}_-(t)\rangle = -\Gamma[2\overline{n}(0)+1]\langle \hat{\sigma}_-(t)\rangle + 2\Gamma\overline{m}(0)\langle \hat{\sigma}_+(t)\rangle, \tag{5.5.70}$$

$$\langle \dot{\hat{\sigma}}_+(t)\rangle = -\Gamma[2\overline{n}(0)+1]\langle \hat{\sigma}_+(t)\rangle + 2\Gamma\overline{m}(0)\langle \hat{\sigma}_-(t)\rangle, \tag{5.5.71}$$

$$\langle \dot{\hat{\sigma}}_3(t)\rangle = -2\Gamma\{[2\overline{n}(0)+1]\langle \hat{\sigma}_3(t)\rangle + 1\}. \tag{5.5.72}$$

The steady-state solution is the same as for a thermal field. However, the components of the atomic dipole, proportional to the expectation values of the operators $\hat{\sigma}_1 = (\hat{\sigma}_+ + \hat{\sigma}_-)$ and $\hat{\sigma}_2 = i(\hat{\sigma}_- - \hat{\sigma}_+)$, decay at different rates since

$$\langle \dot{\hat{\sigma}}_1(t)\rangle = -\Gamma[2\overline{n}(0)+1-2\overline{m}(0)]\langle \hat{\sigma}_1(t)\rangle, \tag{5.5.73}$$

$$\langle \dot{\hat{\sigma}}_2(t)\rangle = -\Gamma[2\overline{n}(0)+1+2\overline{m}(0)]\langle \hat{\sigma}_2(t)\rangle. \tag{5.5.74}$$

The modification of the decay of the atomic dipole by thermal and squeezed fields changes the lineshape of the atomic transition. In order to study this further, we will shortly express the quantum regression theorem in a form suitable for application to a two-level atom.

The normalized spectrum $S(\Omega)$ of light emitted by the atom at long times depends on the two-time correlation function $\langle \hat{\sigma}_+(t+\tau)\hat{\sigma}_-(t)\rangle$ which we will calculate using the quantum regression theorem. Specifically, $S(\Omega)$ is proportional to $C(\Omega)$ given by

$$C(\Omega) = \lim_{T\to\infty}\lim_{t\to\infty}\int_{-T}^{T}\langle \hat{\sigma}_+(t+\tau)\hat{\sigma}_-(t)\rangle \exp(-i\Omega\tau)\,d\tau, \tag{5.5.75}$$

where the correlation function is evaluated in steady state corresponding to $t\to\infty$. This limit is well defined if the correlation function depends only on the difference between the time arguments at long times. This must be checked once the correlation function has been calculated. The limit as $T\to\infty$ is taken after that for t; formally, this omits initial transients. We can write $C(\Omega)$ in a more convenient form by splitting the integration over τ into two parts, as follows:

$$C(\Omega) = \lim_{T\to\infty} \lim_{t\to\infty} \left(\int_0^T \langle \hat{\sigma}_+(t+\tau)\hat{\sigma}_-(t)\rangle \exp(-\mathrm{i}\Omega\tau)\,\mathrm{d}\tau \right.$$
$$\left. + \int_{-T}^0 \langle \hat{\sigma}_+(t+\tau)\hat{\sigma}_-(t)\rangle \exp(-\mathrm{i}\Omega\tau)\,\mathrm{d}\tau \right). \qquad (5.5.76)$$

Assuming that in steady state the correlation function depends only on τ, we can rewrite the correlation function in the second integral in (5.5.76) as $\langle \hat{\sigma}_+(t)\hat{\sigma}_-(t-\tau)\rangle$. Changing the variable of integration in this integral from τ to $-\tau$ gives

$$C(\Omega) = \lim_{T\to\infty} \lim_{t\to\infty} \left(\int_0^T \langle \hat{\sigma}_+(t+\tau)\hat{\sigma}_-(t)\rangle \exp(-\mathrm{i}\Omega\tau)\,\mathrm{d}\Omega \right.$$
$$\left. + \int_0^T \langle \hat{\sigma}_+(t)\hat{\sigma}_-(t+\tau)\rangle \exp(\mathrm{i}\Omega\tau)\,\mathrm{d}\tau \right)$$
$$= \lim_{T\to\infty} \lim_{t\to\infty} 2\,\mathrm{Re} \int_0^T \langle \hat{\sigma}_+(t+\tau)\hat{\sigma}_-(t)\rangle \exp(-\mathrm{i}\Omega\tau)\,\mathrm{d}\tau, \qquad (5.5.77)$$

where Re denotes the real part, since $\langle \hat{\sigma}_+(t)\hat{\sigma}_-(t+\tau)\rangle \exp(\mathrm{i}\Omega\tau)$ is the complex conjugate of $\langle \hat{\sigma}_+(t+\tau)\hat{\sigma}_-(t)\rangle \exp(-\mathrm{i}\Omega\tau)$. It is worth noting from (5.5.77) that $C(\Omega)$ is formally the real part of the Laplace transform of $\lim_{t\to\infty} 2\langle \hat{\sigma}_+(t+\tau)\hat{\sigma}_-(t)\rangle$ evaluated at $s = \mathrm{i}\Omega$. The correlation function in (5.5.77) can be found from the quantum regression theorem; from (5.5.55) its equation of motion is

$$\frac{\mathrm{d}}{\mathrm{d}\tau}\langle \hat{\sigma}_+(t+\tau)\hat{\sigma}_-(t)\rangle = -\Gamma\langle \hat{\sigma}_+(t+\tau)\hat{\sigma}_-(t)\rangle - \left\langle \hat{F}^\dagger(t+\tau)\hat{\sigma}_3(t+\tau)\hat{\sigma}_-(t)\right\rangle. \qquad (5.5.78)$$

Inserting the expression for $\hat{\sigma}_3(t+\tau)$ from (5.5.59), the last term in (5.5.78) becomes

$$\left\langle \hat{F}^\dagger(t+\tau)\hat{\sigma}_3(t+\tau)\hat{\sigma}_-(t)\right\rangle$$
$$= \left\langle \hat{F}^\dagger(t+\tau)\left(-1 + \exp[-2\Gamma(t+\tau)]\left\{ \hat{\sigma}_3(0) + 1 + 2\int_0^{t+\tau} \exp(2\Gamma t') \right.\right.\right.$$
$$\left.\left.\left. \times \left[\hat{\sigma}_+(t')\hat{F}(t') + \hat{F}^\dagger(t')\hat{\sigma}_-(t')\right] \mathrm{d}t' \right\}\right) \hat{\sigma}_-(t) \right\rangle$$
$$= 2\left\langle \hat{F}^\dagger(t+\tau)\left(\int_0^{t+\tau} \exp[-2\Gamma(t+\tau-t')] \right.\right.$$
$$\left.\left. \times \left[\hat{\sigma}_+(t')\hat{F}(t') + \hat{F}^\dagger(t')\hat{\sigma}_-(t')\right] \mathrm{d}t' \right) \hat{\sigma}_-(t) \right\rangle. \qquad (5.5.79)$$

Here we have used the fact that $\hat{\sigma}_-(t)$ and therefore also $\hat{\sigma}_3(0)\hat{\sigma}_-(t)$ only depend on Langevin operators at times prior to t so that $\langle\hat{F}^\dagger(t+\tau)\hat{\sigma}_-(t)\rangle = \langle\hat{F}^\dagger(t+\tau)\hat{\sigma}_3(0)\hat{\sigma}_-(t)\rangle = 0$ as in the derivation of (5.5.38). As in the treatment of the single-time expectation values (5.5.60) to (5.5.62), we neglect higher-order correlation by decorrelating atomic and Langevin operators in (5.5.79) to give

$$\left\langle\hat{F}^\dagger(t+\tau)\hat{\sigma}_3(t+\tau)\hat{\sigma}_-(t)\right\rangle = 2\int_0^{t+\tau}\exp[-2\Gamma(t+\tau-t')] \times\left[\langle\hat{F}^\dagger(t+\tau)\hat{F}(t')\rangle\langle\hat{\sigma}_+(t')\hat{\sigma}_-(t)\rangle + \langle\hat{F}^\dagger(t+\tau)\hat{F}^\dagger(t')\rangle\langle\hat{\sigma}_-(t')\hat{\sigma}_-(t)\rangle\right]\mathrm{d}t'. \tag{5.5.80}$$

We now apply the Markov approximation to the field correlation functions in (5.5.80) to obtain

$$\left\langle\hat{F}^\dagger(t+\tau)\hat{\sigma}_3(t+\tau)\hat{\sigma}_-(t)\right\rangle = 2\Gamma\bar{n}(0)\langle\hat{\sigma}_+(t+\tau)\hat{\sigma}_-(t)\rangle - 2\Gamma\bar{m}(0)\langle\hat{\sigma}_-(t+\tau)\hat{\sigma}_-(t)\rangle, \tag{5.5.81}$$

for a squeezed field, and the same expression with $\bar{m}(0) = 0$ for a thermal field. Substituting (5.5.81) into (5.5.78), we find

$$\frac{\mathrm{d}}{\mathrm{d}\tau}\langle\hat{\sigma}_+(t+\tau)\hat{\sigma}_-(t)\rangle = -\Gamma[2\bar{n}(0)+1]\langle\hat{\sigma}_+(t+\tau)\hat{\sigma}_-(t)\rangle + 2\Gamma\bar{m}(0)\langle\hat{\sigma}_-(t+\tau)\hat{\sigma}_-(t)\rangle. \tag{5.5.82}$$

Comparing this with (5.5.71), we see that the two-time correlation function obeys the same equation as $\langle\hat{\sigma}_+(t+\tau)\rangle$, thus establishing the quantum regression theorem for the two-level atom.

For a thermal field for which $\bar{m}(0) = 0$, the solution of (5.5.82) is

$$\langle\hat{\sigma}_+(t+\tau)\hat{\sigma}_-(t)\rangle = \exp\{-\Gamma[2\bar{n}(0)+1]\tau\}\langle\hat{\sigma}_+(t)\hat{\sigma}_-(t)\rangle = \exp\{-\Gamma[2\bar{n}(0)+1]\tau\}\frac{\bar{n}(0)}{2\bar{n}(0)+1}, \tag{5.5.83}$$

where we have substituted for the steady-state value of $\langle\hat{\sigma}_+(t)\hat{\sigma}_-(t)\rangle$. Inserting this into (5.5.77) and evaluating the transform we have

$$C(\Omega) = \frac{2\bar{n}(0)}{2\bar{n}(0)+1}\,\mathrm{Re}\,\frac{1}{\mathrm{i}\Omega + \Gamma[2\bar{n}(0)+1]}. \tag{5.5.84}$$

Hence the normalized spectrum is

$$S(\Omega) = \frac{[2\bar{n}(0)+1]\Gamma/\pi}{\Omega^2 + \Gamma^2[2\bar{n}(0)+1]^2}. \tag{5.5.85}$$

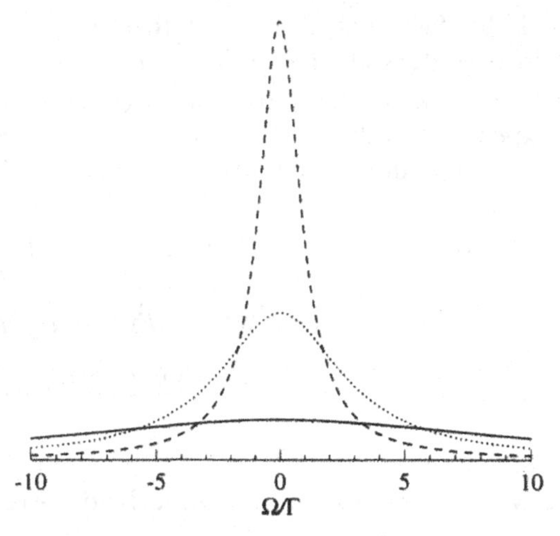

Fig. 5.4 Normalized emission spectrum for an atom interacting with broad-band thermal radiation with $\bar{n}(0) = 0$ (dashed curve), $\bar{n}(0) = 1$ (dotted curve), and $\bar{n}(0) = 5$ (solid curve).

This is a Lorentzian spectrum centred on the transition frequency with a width determined by the temperature of the surrounding field. In Fig. 5.4, we plot $S(\Omega)$ for $\bar{n}(0) = 0, 1, 5$. The increasing width with $\bar{n}(0)$ may be attributed to the action of stimulated processes.

For a squeezed field, the equation of motion of the correlation function $\langle \hat{\sigma}_+(t+\tau)\hat{\sigma}_-(t)\rangle$ also contains the correlation function $\langle \hat{\sigma}_-(t+\tau)\hat{\sigma}_-(t)\rangle$. It is more convenient to express $\langle \hat{\sigma}_+(t+\tau)\hat{\sigma}_-(t)\rangle$ in terms of the operators $\hat{\sigma}_1(t+\tau)$ and $\hat{\sigma}_2(t+\tau)$ as

$$\langle \hat{\sigma}_+(t+\tau)\hat{\sigma}_-(t)\rangle = \tfrac{1}{2}\langle \hat{\sigma}_1(t+\tau)\hat{\sigma}_-(t)\rangle + \tfrac{1}{2}\mathrm{i}\langle \hat{\sigma}_2(t+\tau)\hat{\sigma}_-(t)\rangle \quad (5.5.86)$$

and to apply the quantum regression theorem to each of these terms. From (5.5.73) and (5.5.74), we have

$$\frac{\mathrm{d}}{\mathrm{d}\tau}\langle \hat{\sigma}_1(t+\tau)\hat{\sigma}_-(t)\rangle = -\Gamma[2\bar{n}(0) + 1 - 2\bar{m}(0)]\langle \hat{\sigma}_1(t+\tau)\hat{\sigma}_-(t)\rangle \quad (5.5.87)$$

and

$$\frac{\mathrm{d}}{\mathrm{d}\tau}\langle \hat{\sigma}_2(t+\tau)\hat{\sigma}_-(t)\rangle = -\Gamma[2\bar{n}(0) + 1 + 2\bar{m}(0)]\langle \hat{\sigma}_2(t+\tau)\hat{\sigma}_-(t)\rangle. \quad (5.5.88)$$

The solutions of these equations are

$$\begin{aligned}\langle \hat{\sigma}_1(t+\tau)\hat{\sigma}_-(t)\rangle &= \exp\{-\Gamma[2\bar{n}(0) + 1 - 2\bar{m}(0)]\tau\}\langle \hat{\sigma}_1(t)\hat{\sigma}_-(t)\rangle \\ &= \exp\{-\Gamma[2\bar{n}(0) + 1 - 2\bar{m}(0)]\tau\}\langle \hat{\sigma}_+(t)\hat{\sigma}_-(t)\rangle \quad (5.5.89)\end{aligned}$$

and

$$\begin{aligned}\langle \hat{\sigma}_2(t+\tau)\hat{\sigma}_-(t)\rangle &= \exp\{-\Gamma[2\bar{n}(0) + 1 + 2\bar{m}(0)]\tau\}\langle \hat{\sigma}_2(t)\hat{\sigma}_-(t)\rangle \\ &= -\mathrm{i}\exp\{-\Gamma[2\bar{n}(0) + 1 + 2\bar{m}(0)]\tau\}\langle \hat{\sigma}_+(t)\hat{\sigma}_-(t)\rangle, \quad (5.5.90)\end{aligned}$$

where we have expressed $\hat{\sigma}_1(t)$ and $\hat{\sigma}_2(t)$ in terms of $\hat{\sigma}_+(t)$ and $\hat{\sigma}_-(t)$. Hence from (5.5.86) we find

$$\begin{aligned}\langle \hat{\sigma}_+(t+\tau)\hat{\sigma}_-(t)\rangle &= \tfrac{1}{2}\langle \hat{\sigma}_+(t)\hat{\sigma}_-(t)\rangle[\exp\{-\Gamma[2\bar{n}(0)+1-2\overline{m}(0)]\tau\} \\ &\quad + \exp\{-\Gamma[2\bar{n}(0)+1+2\overline{m}(0)]\tau\}] \\ &= \tfrac{1}{2}\cdot\frac{\bar{n}(0)}{2\bar{n}(0)+1}[\exp\{-\Gamma[2\bar{n}(0)+1-2\overline{m}(0)]\tau\} \\ &\quad + \exp\{-\Gamma[2\bar{n}(0)+1+2\overline{m}(0)]\tau\}], \end{aligned} \tag{5.5.91}$$

where we have again substituted for the steady-state value of $\langle \hat{\sigma}_+(t)\hat{\sigma}_-(t)\rangle$. Inserting this into (5.5.77) and evaluating the transform, we obtain

$$C(\Omega) = \frac{\bar{n}(0)}{2\bar{n}(0)+1} \times \mathrm{Re}\left(\frac{1}{\mathrm{i}\Omega + \Gamma[2\bar{n}(0)+1-2\overline{m}(0)]} + \frac{1}{\mathrm{i}\Omega + \Gamma[2\bar{n}(0)+1+2\overline{m}(0)]}\right). \tag{5.5.92}$$

Hence the normalized spectrum is

$$S(\Omega) = \tfrac{1}{2}\left(\frac{[2\bar{n}(0)+1-2\overline{m}(0)]\Gamma/\pi}{\Omega^2 + \Gamma^2[2\bar{n}(0)+1-2\overline{m}(0)]^2} + \frac{[2\bar{n}(0)+1+2\overline{m}(0)]\Gamma/\pi}{\Omega^2 + \Gamma^2[2\bar{n}(0)+1+2\overline{m}(0)]^2}\right) \tag{5.5.93}$$

which is the sum of two Lorentzians. In Fig. 5.5, we plot $S(\Omega)$ for $\bar{n}(0) = 1$ and $\overline{m}(0) = 0, 1, \surd 2$. When $\overline{m}(0) = 0$, the spectrum is the Lorentzian of the thermal case (5.5.85). For $\overline{m}(0) > 0$, the spectrum consists of two superimposed Lorentzians one of which is broader than that for the thermal field and the other of which is narrower. If $\overline{m}(0) = \bar{n}(0)$, the narrower Lorentzian has the same width as that for a vacuum field, while if $\bar{n}(0) < \overline{m}(0) < \surd\{\bar{n}(0)[\bar{n}(0)+1]\}$, its width is less than that for a vacuum field (the natural linewidth). This subnatural linewidth for $\overline{m}(0) > \bar{n}(0)$ is a non-classical effect arising from the fact that the field is squeezed, that is it exhibits an uncertainty in one of its quadrature components that is less than the vacuum value (see Section 3.7).

5.6 Master equations

In the previous section, the dissipation arising from the interaction between a system and its environment was treated by developing Heisenberg equations of motion for the operators associated with the properties of the system. A parallel analysis is possible working in the Schrödinger picture, as we have already seen in Section 5.3. Usually, however, we only have statistical information about the state of the environment and this is insufficient to construct a total wavefunction for the system and its environment. A description within the Schrödinger picture

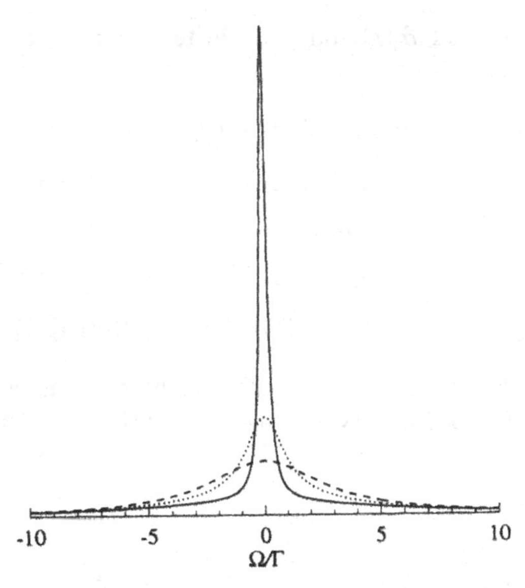

Fig. 5.5 Normalized emission spectrum for an atom interacting with a broad-band squeezed field with $\bar{n}(0)=1$ and $\bar{m}(0)=0$ (dashed curve), $m(0)=1$ (dotted curve), and $\bar{m}(0)=\sqrt{2}$ (solid curve).

of this situation must therefore employ the density matrix. Further, we are primarily interested in the properties of the system *alone* and these may be found from the reduced density matrix obtained by tracing over the degrees of freedom of the environment. The equation for this reduced density matrix is called the master equation, and the treatment of dissipation using master equations is common in quantum optics because of the wide applicability of the technique.

We consider the interaction between a system, with operators $\hat{s}$ and $\hat{s}^\dagger$ and its environment modelled, as before, as a continuum of harmonic oscillators with annihilation and creation operators $\hat{b}(\Delta)$ and $\hat{b}^\dagger(\Delta)$. As we are working in the Schrödinger picture, the operators are time independent and consequently the time dependence of $\hat{b}$ and $\hat{b}^\dagger$ which arose in the previous section is absent here. The operators $\hat{s}$ and $\hat{s}^\dagger$ respectively annihilate and create a quantum of energy $\hbar\omega$ in the system. This implies that if $\hat{H}_0$ is the Schrödinger picture Hamiltonian for the system alone then $[\hat{H}_0, \hat{s}^\dagger] = \hbar\omega\hat{s}^\dagger$. Transforming to an interaction picture in which $\hat{H}_0$ is absent introduces the time dependence $\exp(\mathrm{i}\omega t)$ multiplying $\hat{s}^\dagger$. A further unitary transformation with $\hat{U} = \exp[\mathrm{i}\omega t \int \hat{b}^\dagger(\Delta)\hat{b}(\Delta)\,\mathrm{d}\Delta]$ cancels this time dependence leaving the products $\hat{s}^\dagger\hat{b}(\Delta)$ and $\hat{b}^\dagger(\Delta)\hat{s}$ independent of time. The resulting Hamiltonian for this model consisting of the system and its environment is

$$\hat{H}_\mathrm{I} = \hbar\int \Delta\hat{b}^\dagger(\Delta)\hat{b}(\Delta)\,\mathrm{d}\Delta + \hbar\int W(\Delta) \times\left\{\hat{s}^\dagger\hat{b}(\Delta)\exp[-\mathrm{i}\varphi(\Delta)] + \hat{b}^\dagger(\Delta)\hat{s}\exp[\mathrm{i}\varphi(\Delta)]\right\}\mathrm{d}\Delta. \tag{5.6.1}$$

The form of (5.6.1) is precisely that of (5.5.1) and (5.5.43) and is therefore sufficiently general to describe both the damped cavity mode and the damped two-level atom. The Hamiltonian (5.6.1) is not yet in the most convenient form for deriving the master equation because, in addition to the interaction terms, it contains a term which generates the free time evolution of the environment. We will treat this free evolution exactly but develop approximations for dealing with the interaction terms, and so we first transform $\hat{H}_{\mathrm{I}}$ to remove the free evolution term, as described in Chapter 2. The resulting explicitly time-dependent Hamiltonian is

$$\hat{V}_{\mathrm{I}}(t) = \hbar \int W(\Delta)\{\hat{s}^{\dagger}\hat{b}(\Delta)\exp(-\mathrm{i}\Delta t)\exp[-\mathrm{i}\varphi(\Delta)]$$
$$+\hat{b}^{\dagger}(\Delta)\exp(\mathrm{i}\Delta t)\hat{s}\exp[\mathrm{i}\varphi(\Delta)]\}\,\mathrm{d}\Delta. \tag{5.6.2}$$

Using the definition (5.5.16), and noting that the Heisenberg operator $\hat{b}(\Delta,0)$ is precisely the operator $\hat{b}(\Delta)$ in the Schrödinger picture, the Hamiltonian $\hat{V}_{\mathrm{I}}(t)$ can be written in the compact form

$$\hat{V}_{\mathrm{I}}(t) = \mathrm{i}\hbar[\hat{s}^{\dagger}\hat{F}(t) - \hat{F}^{\dagger}(t)\hat{s}]. \tag{5.6.3}$$

The total density matrix ρ_{T} for the system and its environment evolves according to the Liouville equation

$$\dot{\rho}_{\mathrm{T}}(t) = -\frac{\mathrm{i}}{\hbar}\left[\hat{V}_{\mathrm{I}}(t), \rho_{\mathrm{T}}(t)\right]. \tag{5.6.4}$$

We consider the general class of problems for which the system is initially decorrelated from its environment so that $\rho_{\mathrm{T}}(0) = \rho_{\mathrm{e}}(0) \otimes \rho(0)$ where ρ_{e} is the density matrix for the environment and ρ is the (reduced) density matrix for the system alone. The master equation for $\rho(t)$ is obtained by taking the trace of (5.6.4) over the degrees of freedom of the environment, so that

$$\dot{\rho}(t) = -\frac{\mathrm{i}}{\hbar}\mathrm{Tr}_{\mathrm{e}}\left[\hat{V}_{\mathrm{I}}(t), \rho_{\mathrm{T}}(t)\right]. \tag{5.6.5}$$

In order to evaluate the trace in this equation, we need to know how ρ_{e} evolves under the influence of the coupling. We assume that the environment has a large number of degrees of freedom and therefore is little changed by this coupling. Consequently we will, at a suitable point, use the approximation

$$\rho_{\mathrm{T}}(t) = \rho_{\mathrm{e}}(0) \otimes \rho(t). \tag{5.6.6}$$

If we apply this approximation directly to (5.6.5), we find that $\dot{\rho}(t)$ contains only terms proportional to $\langle\hat{F}(t)\rangle$ and $\langle\hat{F}^{\dagger}(t)\rangle$ and is therefore zero! The problem of when to decorrelate system and environment operators arose in Section 5.5, in the discussion of atom–field interactions, where it was resolved by proceeding to the next order of iteration. We use the same technique in order to deal with (5.6.5). Formally integrating (5.6.3), we obtain

$$\rho_T(t) = \rho_T(0) - \frac{i}{\hbar}\int_0^t \left[\hat{V}_I(t'), \rho_T(t')\right] dt', \tag{5.6.7}$$

which on substitution into (5.6.5) gives

$$\dot{\rho}(t) = -\frac{i}{\hbar}\mathrm{Tr}_e\left[\hat{V}_I(t), \rho_T(0)\right] - \frac{1}{\hbar^2}\int_0^t \mathrm{Tr}_e\left[\hat{V}_I(t), \left[\hat{V}_I(t'), \rho_T(t')\right]\right] dt'. \tag{5.6.8}$$

The first term on the right-hand side of this equation is identically zero since it contains only terms proportional to $\langle \hat{F}(t)\rangle$ and $\langle \hat{F}^\dagger(t)\rangle$. We can now use the approximate form (5.6.6) for $\rho_T(t)$. The resulting equation for $\rho(t)$, using (5.6.3), is

$$\dot{\rho}(t) = \int_0^t \mathrm{Tr}_e\left[(\hat{s}^\dagger\hat{F}(t) - \hat{F}^\dagger(t)\hat{s}), \left[(\hat{s}^\dagger\hat{F}(t') - \hat{F}^\dagger(t')\hat{s}), \rho_e(0)\otimes\rho(t')\right]\right] dt'. \tag{5.6.9}$$

Expanding the commutators gives 16 terms each of which contains a product of two system operators, $\rho(t)$, two Langevin operators, and $\rho_e(0)$. In this decorrelated form, the trace operation acts only on the Langevin operators and $\rho_e(0)$. We use the cyclic property of the trace operation to write, for example, $\mathrm{Tr}_e[\hat{F}(t)\rho_e(0)\hat{F}^\dagger(t')] = \mathrm{Tr}_e[\hat{F}^\dagger(t')\hat{F}(t)\rho_e(0)] = \langle \hat{F}^\dagger(t')\hat{F}(t)\rangle$. The resulting equation for $\rho(t)$ is

$$\begin{aligned}
\dot{\rho}(t) = \int_0^t \Big\{ &\left[\hat{s}\rho(t')\hat{s}^\dagger - \hat{s}^\dagger\hat{s}\rho(t')\right]\langle \hat{F}(t)\hat{F}^\dagger(t')\rangle \\
&+ \left[\hat{s}\rho(t')\hat{s}^\dagger - \rho(t')\hat{s}^\dagger\hat{s}\right]\langle \hat{F}(t')\hat{F}^\dagger(t)\rangle \\
&+ [\hat{s}^\dagger\rho(t')\hat{s} - \hat{s}\hat{s}^\dagger\rho(t')]\langle \hat{F}^\dagger(t)\hat{F}(t')\rangle \\
&+ [\hat{s}^\dagger\rho(t')\hat{s} - \rho(t')\hat{s}\hat{s}^\dagger]\langle \hat{F}^\dagger(t')\hat{F}(t)\rangle \\
&- \left[\hat{s}^\dagger\rho(t')\hat{s}^\dagger - \hat{s}^{\dagger 2}\rho(t')\right]\langle \hat{F}(t)\hat{F}(t')\rangle \\
&- [\hat{s}^\dagger\rho(t')\hat{s}^\dagger - \rho(t')\hat{s}^{\dagger 2}]\langle \hat{F}(t')\hat{F}(t)\rangle \\
&- [\hat{s}\rho(t')\hat{s} - \hat{s}^2\rho(t')]\langle \hat{F}^\dagger(t)\hat{F}^\dagger(t')\rangle \\
&- [\hat{s}\rho(t')\hat{s} - \rho(t')\hat{s}^2]\langle \hat{F}^\dagger(t')\hat{F}^\dagger(t)\rangle \Big\} dt'.
\end{aligned} \tag{5.6.10}$$

The correlation functions of the Langevin operators contain the environment operators and $\rho_e(0)$ and are therefore identical to those calculated within the Heisenberg picture in Section 5.5. The correlation functions in the last four terms of (5.6.10) involving the products of $\hat{F}(t)$ with $\hat{F}(t')$ and $\hat{F}^\dagger(t)$ with $\hat{F}^\dagger(t')$ will be absent in most cases, the notable exception being in the interaction of the system with a rigged reservoir, as discussed in the previous section. Here we restrict our attention to the case of an environment in thermodynamic equilibrium as occurs in most problems. The remaining correlation functions are given by (5.5.24) and (5.5.25) and substituting these into (5.6.10), we find

$$\begin{aligned}\dot{\rho}(t) = \int_0^t \mathrm{d}t' \int W^2(\Delta)\,\mathrm{d}\Delta\{&[\hat{s}\rho(t')\hat{s}^\dagger - \hat{s}^\dagger\hat{s}\rho(t')][\bar{n}(\Delta)+1]\exp[-\mathrm{i}\Delta(t-t')]\\ &+[\hat{s}\rho(t')\hat{s}^\dagger - \rho(t')\hat{s}^\dagger\hat{s}][\bar{n}(\Delta)+1]\exp[\mathrm{i}\Delta(t-t')]\\ &+[\hat{s}^\dagger\rho(t')\hat{s} - \hat{s}\hat{s}^\dagger\rho(t')]\bar{n}(\Delta)\exp[\mathrm{i}\Delta(t-t')]\\ &+[\hat{s}^\dagger\rho(t')\hat{s} - \rho(t')\hat{s}\hat{s}^\dagger]\bar{n}(\Delta)\exp[-\mathrm{i}\Delta(t-t')]\}. \end{aligned} \tag{5.6.11}$$

We now make the Markov approximation to convert (5.6.11) from an integro-differential equation to a differential equation for $\rho(t)$ using the procedure developed in Section 5.3 (which led to (5.3.18)). In this way, we obtain the following *master equation* for the density matrix of the system:

$$\begin{aligned}\dot{\rho}(t) = &-\mathrm{i}\delta\omega\big[\hat{s}^\dagger\hat{s},\rho(t)\big] - \mathrm{i}\delta\omega_{\mathrm{th}}\big[[\hat{s}^\dagger,\hat{s}],\rho(t)\big]\\ &+\Gamma[\bar{n}(0)+1][2\hat{s}\rho(t)\hat{s}^\dagger - \hat{s}^\dagger\hat{s}\rho(t) - \rho(t)\hat{s}^\dagger\hat{s}]\\ &+\Gamma\bar{n}(0)[2\hat{s}^\dagger\rho(t)\hat{s} - \hat{s}\hat{s}^\dagger\rho(t) - \rho(t)\hat{s}\hat{s}^\dagger], \end{aligned} \tag{5.6.12}$$

where $\Gamma = \pi W^2(0)$,

$$\delta\omega = -\mathbb{P}\int \frac{W^2(\Delta)}{\Delta}\,\mathrm{d}\Delta \tag{5.6.13}$$

and

$$\delta\omega_{\mathrm{th}} = -\mathbb{P}\int \frac{\bar{n}(\Delta)W^2(\Delta)}{\Delta}\,\mathrm{d}\Delta. \tag{5.6.14}$$

The first two terms in the master equation (5.6.12) will give rise to frequency shifts, the values of which depend on the nature of the system being damped. As noted before, $\delta\omega$ and $\delta\omega_{\mathrm{th}}$ in (5.6.13) and (5.6.14) must be small compared with Γ and $\Gamma\bar{n}(0)$ respectively in order to give a meaningful value for the frequency shift. For a damped cavity mode with $\hat{s}=\hat{a}$, the annihilation operator for the mode, the second term is zero and the first becomes $-\mathrm{i}\delta\omega[\hat{a}^\dagger\hat{a},\rho(t)]$ corresponding to an increase in the natural frequency by an amount $\delta\omega$. If, on the other hand, we are considering the damping of a two-level atom for which $\hat{s}=\hat{\sigma}_-$, the atomic lowering operator, both terms contribute and combine to give a term $-\mathrm{i}(\delta\omega+2\delta\omega_{\mathrm{th}})[\frac{1}{2}\hat{\sigma}_3,\rho(t)]$ which corresponds to a frequency increase of

$\delta\omega + 2\delta\omega_{th}$. Incorporating these shifts into a redefinition of the natural frequency, as before, gives the master equation

$$\dot{\rho}(t) = \Gamma[\bar{n}(0)+1][2\hat{s}\rho(t)\hat{s}^\dagger - \hat{s}^\dagger\hat{s}\rho(t) - \rho(t)\hat{s}^\dagger\hat{s}] + \Gamma\bar{n}(0)[2\hat{s}^\dagger\rho(t)\hat{s} - \hat{s}\hat{s}^\dagger\rho(t) - \rho(t)\hat{s}\hat{s}^\dagger]. \quad (5.6.15)$$

For a two-level atom, this master equation leads to the familiar Einstein rate equations with Γ being the spontaneous emission rate and $\Gamma\bar{n}(0)$ both the absorption and stimulated emission rates.

Solution of the master equation provides a complete description of the influence of dissipation on the system of interest. If the expectation value of an observable represented by the operator $\hat{A}$ is required then an equation of motion for this quantity, given by $\mathrm{Tr}[\rho(t)\hat{A}]$, is obtainable directly from the master equation. We note that (5.6.15) is consistent with conservation of probability since $\mathrm{Tr}[\dot{\rho}(t)] = 0$, as may be readily verified using the cyclic property of the trace operation. The most obvious way of solving the master equation (5.6.15) is to express $\rho(t)$ as a matrix and solve the resulting coupled differential equations for the components of $\rho(t)$. As an example of this method, we solve the master equation representing a two-level atom coupled to a thermal field for which $\hat{s}$ becomes $\hat{\sigma}_-$ in (5.6.15). With ground and excited states $|1\rangle$ and $|2\rangle$ respectively, we write the elements of the density matrix as $\rho_{ij} = \langle i|\rho(t)|j\rangle$ where $i, j = 1, 2$. The equations for these elements obtained from (5.6.15) are

$$\dot{\rho}_{22} = -\dot{\rho}_{11} = -2\Gamma[\bar{n}(0)+1]\rho_{22} + 2\Gamma\bar{n}(0)\rho_{11}, \quad (5.6.16)$$

$$\dot{\rho}_{21} = \dot{\rho}_{12}^* = -\Gamma[2\bar{n}(0)+1]\rho_{21}. \quad (5.6.17)$$

These equations are equivalent to (5.5.66) and (5.5.68) obtained using the Heisenberg picture. Using the conservation of probability $\rho_{11} + \rho_{22} = 1$, we obtain the solutions

$$\rho_{22} = 1 - \rho_{11} = \rho_{22}(0)\exp\{-2\Gamma[2\bar{n}(0)+1]t\} + \frac{\bar{n}(0)}{2\bar{n}(0)+1}[1 - \exp\{-2\Gamma[2\bar{n}(0)+1]t\}] \quad (5.6.18)$$

and

$$\rho_{21} = \rho_{12}^* = \rho_{21}(0)\exp\{-\Gamma[2\bar{n}(0)+1]t\}. \quad (5.6.19)$$

As expected, the atom evolves towards thermodynamic equilibrium with the field.

When the system consists of a large number of states, direct solution for the elements of the density matrix becomes impractical and we must seek alternative methods of solution. We shall describe some of these in the next section and here we give one simple example. Consider a cavity mode coupled to a zero-temperature environment. From the master equation (5.6.15) with $\hat{s} = \hat{a}$, the annihilation operator for the mode, we find

$$\dot{\rho}(t) = \Gamma\left[2\hat{a}\rho(t)\hat{a}^\dagger - \hat{a}^\dagger\hat{a}\rho(t) - \rho(t)\hat{a}^\dagger\hat{a}\right]. \quad (5.6.20)$$

This can be written in a more compact form using the superoperators $\hat{J}$ and $\hat{L}$ defined as follows by their action on the density matrix $\rho(t)$:

$$\hat{J}\rho(t) = \hat{a}\rho(t)\hat{a}^\dagger, \tag{5.6.21}$$

$$\hat{L}\rho(t) = -\tfrac{1}{2}\left[\hat{a}^\dagger\hat{a}\rho(t) + \rho(t)\hat{a}^\dagger\hat{a}\right]. \tag{5.6.22}$$

Superoperators differ from conventional operators in that they only act to the right and only on each other and the density matrix, their action on a wave-function being undefined. Using the definitions (5.6.21) and (5.6.22), the master equation (5.6.20) becomes

$$\dot{\rho}(t) = 2\Gamma(\hat{J}+\hat{L})\rho(t), \tag{5.6.23}$$

the formal solution of which is

$$\rho(t) = \exp[2\Gamma t(\hat{J}+\hat{L})]\,\rho(0). \tag{5.6.24}$$

The utility of this solution lies in the fact that it can be simplified by dis-entangling the exponential operator. The commutator of the superoperators is given by

$$\begin{aligned}[\hat{J},\hat{L}]\rho &= \hat{J}\hat{L}\rho - \hat{L}\hat{J}\rho \\ &= -\tfrac{1}{2}\left(\hat{a}\hat{a}^\dagger\hat{a}\rho\hat{a}^\dagger + \hat{a}\rho\hat{a}^\dagger\hat{a}\hat{a}^\dagger - \hat{a}^\dagger\hat{a}\hat{a}\rho\hat{a}^\dagger - \hat{a}\rho\hat{a}^\dagger\hat{a}^\dagger\hat{a}\right) \\ &= -\hat{J}\rho. \end{aligned} \tag{5.6.25}$$

We saw in Section 3.3 that an exponential containing the sum of two operators, the commutator of which is of the form (5.6.25), may be disentangled as in (3.3.41) and (3.3.42) to give either

$$\rho(t) = \exp(2\Gamma t\hat{L})\exp\{[1-\exp(-2\Gamma t)]\hat{J}\}\rho(0) \tag{5.6.26}$$

or

$$\rho(t) = \exp\{[\exp(2\Gamma t)-1]\hat{J}\}\exp(2\Gamma t\hat{L})\rho(0). \tag{5.6.27}$$

These solutions lead to simple closed-form expressions for $\rho(t)$ if we can express $\rho(0)$ in the coherent state basis or the number state basis. As examples, we consider the damping of a cavity mode prepared initially in a coherent state $|\alpha\rangle$ or a number state $|n\rangle$. In the former, $\rho(0) = |\alpha\rangle\langle\alpha|$ and we choose the solution (5.6.26) because $\rho(0)$ is an eigenmatrix of the superoperator $\hat{J}$, since

$$\hat{J}\rho(0) = \hat{a}\rho(0)\hat{a}^\dagger = \hat{a}|\alpha\rangle\langle\alpha|\hat{a}^\dagger = |\alpha|^2\,\rho(0). \tag{5.6.28}$$

From (5.6.26), we then obtain

$$\rho(t) = \exp\{[1-\exp(-2\Gamma t)]|\alpha|^2\}\exp(2\Gamma t\hat{L})\rho(0). \tag{5.6.29}$$

In order to evaluate $\exp(2\Gamma t\hat{L})\rho(0)$, we consider

$$\exp(2k\hat{L})\rho(0) = \exp(-k\hat{a}^\dagger\hat{a})|\alpha\rangle\langle\alpha|\exp(-k\hat{a}^\dagger\hat{a}). \tag{5.6.30}$$

Using the number state expansion (3.6.4) for the coherent state $|\alpha\rangle$, we find

$$\begin{aligned}\exp(-k\hat{a}^\dagger\hat{a})|\alpha\rangle &= \exp(-|\alpha|^2/2)\sum_{n=0}^{\infty}\frac{[\alpha\exp(-k)]^n}{\sqrt{(n!)}}|n\rangle\\ &= \exp\left\{-\tfrac{1}{2}|\alpha|^2[1-\exp(-2k)]\right\}|\alpha\exp(-k)\rangle,\end{aligned} \tag{5.6.31}$$

where k is real. Finally, we find from (5.6.29) with $k = \Gamma t$ that

$$\rho(t) = |\alpha\exp(-\Gamma t)\rangle\langle\alpha\exp(-\Gamma t)|, \tag{5.6.32}$$

corresponding to a coherent state, the amplitude of which decays at the rate Γ. In the second example, $\rho(0) = |n\rangle\langle n|$ and we choose the solution (5.6.27) because $\rho(0)$ is an eigenmatrix of the superoperator $\hat{L}$, since

$$\begin{aligned}\hat{L}\rho(0) &= -\tfrac{1}{2}\left(\hat{a}^\dagger\hat{a}|n\rangle\langle n| + |n\rangle\langle n|\hat{a}^\dagger\hat{a}\right)\\ &= -n|n\rangle\langle n|.\end{aligned} \tag{5.6.33}$$

From (5.6.27), we obtain

$$\rho(t) = \exp(-2\Gamma nt)\exp\left\{[\exp(2\Gamma t)-1]\hat{J}\right\}\rho(0). \tag{5.6.34}$$

In order to evaluate (5.6.34), we consider $\exp(k\hat{J})\rho(0)$ as a power series in k giving

$$\begin{aligned}\exp(k\hat{J})\rho(0) &= \sum_{l=0}^{\infty}\frac{k^l}{l!}\hat{a}^l|n\rangle\langle n|\hat{a}^{\dagger l}\\ &= \sum_{l=0}^{n}\frac{k^l}{l!}\frac{n!}{(n-l)!}|n-l\rangle\langle n-l|.\end{aligned} \tag{5.6.35}$$

Finally we see from (5.6.34) with $k = \exp(2\Gamma \mathrm{t}) - 1$ that

$$\rho(t) = \sum_{l=0}^{n}[\exp(-2\Gamma t)]^{n-l}[1-\exp(-2\Gamma t)]^l\frac{n!}{l!(n-l)!}|n-l\rangle\langle n-l|, \tag{5.6.36}$$

corresponding to a mixed state with a binomial photon number probability distribution.

Master equations can also be obtained to describe more complicated dynamics in which there are strong couplings in addition to the weak interaction with the environment responsible for the dissipation. Consider a Hamiltonian in the interaction picture of the form

$$\hat{V}_{\mathrm{I}}(t) = \mathrm{i}\hbar[\hat{s}^\dagger\hat{F}(t) - \hat{F}^\dagger(t)\hat{s}] + \hat{W}_{\mathrm{I}}(t), \tag{5.6.37}$$

where the first part is identical to (5.6.3) and describes the dissipative interaction with the environment and $\hat{W}_{\mathrm{I}}(t)$ is the strong coupling. The strategy for deriving the corresponding master equation involves treating the effect of the strong coupling exactly whilst applying the now familiar approximations for dealing with the coupling to the environment. The Liouville equation (5.6.4) now contains the term $-\mathrm{i}[\hat{W}_{\mathrm{I}}(t), \rho_{\mathrm{T}}(t)]/\hbar$ in addition to the term previously studied. Taking the trace over the degrees of freedom of the environment and employing the decorrelation approximation (5.6.6) gives the following master equation for the system density matrix:

$$\dot{\rho}(t) = -\frac{\mathrm{i}}{\hbar}\left[\hat{W}_{\mathrm{I}}(t), \rho(t)\right] + \hat{\Lambda}\rho(t), \tag{5.6.38}$$

where $\hat{\Lambda}$ is a superoperator containing the effects of the dissipative couplings. The term $\hat{\Lambda}\rho(t)$ in (5.6.38) has the form, for example, of the right-hand side of (5.6.12) which was derived in the absence of the strong coupling. We see that the master equation is comprised of two terms: the strong coupling term having precisely the same form as it would have in the absence of dissipation, together with the dissipative term having precisely the same form as it would have in the absence of the strong coupling. It is important to realize that the simple form of (5.6.38) relies on the assumption that the properties of the environment at the eigenfrequencies of the strongly coupled system are not significantly different from those at the natural frequencies of the system in the absence of strong coupling. Any differences have been neglected in deriving (5.6.38).

As an example of a problem of this type, we consider a two-level atom driven by a resonant CW laser and damped by spontaneous emission induced by coupling to the surrounding field initially in its vacuum state. In (5.6.38), the strong coupling $\hat{W}_{\mathrm{I}}$ is the interaction picture Hamiltonian (2.2.19) with $\Delta = 0$ and $\varphi = 0$. The dissipative term $\hat{\Lambda}\rho(t)$ is given by the right-hand side of (5.6.15) with $\hat{s}$ becoming $\hat{\sigma}_-$ and $\bar{n}(0) = 0$. The resulting master equation is

$$\dot{\rho}(t) = -\mathrm{i}[\tfrac{1}{2}V(\hat{\sigma}_+ + \hat{\sigma}_-), \rho(t)] + \Gamma\left[2\hat{\sigma}_-\rho(t)\hat{\sigma}_+ - \hat{\sigma}_+\hat{\sigma}_-\rho(t) - \rho(t)\hat{\sigma}_+\hat{\sigma}_-\right], \tag{5.6.39}$$

where Γ is one-half of the Einstein A-coefficient. This equation can be solved by a variety of methods, including direct solution for the elements of the density matrix. Here we proceed by recalling from Section 1.3 that the density matrix may be expressed in the form

$$\rho(t) = \tfrac{1}{2}[\mathbf{1} + u(t)\hat{\sigma}_1 + v(t)\hat{\sigma}_2 + w(t)\hat{\sigma}_3], \tag{5.6.40}$$

where $\mathbf{1}$ is the 2×2 identity matrix and u, v, and w are the expectation values of

$\hat{\sigma}_1$, $\hat{\sigma}_2$, and $\hat{\sigma}_3$, respectively. We find the differential equation for w by multiplying the master equation (5.6.38) by $\hat{\sigma}_3$ and taking the trace to give

$$\begin{aligned}\dot{w}(t) &= \mathrm{Tr}[\hat{\sigma}_3\,\dot{\rho}(t)]\\ &= -\mathrm{i}\tfrac{1}{2}V\,\mathrm{Tr}[\hat{\sigma}_3\hat{\sigma}_1\,\rho(t) - \hat{\sigma}_3\,\rho(t)\hat{\sigma}_1]\\ &\quad +\Gamma\,\mathrm{Tr}\Big[2\hat{\sigma}_3\hat{\sigma}_-\rho(t)\hat{\sigma}_+ - \hat{\sigma}_3\hat{\sigma}_+\hat{\sigma}_-\rho(t) - \hat{\sigma}_3\,\rho(t)\hat{\sigma}_+\hat{\sigma}_-\Big].\end{aligned} \tag{5.6.41}$$

Using the properties $\hat{\sigma}_3\hat{\sigma}_\pm = \pm\hat{\sigma}_\pm$ and $\hat{\sigma}_\pm\hat{\sigma}_3 = \mp\hat{\sigma}_\pm$, and the cyclic property of the trace operation, we obtain

$$\begin{aligned}\dot{w}(t) &= -\mathrm{i}\tfrac{1}{2}V\,\mathrm{Tr}\{[\hat{\sigma}_3,\hat{\sigma}_1]\,\rho(t)\} - 4\Gamma\,\mathrm{Tr}[\hat{\sigma}_+\hat{\sigma}_-\rho(t)]\\ &= Vv(t) - 2\Gamma[w(t)+1].\end{aligned} \tag{5.6.42}$$

Similarly, we find

$$\dot{u}(t) = -\Gamma u(t) \tag{5.6.43}$$

and

$$\dot{v}(t) = -Vw(t) - \Gamma v(t). \tag{5.6.44}$$

Equations (5.6.42) to (5.6.44) are known as the optical Bloch equations, and their solution enables the density matrix $\rho(t)$ to be reconstructed using (5.6.40). The solution for $u(t)$ is given simply by

$$u(t) = u(0)\exp(-\Gamma t), \tag{5.6.45}$$

while (5.6.42) and (5.6.44) can be solved, for example, by using Laplace transforms. We find

$$\begin{aligned}v(t) = \frac{1}{(V^2+2\Gamma^2)}\Bigg[\!\!\Bigg[2\Gamma V + \exp(-3\Gamma t/2)\Bigg\{[(V^2+2\Gamma^2)v(0) - 2\Gamma V]\cos\Omega_{\mathrm{R}}t\\ + \frac{1}{\Omega_{\mathrm{R}}}\Big[-3\Gamma^2 V + (V^2+2\Gamma^2)(-Vw(0) + \tfrac{1}{2}\Gamma v(0))\Big]\sin\Omega_{\mathrm{R}}t\Bigg\}\Bigg]\!\!\Bigg]\end{aligned} \tag{5.6.46}$$

and

$$\begin{aligned}w(t) = \frac{1}{(V^2+2\Gamma^2)}\Bigg[\!\!\Bigg[-2\Gamma^2 + \exp(-3\Gamma t/2)\Bigg\{[2\Gamma^2 + (V^2+2\Gamma^2)w(0)]\cos\Omega_{\mathrm{R}}t\\ + \frac{1}{\Omega_{\mathrm{R}}}\Big[-2\Gamma(V^2+\tfrac{1}{2}\Gamma^2) + (V^2+2\Gamma^2)(Vv(0) - \tfrac{1}{2}\Gamma w(0))\Big]\sin\Omega_{\mathrm{R}}t\Bigg\}\Bigg]\!\!\Bigg],\end{aligned} \tag{5.6.47}$$

where $v(0)$ and $w(0)$ are the initial values of v and w and Ω_{R} is the Rabi frequency as modified by the spontaneous emission, given by

$$\Omega_{\mathrm{R}} = (V^2 - \Gamma^2/4)^{1/2}. \tag{5.6.48}$$

In the long-time limit, the steady-state values of u, v, and w are

$$u(\infty) = 0, \tag{5.6.49}$$

$$v(\infty) = \frac{2\Gamma V}{(V^2 + 2\Gamma^2)}, \tag{5.6.50}$$

$$w(\infty) = \frac{-2\Gamma^2}{(V^2 + 2\Gamma^2)}, \tag{5.6.51}$$

corresponding to a state with a non-zero probability for being in the upper level since $\langle \hat{\sigma}_3 \rangle \neq -1$, and having a residual dipole moment, $\langle \hat{\sigma}_2 \rangle \neq 0$. Substituting these steady-state values into (5.6.40), we find that the steady-state density matrix is

$$\rho(\infty) = \tfrac{1}{2}\left(1 + \frac{2\Gamma}{(V^2 + 2\Gamma^2)}(V\hat{\sigma}_2 - \Gamma\hat{\sigma}_3)\right), \tag{5.6.52}$$

representing a mixture of the eigenstates of the operator $V\hat{\sigma}_2 - \Gamma\hat{\sigma}_3$. The eigenvalues of $\rho(\infty)$ are the steady-state occupation probabilities of these two eigenstates.

As a special case, consider the atom to be initially in the lower level so that $w(0) = -1$ and $u(0) = v(0) = 0$. With these initial conditions, $u(t) \equiv 0$ while $v(t)$ and $w(t)$ are found from (5.6.46) and (5.6.47) to be

$$v(t) = \frac{V}{(V^2 + 2\Gamma^2)} \times \left[2\Gamma + \exp(-3\Gamma t/2)\left(-2\Gamma \cos \Omega_R t + \frac{1}{\Omega_R}(V^2 - \Gamma^2)\sin \Omega_R t\right)\right] \tag{5.6.53}$$

and

$$w(t) = \frac{-1}{(V^2 + 2\Gamma^2)}\left[2\Gamma^2 + V^2 \exp(-3\Gamma t/2)\left(\cos \Omega_R t + \frac{3\Gamma}{2\Omega_R}\sin \Omega_R t\right)\right]. \tag{5.6.54}$$

The inversion $w(t)$ and the dipole $v(t)$ will exhibit damped Rabi oscillations if Ω_R is real, that is if $2V > \Gamma$. This is illustrated in Fig. 5.6 where we plot $w(t)$ and $v(t)$ as functions of Γt for $V = 20\Gamma$.

Now that we have obtained the general solution for $u(t)$, $v(t)$, and $w(t)$ given in (5.6.45) to (5.6.47), we can calculate the expectation value of any observable at any time using the density matrix (5.6.40). However, not all properties of interest may be expressed in terms of single-time averages. In particular, the fluorescence spectrum is obtained from the two-time correlation function $\langle \hat{\sigma}_+(t+\tau)\hat{\sigma}_-(t) \rangle$, as given by (5.5.77) in the previous section. There we established the quantum regression theorem for the two-level atom within the Heisenberg picture. This theorem gives the equations of motion for multiple-time

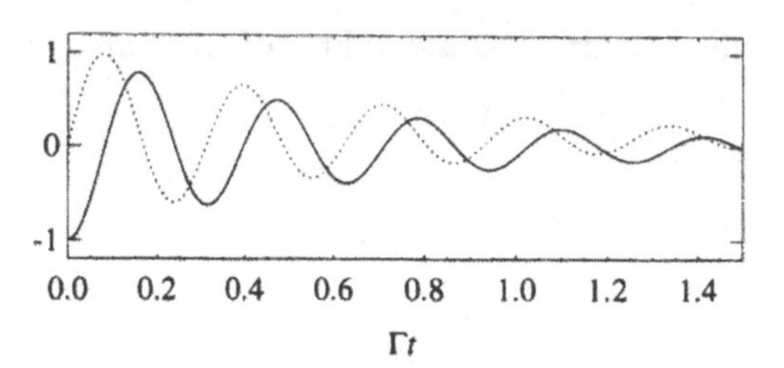

Fig. 5.6 Evolution of the inversion $w(t)$ (solid curve) and dipole $v(t)$ (dotted curve) as functions of the scaled time for $V = 20\Gamma$.

correlation functions from the equations of motion for single-time averages. Since these quantities are independent of the picture in which they are calculated, the quantum regression theorem enables us to find the equations of motion for two-time correlation functions from those for $u(t)$, $v(t)$, and $w(t)$, obtained from the master equation (5.6.41) within the Schrödinger picture. Since u, v, and w are the expectation values $\langle\hat{\sigma}_1\rangle$, $\langle\hat{\sigma}_2\rangle$, and $\langle\hat{\sigma}_3\rangle$, respectively, the quantum regression theorem applied to (5.6.42) to (5.6.44) gives, for $\tau > 0$,

$$\frac{\mathrm{d}}{\mathrm{d}\tau}\langle\hat{\sigma}_1(t+\tau)\hat{\sigma}_-(t)\rangle = -\Gamma\langle\hat{\sigma}_1(t+\tau)\hat{\sigma}_-(t)\rangle, \tag{5.6.55}$$

$$\frac{\mathrm{d}}{\mathrm{d}\tau}\langle\hat{\sigma}_2(t+\tau)\hat{\sigma}_-(t)\rangle = -V\langle\hat{\sigma}_3(t+\tau)\hat{\sigma}_-(t)\rangle - \Gamma\langle\hat{\sigma}_2(t+\tau)\hat{\sigma}_-(t)\rangle, \tag{5.6.56}$$

$$\frac{\mathrm{d}}{\mathrm{d}\tau}\langle\hat{\sigma}_3(t+\tau)\hat{\sigma}_-(t)\rangle = V\langle\hat{\sigma}_2(t+\tau)\hat{\sigma}_-(t)\rangle - 2\Gamma[\langle\hat{\sigma}_3(t+\tau)\hat{\sigma}_-(t)\rangle + \langle\hat{\sigma}_-(t)\rangle]. \tag{5.6.57}$$

The solution of (5.6.55) is

$$\langle\hat{\sigma}_1(t+\tau)\hat{\sigma}_-(t)\rangle = \langle\hat{\sigma}_+(t)\hat{\sigma}_-(t)\rangle\exp(-\Gamma\tau) = \tfrac{1}{2}(1+w(t))\exp(-\Gamma\tau), \tag{5.6.58}$$

where we have used the fact that $\hat{\sigma}_1 = \hat{\sigma}_+ + \hat{\sigma}_-$ and $\hat{\sigma}_3 = 2\hat{\sigma}_+\hat{\sigma}_- - 1$. The solution for $\langle\hat{\sigma}_2(t+\tau)\hat{\sigma}_-(t)\rangle$ can be obtained directly from the solution for $v(t)$ given in (5.6.46). We note that the constant term -2Γ in the equation of motion (5.6.42) for $w(t)$ has become $-2\Gamma\langle\hat{\sigma}_-(t)\rangle$ in (5.6.57). Consequently, a corresponding replacement must be made in (5.6.46) to obtain the solution for $\langle\hat{\sigma}_2(t+\tau)\hat{\sigma}_-(t)\rangle$. We find

$$\begin{aligned}\langle\hat{\sigma}_2(t+\tau)\hat{\sigma}_-(t)\rangle = \frac{1}{(V^2+2\Gamma^2)}\Bigg[\!\!\Bigg[& 2\Gamma V\langle\hat{\sigma}_-(t)\rangle + \exp(-3\Gamma\tau/2) \\ & \times\Bigg\{\Big[(V^2+2\Gamma^2)\langle\hat{\sigma}_2(t)\hat{\sigma}_-(t)\rangle - 2\Gamma V\langle\hat{\sigma}_-(t)\rangle\Big]\cos\Omega_R\tau \\ & + \frac{1}{\Omega_R}\Big[-3\Gamma^2 V\langle\hat{\sigma}_-(t)\rangle + (V^2+2\Gamma^2)(-V\langle\hat{\sigma}_3(t)\hat{\sigma}_-(t)\rangle \\ & + \tfrac{1}{2}\Gamma\langle\hat{\sigma}_2(t)\hat{\sigma}_-(t)\rangle)\Big]\sin\Omega_R\tau\Bigg\}\Bigg]\!\!\Bigg].\end{aligned} \tag{5.6.59}$$

The expectation values on the right-hand side of this equation can be calculated using $\hat{\sigma}_- = (\hat{\sigma}_1 - i\hat{\sigma}_2)/2$, $\hat{\sigma}_2\hat{\sigma}_- = -i\hat{\sigma}_+\hat{\sigma}_-$, and $\hat{\sigma}_3\hat{\sigma}_- = -\hat{\sigma}_-$, so that $\langle\hat{\sigma}_-(t)\rangle = [u(t) - iv(t)]/2$, $\langle\hat{\sigma}_2(t)\hat{\sigma}_-(t)\rangle = -i[1 + w(t)]/2$, and $\langle\hat{\sigma}_3(t)\hat{\sigma}_-(t)\rangle = -[u(t) - iv(t)]/2$. In order to calculate the spectrum, we require the limit of (5.6.58) and (5.6.59) as $t \to \infty$. In this limit, the above expectation values may all be expressed in terms of $u(\infty)$, $v(\infty)$, and $w(\infty)$ given by (5.6.49) to (5.6.51). With these substitutions, the limiting forms of the two-time correlation functions are

$$\lim_{t\to\infty} \langle\hat{\sigma}_1(t+\tau)\hat{\sigma}_-(t)\rangle = \frac{V^2}{2(V^2+2\Gamma^2)} \exp(-\Gamma\tau) \tag{5.6.60}$$

and

$$\lim_{t\to\infty} \langle\hat{\sigma}_2(t+\tau)\hat{\sigma}_-(t)\rangle = \frac{-iV^2}{2(V^2+2\Gamma^2)^2}\Bigg[4\Gamma^2 + \exp(-3\Gamma\tau/2) \times\left((V^2-2\Gamma^2)\cos\Omega_R\tau + \frac{\Gamma}{2\Omega_R}(5V^2-2\Gamma^2)\sin\Omega_R\tau\right)\Bigg]. \tag{5.6.61}$$

Since $\langle\hat{\sigma}_+(t+\tau)\hat{\sigma}_-(t)\rangle = [\langle\hat{\sigma}_1(t+\tau)\hat{\sigma}_-(t)\rangle + i\langle\hat{\sigma}_2(t+\tau)\hat{\sigma}_-(t)\rangle]/2$, $C(\Omega)$ given by (5.5.77), and proportional to the spectrum, may be found using (5.6.60) and (5.5.61). It is important to note that Ω_R is imaginary if $2V < \Gamma$ and that, in this case, care must be taken when evaluating the real part in (5.5.77). The general expression for $C(\Omega)$, valid for all V, is

$$C(\Omega) = \frac{\Gamma V^2}{2(V^2+2\Gamma^2)} \times\left(\frac{4\pi\Gamma\delta(\Omega)}{(V^2+2\Gamma^2)} + \frac{1}{(\Gamma^2+\Omega^2)} + \frac{4(V^2-\Gamma^2)-\Omega^2}{(V^2+2\Gamma^2-\Omega^2)^2+9\Gamma^2\Omega^2}\right). \tag{5.6.62}$$

This is the Mollow spectrum of the fluorescence emitted by a coherently driven two-level atom. The delta-function term centred on the laser frequency (which in this resonant case coincides with the atomic transition frequency) is the contribution due to elastic scattering. In Fig. 5.7, we plot the inelastic part of $C(\Omega)$ as a function of Ω/Γ for (a) $V = \Gamma$, (b) $V = 3\Gamma$, and (c) $V = 9\Gamma$. For small values of V, the spectrum has a single peak at $\Omega = 0$. As V increases, two further peaks begin to form and separate from the central peak to give a three-peaked spectrum. For $V \gg \Gamma$, the spectrum exhibits three distinct, nearly Lorentzian peaks centred at $\Omega = 0$ and $\Omega = \pm\Omega_R$. Near $\Omega = 0$, $C(\Omega)$ has the approximate form

$$C(\Omega \sim 0) \simeq \frac{\Gamma}{2V^2}\left(4\pi\Gamma\delta(\Omega) + \frac{V^2}{(\Gamma^2+\Omega^2)}\right), \tag{5.6.63}$$

while near the two side peaks at $\Omega = \pm\Omega_R$, $C(\Omega)$ has the approximate form

$$C(\Omega \sim \pm\Omega_R) \simeq \frac{3\Gamma}{8}\frac{1}{9\Gamma^2/4 + (\Omega \mp \Omega_R)^2}. \tag{5.6.64}$$

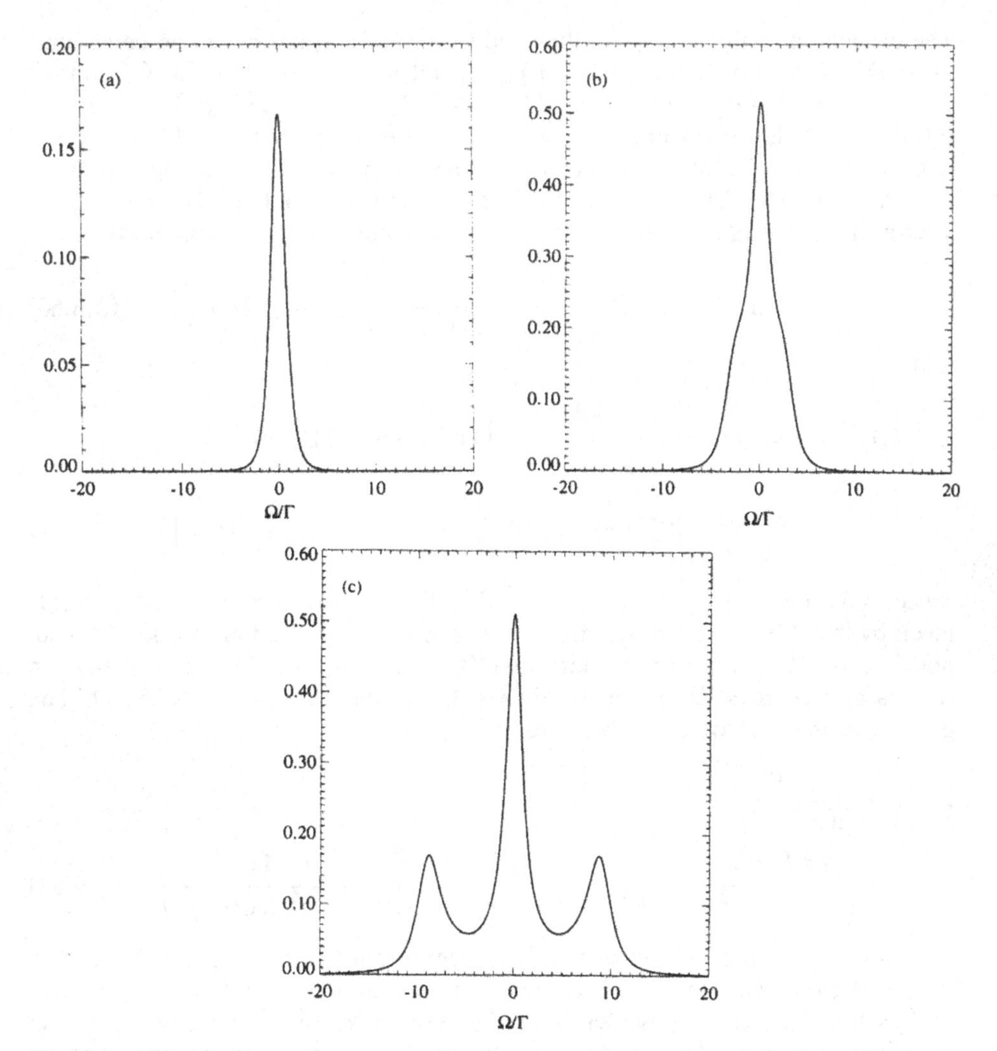

Fig. 5.7 The spectrum of resonance fluorescence as a function of Ω/Γ for (a) $V=\Gamma$, (b) $V=3\Gamma$, and (c) $V=9\Gamma$.

These side peaks therefore have heights which are one-third that of the central peak and widths which are $3/2$ times that of the central peak.

It is usually the case that the density matrix for an initially pure state evolves into a mixed state. This is because of the loss of information when tracing over the degrees of freedom of the environment. It is possible, however, to monitor the environment in such a way as to retain enough information to maintain a pure state description. The state then evolves in a manner conditioned by this monitoring. Any master equation can be written in the form

$$\dot{\rho} = \hat{K}\rho, \tag{5.6.65}$$

where $\hat{K}$ is a superoperator. The formal solution

$$\begin{aligned}\rho(t) &= \exp(\hat{K}t)\rho(0)\\ &= \exp[(\hat{K}-\hat{S})t+\hat{S}t]\,\rho(0),\end{aligned} \tag{5.6.66}$$

for any superoperator $\hat{S}$, can be expanded as

$$\begin{aligned}\rho(t) = \sum_{m=0}^{\infty}\int_0^t \mathrm{d}t_m \int_0^{t_m}\mathrm{d}t_{m-1}\cdots\int_0^{t_2}\mathrm{d}t_1 \exp\big[(\hat{K}-\hat{S})(t-t_m)\big]\hat{S}\\ \times\exp\big[(\hat{K}-\hat{S})(t_m-t_{m-1})\big]\hat{S}\cdots\exp\big[(\hat{K}-\hat{S})(t_2-t_1)\big]\hat{S}\\ \times\exp\big[(\hat{K}-\hat{S})t_1\big]\,\rho(0).\end{aligned} \tag{5.6.67}$$

It can be verified that (5.6.67) is indeed the general solution of (5.6.65) by direct substitution, and suitable choices of $\hat{K}$ and $\hat{S}$ allow for an interpretation in terms of conditioned evolution. A simple example is the master equation (5.6.23) for a damped cavity mode with the identifications $\hat{S}=2\Gamma\hat{J}$ and $\hat{K}-\hat{S}=2\Gamma\hat{L}$. Each occurrence of $\hat{J}$ in the formal solution (5.6.67) acts to the right with the annihilation operator $\hat{a}$ and to the left with $\hat{a}^{\dagger}$ and corresponds to the loss of one photon to the environment. It follows that the integrand in (5.6.67) may be interpreted as the evolved density matrix conditioned on the loss of m photons to the environment at the sequence of times $t_1,t_2,\ldots,t_m$. The integrations and the summation account for the loss of all possible numbers of photons at all possible times. If we know that precisely N photons were lost, one at each of the sequence of times $t_1,t_2,\ldots,t_N$, then the conditioned density matrix is given by

$$\rho_c(t)=\frac{\bar{\rho}_c(t)}{\mathrm{Tr}[\,\bar{\rho}_c(t)]}, \tag{5.6.68}$$

where $\bar{\rho}_c(t)$ has the unnormalized form

$$\bar{\rho}_c(t)=\exp\big[2\Gamma\hat{L}(t-t_N)\big]2\Gamma\hat{J}\exp\big[2\Gamma\hat{L}(t_N-t_{N-1})\big]\cdots2\Gamma\hat{J}\exp\big(2\Gamma\hat{L}t_1\big)\rho(0). \tag{5.6.69}$$

In general, both $\hat{J}$ and $\hat{L}$ contribute to the decay of the field although the precise effect of each of these is determined by the initial state $\rho(0)$. Extreme examples occur when $\rho(0)=|n\rangle\langle n|$ and $\rho(0)=|\alpha\rangle\langle\alpha|$. For the number state, the conditioned density matrix is $\rho_c(t)=|n-m\rangle\langle n-m|$ which is simply the normalized form of $\hat{J}^m\,|n\rangle\langle n|$. The interpretation is that, with the loss of each photon, the number is reduced by one. In this case, $\hat{L}$ does not contribute to the decay because $\hat{J}$ transforms one number state into another and $|n'\rangle\langle n'|$ is an eigenstate of $\hat{L}$. The exponential factors in (5.6.69) lead to time-dependent factors which cancel in the normalization process. For the coherent state, the conditioned density matrix is $\rho_c(t) = |\alpha \exp(-\Gamma t)\rangle \langle \alpha \exp(-\Gamma t)|$ which is also the unconditioned density matrix (5.6.32). In this case, it is $\hat{J}$ which does not contribute to the decay because the exponential factors transform one coherent state into another and $|\alpha'\rangle\langle\alpha'|$ is an eigenstate of $\hat{J}$. The decay of a mode prepared in a coherent state changes the amplitude but does not produce a mixed state. No information is lost in this case and monitoring the environment adds nothing to our knowledge of the state. It is for this reason that the conditioned and unconditioned density matrices are the same.

Conditioned evolutions can be obtained from more general master equations. Consider, for example,

$$\dot{\rho}(t) = -\frac{i}{\hbar}\left[\hat{W}, \rho(t)\right] + \Gamma\left[2\hat{s}\rho(t)\hat{s}^\dagger - \hat{s}^\dagger\hat{s}\rho(t) - \rho(t)\hat{s}^\dagger\hat{s}\right], \qquad (5.6.70)$$

where $\hat{W}$ is a time-independent operator associated with strong coupling or free evolution. It is useful to write the solution of (5.6.70) in the form (5.6.67) with

$$\hat{S}\rho(t) = 2\Gamma\hat{s}\rho(t)\hat{s}^\dagger \qquad (5.6.71)$$

and

$$(\hat{K} - \hat{S})\rho(t) = -\frac{i}{\hbar}\left[\hat{W}, \rho(t)\right] - \Gamma\left[\hat{s}^\dagger\hat{s}\rho(t) + \rho(t)\hat{s}^\dagger\hat{s}\right]. \qquad (5.6.72)$$

As with the damped mode, we can construct from the integrand of (5.6.67) a conditioned density matrix. Each occurrence of (5.6.71) in this density matrix corresponds to the observation of the loss of or addition to the system of one quantum, depending on the nature of $\hat{s}$. For example, if $\hat{s} = \hat{\sigma}_-$, then the process described by (5.6.71) is the spontaneous emission of one photon.

If the initial state is pure, then the conditioned density matrix also corresponds to a pure state. Hence we can dispense with the density matrix and adopt a conditioned *wavefunction* for the system. The evolution of this wavefunction depends on the times of loss to or absorption from the environment of a quantum but we only know the probability of this to occur at any particular time. This is the basis for the recent development of stochastic wavefunction simulation. The principle is that a conditioned wavefunction is obtained using randomly generated numbers to determine whether a quantum is lost or gained at each integration time step. Averaging over a sufficiently large number of conditioned wavefunctions reproduces the full unconditioned density matrix.

5.7 Evolution of field statistics

The solution of a master equation gives the density matrix and hence the fullest possible description of the evolved state at a given time. Often, however, it is sufficient to obtain selected statistical information. If this is the case then we can combine the methods introduced in Chapter 4 with a master equation to obtain a partial differential equation governing the evolution of the field statistics. In this section, we present three simple examples using the moment generating function $M(\mu)$, the characteristic function $\chi(\xi, p)$, and the quasi-probability distribution $W(\alpha, p)$.

If the master equation only couples diagonal elements in the number state basis to other diagonal elements then the evolution of a given photon number probability $P(n)$ only depends on the other probabilities $P(n')$. In such cases we can transform the master equation into a partial differential equation for the moment generating function $M(\mu)$. As an example consider the evolution of a single field mode coupled to an environment at a finite temperature. The evolution of the density matrix for the mode is described by the master equation (see (5.6.15))

$$\dot{\rho}(t) = \Gamma(\bar{n}+1)\left[2\hat{a}\rho(t)\hat{a}^\dagger - \hat{a}^\dagger\hat{a}\rho(t) - \rho(t)\hat{a}^\dagger\hat{a}\right] + \Gamma\bar{n}[2\hat{a}^\dagger\rho(t)\hat{a} - \hat{a}\hat{a}^\dagger\rho(t) - \rho(t)\hat{a}\hat{a}^\dagger]. \tag{5.7.1}$$

The probability for being in the number state $|n\rangle$ is $P(n) = \langle n|\,\rho\,|n\rangle$ and, from (5.7.1), this satisfies the differential equation

$$\dot{P}(n) = -2\Gamma[n(2\bar{n}+1)+\bar{n}]P(n) + 2\Gamma(\bar{n}+1)(n+1)P(n+1) + 2\Gamma\bar{n}nP(n-1). \tag{5.7.2}$$

The moment generating function is given by (4.2.1) as

$$M(\mu, t) = \sum_{n=0}^{\infty} P(n)(1-\mu)^n, \tag{5.7.3}$$

and we can find an equation for the evolution of $M(\mu, t)$ by multiplying (5.7.2) by $(1-\mu)^n$ and summing over n. Terms containing factors of n can be obtained by differentiation with respect to μ so that, for example,

$$\sum_{n=0}^{\infty} n(1-\mu)^n P(n) = (1-\mu)\left(-\frac{\partial}{\partial\mu}\right)M(\mu, t). \tag{5.7.4}$$

We find

$$\begin{aligned}\frac{\partial M}{\partial t}(\mu, t) &= -2\Gamma\bar{n}M(\mu, t) + 2\Gamma(2\bar{n}+1)(1-\mu)\frac{\partial M}{\partial\mu}(\mu, t) \\ &\quad - 2\Gamma(\bar{n}+1)\frac{\partial M}{\partial\mu}(\mu, t) - 2\Gamma\bar{n}(1-\mu)\frac{\partial}{\partial\mu}[(1-\mu)M(\mu, t)] \\ &= -2\Gamma\mu\frac{\partial}{\partial\mu}[(1+\mu\bar{n})M(\mu, t)].\end{aligned} \tag{5.7.5}$$

The steady-state solution $M(\mu, \infty)$ of (5.7.5), obtained by setting $\dot{M} = 0$ and imposing the condition that $M(0, t) = 1$, is $M(\mu, \infty) = (1+\mu\bar{n})^{-1}$ which is that for the thermal distribution with mean photon number $\bar{n}$, as in (4.2.22). In

Appendix 11, we solve the partial differential equation (5.7.5) by the method of characteristics and obtain the general solution

$$M(\mu,t)=\frac{1}{1+\mu\bar{n}[1-\exp(-2\Gamma t)]}M\left(\frac{\mu\exp(-2\Gamma t)}{1+\mu\bar{n}[1-\exp(-2\Gamma t)]},0\right). \quad (5.7.6)$$

As an example of the use of this solution, consider a mode initially prepared in the coherent state $|\alpha\rangle$ for which $M(\mu,0)=\exp(-\mu|\alpha|^2)$ (see (4.2.18)). The time-evolved moment generating function is then, from (5.7.6),

$$M(\mu,t)=\frac{1}{1+\mu\bar{n}[1-\exp(-2\Gamma t)]}\exp\left(-\frac{\mu\exp(-2\Gamma t)}{1+\mu\bar{n}[1-\exp(-2\Gamma t)]}|\alpha|^2\right). \quad (5.7.7)$$

All the information about the photon statistics at time t is contained within this function. In particular, the first two factorial moments are, from (4.2.10),

$$n^{(1)}=\langle\hat{n}\rangle=|\alpha|^2\exp(-2\Gamma t)+\bar{n}[1-\exp(-2\Gamma t)] \quad (5.7.8)$$

and

$$n^{(2)}=\langle:\hat{n}^2:\rangle=|\alpha|^4\exp(-4\Gamma t)+4|\alpha|^2\bar{n}[1-\exp(-2\Gamma t)]\exp(-2\Gamma t) + 2\bar{n}^2[1-\exp(-2\Gamma t)]^2. \quad (5.7.9)$$

The normal ordered variance is

$$\begin{aligned}:\Delta n^2: &= \langle:\hat{n}^2:\rangle-\langle\hat{n}\rangle^2\\ &=\{2|\alpha|^2\exp(-2\Gamma t)+\bar{n}[1-\exp(-2\Gamma t)]\}\bar{n}[1-\exp(-2\Gamma t)],\end{aligned} \quad (5.7.10)$$

which evolves from the zero value at $t=0$, characteristic of a coherent state, to $\bar{n}^2$ as $t\to\infty$, characteristic of a thermal state.

The master equation for a field mode may be transformed into a partial differential equation for either the characteristic function $\chi(\xi,p)$ or the quasi-probability distribution $W(\alpha,p)$. This is done by identifying the actions on ρ of $\hat{a}$ and $\hat{a}^\dagger$ with linear partial differential operators acting on either $\chi(\xi,p)$ or $W(\alpha,p)$. These identities are derived in Appendix 12. As an example, consider the master equation

$$\begin{aligned}\dot{\rho}(t)= &-\mathrm{i}[W(\hat{a}+\hat{a}^\dagger),\rho(t)]+\Gamma(\bar{n}+1)\left[2\hat{a}\rho(t)\hat{a}^\dagger-\hat{a}^\dagger\hat{a}\rho(t)-\rho(t)\hat{a}^\dagger\hat{a}\right]\\ &+\Gamma\bar{n}[2\hat{a}^\dagger\rho(t)\hat{a}-\hat{a}\hat{a}^\dagger\rho(t)-\rho(t)\hat{a}\hat{a}^\dagger]\end{aligned} \quad (5.7.11)$$

which describes a mode driven by a resonant classical current and coupled to a reservoir at finite temperature. For $\Gamma=0$, (5.7.11) will generate the coherent state having $\alpha=-\mathrm{i}Wt$ from the vacuum. Applying the replacements (A12.6) to (A12.9) to (5.7.11) gives the corresponding equation

$$\begin{aligned}\frac{\partial}{\partial t}\chi(\xi,p,t)= &\,\mathrm{i}W(\xi^*+\xi)\chi(\xi,p,t)-\Gamma\left(\xi\frac{\partial}{\partial\xi}+\xi^*\frac{\partial}{\partial\xi^*}\right)\chi(\xi,p,t)\\ &-\Gamma(2\bar{n}+1-p)|\xi|^2\chi(\xi,p,t)\end{aligned} \quad (5.7.12)$$

for the time-evolved characteristic function $\chi(\xi, p, t)$. In (4.4.1), we see that characteristic functions for different values of p are simply related to each other and it is sensible, therefore, to choose the value of p in (5.7.12) for which this equation can most easily be solved. Setting $p = 2\bar{n} + 1$ removes the last term on the right-hand side of (5.7.12) and we show in Appendix 11 (where we have written $\chi(\xi, 2\bar{n}+1, t) = \chi(\xi, t)$) that the resulting equation has the solution

$$\chi(\xi, 2\bar{n}+1, t) = \exp\left(\frac{\mathrm{i}W}{\Gamma}(\xi + \xi^*)[1 - \exp(-\Gamma t)]\right)\chi(\xi\exp(-\Gamma t), 2\bar{n}+1, 0). \tag{5.7.13}$$

The characteristic function for arbitrary p can be found using (4.4.1) to relate $\chi(\xi, 2\bar{n}+1, t)$ to $\chi(\xi, p, t)$ and $\chi(\xi\exp(-\Gamma t), 2\bar{n}+1, 0)$ to $\chi(\xi\exp(-\Gamma t), p, 0)$ with the result that

$$\chi(\xi, p, t) = \exp\{[p - (2\bar{n}+1)]|\xi|^2[1 - \exp(-2\Gamma t)]\}$$
$$\times \exp\left(\frac{\mathrm{i}W}{\Gamma}(\xi + \xi^*)(1 - \exp(-\Gamma t))\right)\chi(\xi\exp(-\Gamma t), p, 0). \tag{5.7.14}$$

The expectation value of the p-ordered product of $\hat{a}^{\dagger m}$ and $\hat{a}^n$ can be found from (5.7.14) by differentiation, as shown in Section 4.4.

As a final example, we consider the evolution of the quasi-probability distribution $W(\alpha, p)$ associated with the master equation

$$\dot{\rho}(t) = -\frac{g}{2}\left[(\hat{a}^{\dagger 2} - \hat{a}^2), \rho(t)\right] + \Gamma\left[2\hat{a}\rho(t)\hat{a}^\dagger - \hat{a}^\dagger\hat{a}\rho(t) - \rho(t)\hat{a}^\dagger\hat{a}\right]. \tag{5.7.15}$$

This equation arises from a simple model of a degenerate parametric amplifier, or similar non-linear optical device. For $\Gamma = 0$, (5.7.15) will generate the squeezed state with $\zeta = gt$ from the vacuum. Applying the replacements (A12.8) to (A12.21) to (5.7.15) gives the corresponding partial differential equation

$$\frac{\partial}{\partial t}W(\alpha, p, t) = g\left[\frac{\partial}{\partial\alpha}\alpha^* + \frac{\partial}{\partial\alpha^*}\alpha - \frac{p}{2}\left(\frac{\partial^2}{\partial\alpha^2} + \frac{\partial^2}{\partial\alpha^{*2}}\right)\right]W(\alpha, p, t)$$
$$+\Gamma\left(\frac{\partial}{\partial\alpha^*}\alpha^* + \frac{\partial}{\partial\alpha}\alpha + (1-p)\frac{\partial^2}{\partial\alpha\,\partial\alpha^*}\right)W(\alpha, p, t) \tag{5.7.16}$$

for the time-evolved quasi-probability distribution $W(\alpha, p, t)$. This equation is of the Fokker–Planck type, the properties of which are described in Appendix 13. The first-order derivatives are drift terms which cause the expectation value of $\hat{a}$ to change. The second-order derivatives are diffusion terms which cause $W(\alpha, p, t)$ to spread. An important requirement for a well-behaved quasi-probability is that the diffusion is positive. If this is not the case then $W(\alpha, p, t)$ evolves towards a form which is only expressible in terms of generalized functions. Comparing (5.7.16) with the more general form (A13.55), we find that the condition $D_{\alpha\alpha^*} > |D_{\alpha\alpha}|$ for positive diffusion is satisfied if $p < \Gamma/(\Gamma + g)$, where we have set g to be positive. The general solution of (5.7.16) may be expressed in terms of a Green function and is given by (A13.14).

The problem can be reformulated in terms of a pair of stochastic differential (or c-number Langevin) equations for the amplitudes α and α^*. Comparing (5.7.16) with the general form (A13.55) and using (A13.57) and (A13.58), we find

$$\dot{\alpha} = -\Gamma\alpha - g\alpha^* + f_\alpha(t), \tag{5.7.17}$$

$$\dot{\alpha}^* = -\Gamma\alpha^* - g\alpha + f_{\alpha^*}(t), \tag{5.7.18}$$

where, using (A13.59) to (A13.61), f_α and f_{α^*} are Langevin terms having the correlation functions

$$\langle f_\alpha(t)\rangle = \langle f_{\alpha^*}(t)\rangle = 0, \tag{5.7.19}$$

$$\langle f_\alpha(t) f_\alpha(t')\rangle = \langle f_{\alpha^*}(t) f_{\alpha^*}(t')\rangle = -gp\delta(t-t') \tag{5.7.20}$$

and

$$\langle f_\alpha(t) f_{\alpha^*}(t')\rangle = \Gamma(1-p)\delta(t-t'). \tag{5.7.21}$$

The stochastic differential equations (5.7.17) and (5.7.18) are analogous to the operator Langevin equations discussed in Section 5.5. Indeed, had we chosen to analyse this problem in the Heisenberg picture, we would have obtained the Langevin equations

$$\dot{\hat{a}}(t) = -\Gamma\hat{a}(t) - g\hat{a}^\dagger(t) + \hat{F}(t), \tag{5.7.22}$$

$$\dot{\hat{a}}^\dagger(t) = -\Gamma\hat{a}^\dagger(t) - g\hat{a}(t) + \hat{F}^\dagger(t). \tag{5.7.23}$$

Since these are operator equations, care must be taken to preserve ordering when calculating moments. The stochastic differential equations (5.7.17) and (5.7.18) are for the c-numbers α and α^* but there is an implicit ordering associated with the parameter p. We solve (5.7.17) and (5.7.18) by introducing the real variables

$$x_0 = \frac{1}{\sqrt{2}}(\alpha + \alpha^*), \tag{5.7.24}$$

$$x_{\pi/2} = -\frac{\mathrm{i}}{\sqrt{2}}(\alpha - \alpha^*), \tag{5.7.25}$$

which are analogous to the quadrature operators $\hat{x}_0$ and $\hat{x}_{\pi/2}$. In terms of these variables (5.7.17) and (5.7.18) become

$$\dot{x}_0 = -(\Gamma + g)x_0 + \frac{1}{\sqrt{2}}[f_\alpha(t) + f_{\alpha^*}(t)], \tag{5.7.26}$$

$$\dot{x}_{\pi/2} = -(\Gamma - g)x_{\pi/2} - \frac{\mathrm{i}}{\sqrt{2}}[f_\alpha(t) - f_{\alpha^*}(t)]. \tag{5.7.27}$$

The solutions of this pair of uncoupled equations are

$$x_0(t) = x_0(0)\exp[-(\Gamma+g)t] + \frac{1}{\sqrt{2}}\int_0^t \exp[-(\Gamma+g)(t-t')] \times [f_\alpha(t') + f_{\alpha^*}(t')]\,\mathrm{d}t', \tag{5.7.28}$$

$$x_{\pi/2}(t) = x_{\pi/2}(0)\exp[-(\Gamma-g)t] - \frac{\mathrm{i}}{\sqrt{2}}\int_0^t \exp[-(\Gamma-g)(t-t')] \times [f_\alpha(t') - f_{\alpha^*}(t')]\,\mathrm{d}t'. \tag{5.7.29}$$

We can use these solutions together with the correlation functions (5.7.19) to (5.7.21) to calculate the moments of x_0 and $x_{\pi/2}$. The simplest of these are

$$\langle x_0(t)\rangle = \langle x_0(0)\rangle \exp[-(\Gamma+g)t], \tag{5.7.30}$$

$$\langle x_{\pi/2}(t)\rangle = \langle x_{\pi/2}(0)\rangle \exp[-(\Gamma-g)t], \tag{5.7.31}$$

$$\langle x_0^2(t)\rangle = \langle x_0^2(0)\rangle \exp[-2(\Gamma+g)t] + \frac{[\Gamma(1-p)-gp]}{2(\Gamma+g)}\{1-\exp[-2(\Gamma+g)t]\}, \tag{5.7.32}$$

$$\langle x_{\pi/2}^2(t)\rangle = \langle x_{\pi/2}^2(0)\rangle \exp[-2(\Gamma-g)t] + \frac{[\Gamma(1-p)+gp]}{2(\Gamma-g)}\{1-\exp[-2(\Gamma-g)t]\}, \tag{5.7.33}$$

and

$$\langle x_0(t)x_{\pi/2}(t)\rangle = \langle x_0(0)x_{\pi/2}(0)\rangle \exp(-2\Gamma t). \tag{5.7.34}$$

It is also straightforward to obtain two-time correlation functions should these be required. The moments (5.7.32) to (5.7.34) are the p-ordered moments of the corresponding quadrature operators so that $\langle x_0^2\rangle = \langle \hat{x}_0^2\rangle_p$, $\langle x_{\pi/2}^2\rangle = \langle \hat{x}_{\pi/2}^2\rangle_p$, and $\langle x_0 x_{\pi/2}\rangle = \langle \hat{x}_0\hat{x}_{\pi/2}\rangle_p$. It may be surprising that the last of these tends to zero for all possible orderings since $\hat{x}_0$ and $\hat{x}_{\pi/2}$ do not commute. The resolution of this apparent difficulty lies in the form of the p-ordered moments since

$$\begin{aligned}\langle \hat{x}_0\hat{x}_{\pi/2}\rangle_p &= -\frac{\mathrm{i}}{2}\int \mathrm{d}^2\alpha\, W(\alpha,p)(\alpha+\alpha^*)(\alpha-\alpha^*) \\ &= -\frac{\mathrm{i}}{2}\int \mathrm{d}^2\alpha\, W(\alpha,p)(\alpha^2-\alpha^{*2}) \\ &= -\mathrm{i}\langle \hat{a}^2 - \hat{a}^{\dagger 2}\rangle,\end{aligned} \tag{5.7.35}$$

which is clearly independent of p and can take the value 0.

6
DRESSED STATES

6.1 Introduction

In describing a dynamical system in quantum optics, we usually begin with a set of bare states. These are eigenstates of a time-independent Hamiltonian $\hat{H}_0$, say, which contains no interactions. The bare states are coupled together by the addition of interaction terms to $\hat{H}_0$, producing the total Hamiltonian for the coupled system. We define a dressed state as an eigenstate of the time-independent form of the total Hamiltonian, including interactions. The reason we want to find these dressed states is that once they and their energies E_n are known, the dynamics of the system is simple: the total state is a superposition of these dressed states (with phases $\exp(-iE_nt/\hbar)$), the amplitudes being *constants*. The probability for being in any one of the dressed states is time independent.

Sometimes it is difficult or impractical to find the dressed states. However, in such cases, it may be convenient or instructive to diagonalize part of the Hamiltonian. The eigenstates of the part of the Hamiltonian which has been diagonalized we refer to as partly dressed states. These are not true dressed states because the remainder of the Hamiltonian will couple them together, but their dynamics may be simpler than that of the original bare states.

6.2 Two-state systems

In Section 2.3, we considered the dynamics of the two-level atom driven by a classical electromagnetic field and found Rabi oscillations in which the probability for being inverted varies sinusoidally. This is the simplest dynamical system in quantum optics and we use it to illustrate the dressed state technique. In Section 1.3, we showed that the two-state atom can be conveniently represented in terms of Pauli operators and that the time-independent Hamiltonian in the rotating-wave approximation can be written (see (2.2.19) with $\delta = \Delta/2$)

$$\hat{H}_\mathrm{I} = \frac{\hbar\Delta}{2}\hat{\sigma}_3 - \frac{\hbar V}{2}[\hat{\sigma}_+ \exp(-i\varphi) + \hat{\sigma}_- \exp(i\varphi)]. \qquad (6.2.1)$$

The dressed states are the eigenstates of this Hamiltonian. Since we are considering a two-state system, there are two dressed states $|+\rangle$ and $|-\rangle$. We express these as superpositions of the bare states $|1\rangle$ and $|2\rangle$, so that

$$|+\rangle = \alpha_+ |1\rangle + \beta_+ |2\rangle, \qquad (6.2.2)$$

$$|-\rangle = \alpha_- |1\rangle + \beta_- |2\rangle, \qquad (6.2.3)$$

where $\alpha_\pm$ and $\beta_\pm$ are complex constants. We require these states to be eigenstates satisfying the equations

$$\hat{H}_\text{I}\,|\pm\rangle = \hbar\omega_\pm\,|\pm\rangle. \tag{6.2.4}$$

As we showed in Section 1.3, the representation of the Pauli operators in the bare state basis is

$$\hat{\sigma}_3 = |2\rangle\langle 2| - |1\rangle\langle 1|, \tag{6.2.5}$$

$$\hat{\sigma}_+ = |2\rangle\langle 1|, \tag{6.2.6}$$

$$\hat{\sigma}_- = |1\rangle\langle 2|. \tag{6.2.7}$$

Using this representation, and (6.2.1) to (6.2.3), the eigenvalue equation (6.2.4) becomes

$$\left\{\frac{\Delta}{2}(|2\rangle\langle 2| - |1\rangle\langle 1|) - \frac{V}{2}[|2\rangle\langle 1|\exp(-\mathrm{i}\varphi) + |1\rangle\langle 2|\exp(\mathrm{i}\varphi)]\right\}(\alpha_\pm\,|1\rangle + \beta_\pm\,|2\rangle) = \omega_\pm(\alpha_\pm\,|1\rangle + \beta_\pm\,|2\rangle). \tag{6.2.8}$$

Evaluating the inner products gives

$$\tfrac{1}{2}\{[-\Delta\alpha_\pm - V\exp(\mathrm{i}\varphi)\beta_\pm]|1\rangle + [\Delta\beta_\pm - V\exp(-\mathrm{i}\varphi)\alpha_\pm]|2\rangle\} = \omega_\pm(\alpha_\pm\,|1\rangle + \beta_\pm\,|2\rangle). \tag{6.2.9}$$

Comparing the coefficients of the bare states in (6.2.9), we find two simultaneous equations for $\alpha_\pm$ and $\beta_\pm$ which may be written in matrix form as

$$\begin{pmatrix} \omega_\pm + \dfrac{\Delta}{2} & \dfrac{V}{2}\exp(\mathrm{i}\varphi) \\ \dfrac{V}{2}\exp(-\mathrm{i}\varphi) & \omega_\pm - \dfrac{\Delta}{2} \end{pmatrix}\begin{pmatrix} \alpha_\pm \\ \beta_\pm \end{pmatrix} = 0. \tag{6.2.10}$$

The condition for a non-trivial solution for $\alpha_\pm$ and $\beta_\pm$ is that the determinant of the matrix in (6.2.10) is zero. This condition leads to a quadratic equation for the eigenfrequencies $\omega_\pm$, the solutions of which are

$$\omega_\pm = \pm\tfrac{1}{2}(\Delta^2 + V^2)^{1/2} = \pm\tfrac{1}{2}\Omega_\text{R}, \tag{6.2.11}$$

where we recognize Ω_R as the Rabi frequency obtained in Section 2.3. With these eigenfrequencies, the normalized eigenvectors of the matrix in (6.2.10) have the elements

$$\alpha_+ = \beta_- = \left(\frac{\Omega_R - \Delta}{2\Omega_R} \right)^{1/2}, \tag{6.2.12}$$

$$\alpha_- = -\beta_+^* = \frac{V \exp(i\varphi)}{[2\Omega_R(\Omega_R - \Delta)]^{1/2}}, \tag{6.2.13}$$

where we have chosen the arbitrary phase of the eigenvectors so that α_+ and β_- are real and positive. We note that the form of the dressed states, obtained by substituting (6.2.12) and (6.2.13) into (6.2.2) and (6.2.3), depends on the phase of the driving field as well as its amplitude.

In terms of the dressed states, the evolved state of the atom can be written as

$$|\psi_I(t)\rangle = k_+ \exp(-i\omega_+ t)|+\rangle + k_- \exp(-i\omega_- t)|-\rangle \tag{6.2.14}$$

$$= \exp(-i\Omega_R t/2)[k_+ |+\rangle + k_- \exp(i\Omega_R t)|-\rangle], \tag{6.2.15}$$

where k_+ and k_- are complex *constants* determined by the initial conditions. Once these are known, the dynamics is completely determined. Consider for example the case where the atom is initially in the lower of the bare states, so that $|\psi_I(0)\rangle = |\psi(0)\rangle = |1\rangle$. From (6.2.2), (6.2.3), (6.2.12), and (6.2.13) we can express the initial state in terms of the dressed states as

$$|\psi_I(0)\rangle = |1\rangle = \frac{\beta_- |+\rangle - \beta_+ |-\rangle}{\alpha_+\beta_- - \alpha_-\beta_+} = \beta_- |+\rangle - \beta_+ |-\rangle. \tag{6.2.16}$$

Comparing this with the general form of the state vector given in (6.2.15) evaluated at $t = 0$, we see that $k_+ = \beta_-$ and $k_- = -\beta_+$, giving

$$|\psi_I(t)\rangle = \exp(-i\Omega_R t/2)[\beta_- |+\rangle - \beta_+ \exp(i\Omega_R t)|-\rangle]. \tag{6.2.17}$$

Any physical property of the two-level atom can now be calculated from this state. In particular, the probability that the atom is in state $|1\rangle$ at time t is

$$\begin{aligned} P_1(t) &= |\langle 1 | \psi(t)\rangle|^2 \\ &= |\beta_- \langle 1 | +\rangle - \beta_+ \exp(i\Omega_R t)\langle 1 | -\rangle|^2 \\ &= |\beta_-\alpha_+ - \beta_+ \exp(i\Omega_R t)\alpha_-|^2 \\ &= \left| \frac{\Omega_R - \Delta}{2\Omega_R} + \frac{V^2 \exp(i\Omega_R t)}{2\Omega_R(\Omega_R - \Delta)} \right|^2. \end{aligned} \tag{6.2.18}$$

Evaluating the square of the modulus gives

$$P_1(t) = 1 - \frac{V^2}{2\Omega_R^2}(1 - \cos \Omega_R t), \tag{6.2.19}$$

which is, once again, the Rabi solution (2.3.12). In the dressed state treatment, the Rabi oscillations appear as a consequence of beating between the two dressed state amplitudes. This contrasts with the previous interpretation in which Rabi oscillations arose directly from periodic transitions between the bare states.

It is not necessary that the atom is initially in one of its bare states; it could have been prepared in any superposition of these states. Consider an initial state which is an equal superposition of $|1\rangle$ and $|2\rangle$, so that $|\psi_{\mathrm{I}}(0)\rangle = (|1\rangle + |2\rangle)/\sqrt{2}$. Proceeding as before, we find

$$|\psi_{\mathrm{I}}(0)\rangle = \frac{1}{\sqrt{2}}(|1\rangle + |2\rangle) = \frac{1}{\sqrt{2}}[(\beta_- - \alpha_-)|+\rangle + (\alpha_+ - \beta_+)|-\rangle], \quad (6.2.20)$$

and comparing this with the general form (6.2.15) evaluated at $t = 0$, we see that $k_+ = (\beta_- - \alpha_-)/\sqrt{2}$ and $k_- = (\alpha_+ - \beta_+)/\sqrt{2}$, giving

$$|\psi_{\mathrm{I}}(t)\rangle = \frac{\exp(-\mathrm{i}\Omega_{\mathrm{R}}t/2)}{\sqrt{2}}[(\beta_- - \alpha_-)|+\rangle + (\alpha_+ - \beta_+)\exp(\mathrm{i}\Omega_{\mathrm{R}}t)|-\rangle]. \quad (6.2.21)$$

As before, we calculate the probability that the atom is in state $|1\rangle$ at time t, and find

$$P_1(t) = \tfrac{1}{2}\left[1 + \frac{V}{\Omega_{\mathrm{R}}}\left(\frac{\Delta}{\Omega_{\mathrm{R}}}\cos\varphi(1 - \cos\Omega_{\mathrm{R}}t) - \sin\varphi\sin\Omega_{\mathrm{R}}t\right)\right]. \quad (6.2.22)$$

We note that this probability depends on the phase of the field. On resonance, that is when $\Delta = 0$, it is possible that $P_1(t)$ is *time independent*. This occurs when $\sin\varphi = 0$. On the other hand, if $\Delta = 0$ and $\sin\varphi = \pm 1$, then $P_1(t)$ oscillates over the full range between 0 and 1. In order to see the origin of this dependence on the phase of the field, we evaluate $|\psi_{\mathrm{I}}(t)\rangle$ in the dressed state basis given in (6.2.21) when $\Delta = 0$ and find

$$|\psi_{\mathrm{I}}(t)\rangle = \frac{\exp(-\mathrm{i}\Omega_{\mathrm{R}}t/2)}{2}\{[1 + \exp(\mathrm{i}\varphi)]|+\rangle + \exp(\mathrm{i}\Omega_{\mathrm{R}}t)[1 - \exp(-\mathrm{i}\varphi)]|-\rangle\}. \quad (6.2.23)$$

We see that when $\varphi = 0$, the coefficient of the dressed state $|-\rangle$ is zero and the initial condition corresponds to preparing the atom in the dressed state $|+\rangle$. At subsequent times

$$|\psi_{\mathrm{I}}(t)\rangle = \exp(-\mathrm{i}\Omega_{\mathrm{R}}t/2)|+\rangle, \quad (6.2.24)$$

and as this is the dressed state $|+\rangle$ multiplied by a phase factor, there is no beating between dressed state amplitudes and hence no Rabi oscillations.

Similarly when $\varphi = \pi$, the initial condition corresponds to preparing the atom in the dressed state $|-\rangle$ and again there are no Rabi oscillations. If, however, $\varphi = \pm\pi/2$, we find

$$|\psi_I(t)\rangle = \exp(-i\Omega_R t/2)\frac{\exp(\pm i\pi/4)}{\sqrt{2}}[|+\rangle + \exp(i\Omega_R t)|-\rangle], \quad (6.2.25)$$

and the probabilities for being in the states $|+\rangle$ and $|-\rangle$ are both equal to 1/2 leading to on-resonance Rabi oscillations. It is worth emphasizing that this phase dependence is a characteristic feature of the coherent interaction between the atom and the field.

In the above discussion, we have found the dressed states for an isolated atom interacting coherently with a classical monochromatic field. There will usually be additional interactions present including possible couplings between the bare states, in particular spontaneous emission from $|2\rangle$ to $|1\rangle$, and coupling to other states. In such cases the states $|+\rangle$ and $|-\rangle$ will only be partly dressed states of the total Hamiltonian. Nevertheless, they may still be a useful aid to calculation and to interpretation. In practice it is often useful to dress that part of the Hamiltonian containing the strongest interaction. The partly dressed states so obtained will evolve under the influence of the remaining weaker couplings. The effect of this dressing is to replace two strongly coupled bare states by the states $|+\rangle$ and $|-\rangle$ separated in energy by an amount $\hbar\Omega_R$. Such a splitting of levels is a common feature in the spectroscopy of strongly driven atoms. We briefly discuss two important examples: the Autler–Townes doublet observed in the absorption spectrum of light from a weak probe laser by a pair of strongly coupled levels, and the Mollow triplet observed in resonance fluorescence and analysed in Section 5.6.

In the first of these, the bare states $|1\rangle$ and $|2\rangle$ are two excited states of the atom strongly driven by a near-resonant laser. A second, weaker laser is tuned to near resonance between the ground state $|0\rangle$ and the state $|1\rangle$ and acts as a probe (see Fig. 6.1). As the frequency of the probe laser is swept through resonance with the bare transition frequency ω_0 between states $|0\rangle$ and $|1\rangle$, the absorption spectrum obtained exhibits two peaks separated by the Rabi frequency Ω_R associated with transitions between the bare states $|1\rangle$ and $|2\rangle$ (see Fig. 6.1). The positions of these peaks differ from the bare transition frequency ω_0 by $\pm\Omega_R/2$. The reason for this splitting is made clear by dressing the strong transition between $|1\rangle$ and $|2\rangle$ to obtain a pair of partly dressed states $|+\rangle$ and $|-\rangle$ separated in energy by $\hbar\Omega_R$. In this description the two peaks of the Autler–Townes spectrum correspond to the excitation of one or other of the two partly dressed states $|+\rangle$ and $|-\rangle$ by absorption of energy from the probe laser. The exact form of the spectrum, including the widths of the peaks, requires the inclusion of other couplings, in particular spontaneous emission. However, the principal feature of the spectrum, that is the appearance of two peaks separated by the Rabi frequency, is easily understood in terms of the partly dressed states. A related phenomenon occurs in the ionization of two

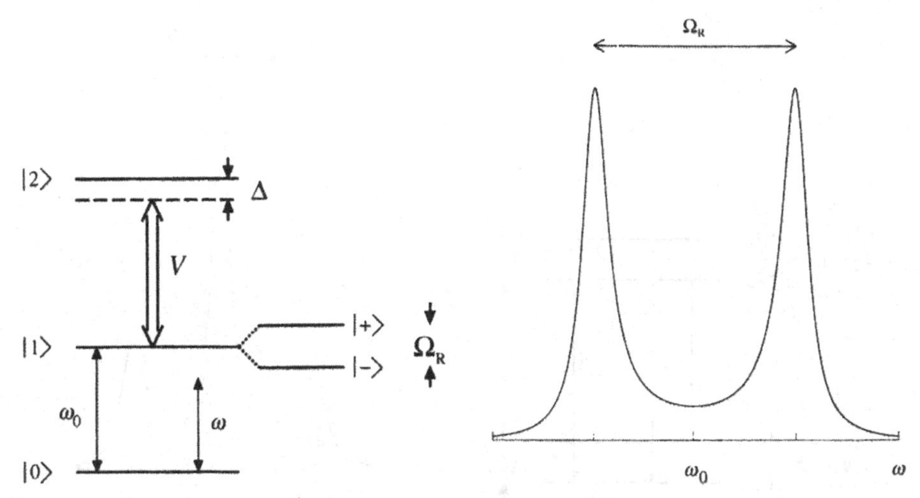

Fig. 6.1 Level scheme for a pump–probe experiment showing the Rabi splitting and the associated Autler–Townes doublet.

strongly coupled levels. The resulting photoelectron spectrum again exhibits two peaks separated in energy by $\hbar\Omega_R$.

In the second example, we add spontaneous emission between a pair of resonantly coupled bare states. Although the spontaneous emission process is weak compared with the coherent coupling, the repeated excitation of state $|2\rangle$ by the field produces a large number of spontaneously emitted fluorescence photons from the atom. The fluorescence spectrum exhibits three peaks (see Fig. 5.7). The central peak coincides in frequency with the bare transition frequency ω_0 between the states $|1\rangle$ and $|2\rangle$. The smaller peaks are displaced from ω_0 by $\pm\Omega_R$. Dressing only the strong transition produces partly dressed states $|+\rangle$ and $|-\rangle$ separated in energy, as before, by $\hbar\Omega_R$. The spontaneous emission of a photon with frequency ω removes an amount of energy $\hbar\omega$ from the coupled atom–field system. This energy is small compared with the total energy and does not affect the structure of the partly dressed states. Prior to emission of a fluorescence photon, it is helpful to picture the system as having $N+1$ quanta shared between the field and the atom. In the partly dressed description, the two states $|+\rangle_{N+1}$ and $|-\rangle_{N+1}$ have energies $(N+1)\hbar\omega_0 \pm \hbar\Omega_R/2$. The complete energy-level diagram consists of pairs of partly dressed states, each pair separated from the next by energy $\hbar\omega_0$. Peaks in the fluorescence spectrum coincide with resonant transitions between adjacent pairs of partly dressed states (see Fig. 6.2). These transitions correspond to frequencies ω_0, $\omega_0+\Omega_R$, and $\omega_0-\Omega_R$. The central peak at the bare transition frequency ω_0 is due to two possible transitions: from $|+\rangle_{N+1}$ to $|+\rangle_N$, and from $|-\rangle_{N+1}$ to $|-\rangle_N$. The peaks at the lower and higher frequencies are due to the transitions between $|-\rangle_{N+1}$ and $|+\rangle_N$, and between $|+\rangle_{N+1}$ and $|-\rangle_N$, respectively. The fact that there are two contributions to the central peak is part of the explanation for this peak being the most prominent. As with the Autler–Townes

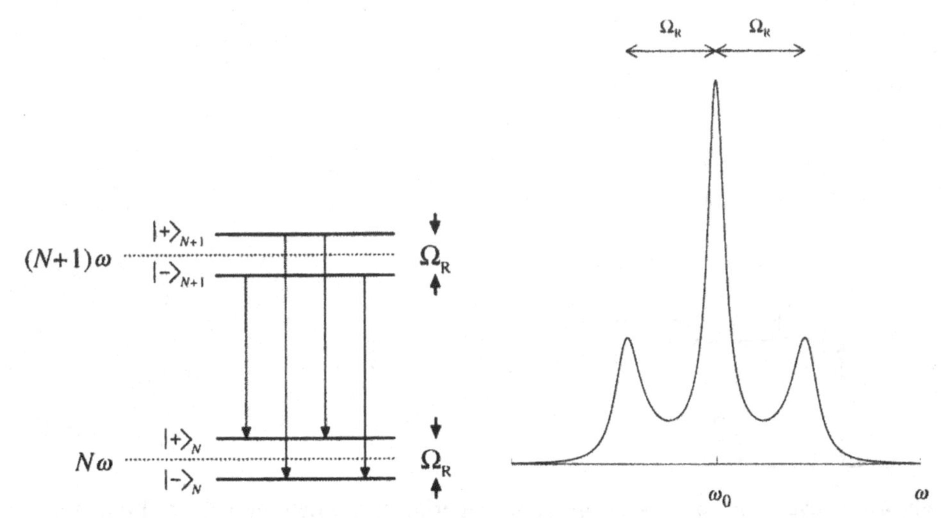

Fig. 6.2 Partly dressed states for resonance fluorescence and the associated Mollow triplet.

spectrum, a full description of the spectrum requires knowledge of the details of the spontaneous emission process. However, the origin of the three peaks is easily understood in terms of the partly dressed states.

In this section, we have calculated the dressed states and their energies for a driven two-state system, using the specific example of an atom interacting with a classical field. This approach may be applied to any coupled two-state system and we illustrate the generality of the method by finding the dressed states for the Jaynes–Cummings model (see Section 2.4) in which the field is quantized. The Hamiltonian in the interaction picture for this system is

$$\hat{H}_{\mathrm{I}} = \frac{\hbar\Delta}{2}\hat{\sigma}_3 - \mathrm{i}\hbar\lambda\left(\hat{\sigma}_+\hat{a} - \hat{a}^\dagger\hat{\sigma}_-\right), \tag{6.2.26}$$

as in (2.4.3). We saw in Section 2.4 that $\hat{N} = \hat{a}^\dagger\hat{a} + \hat{\sigma}_+\hat{\sigma}_-$ is a constant of the motion and hence that the interaction preserves the total number of excitations. The dressed states are found by considering pairs of bare states having the same number of excitations, for example the pair $|1\rangle|N\rangle$ and $|2\rangle|N-1\rangle$ each having N quanta. These states are only coupled to each other and therefore form a two-state system which can be dressed. The exception is when there are no quanta and a single uncoupled state $|1\rangle|0\rangle$.

For a pair of bare states with N excitations, there are two dressed states $|+,N\rangle$ and $|-,N\rangle$. We express these as superpositions of the bare states $|1\rangle|N\rangle$ and $|2\rangle|N-1\rangle$ so that

$$|+,N\rangle = \alpha_+(N)|1\rangle|N\rangle + \beta_+(N)|2\rangle|N-1\rangle, \tag{6.2.27}$$

$$|-,N\rangle = \alpha_-(N)|1\rangle|N\rangle + \beta_-(N)|2\rangle|N-1\rangle. \tag{6.2.28}$$

The eigenvalue equation $\hat{H}_{\mathrm{I}}\,|\pm, N\rangle = \hbar\omega_{\pm}(N)|\pm, N\rangle$ is

$$\left(\frac{\Delta}{2}\hat{\sigma}_3 - \mathrm{i}\lambda\left(\hat{\sigma}_+\hat{a} - \hat{a}^\dagger\hat{\sigma}_-\right)\right)[\alpha_{\pm}(N)|1\rangle|N\rangle + \beta_{\pm}(N)|2\rangle|N-1\rangle]$$
$$= \omega_{\pm}(N)[\alpha_{\pm}(N)|1\rangle|N\rangle + \beta_{\pm}(N)|2\rangle|N-1\rangle]. \tag{6.2.29}$$

Taking matrix elements of this equation with the bare states, we find two simultaneous equations for $\alpha_{\pm}(N)$ and $\beta_{\pm}(N)$ which may be written in matrix form as

$$\begin{pmatrix} \omega_{\pm}(N) + \dfrac{\Delta}{2} & \mathrm{i}\lambda N^{1/2} \\ -\mathrm{i}\lambda N^{1/2} & \omega_{\pm}(N) - \dfrac{\Delta}{2} \end{pmatrix}\begin{pmatrix} \alpha_{\pm}(N) \\ \beta_{\pm}(N) \end{pmatrix} = 0. \tag{6.2.30}$$

This is formally identical to (6.2.10), where $\omega_{\pm}$, $\alpha_{\pm}$, and $\beta_{\pm}$ are now N dependent, $\varphi = \pi/2$, and we replace $V/2$ by $\lambda N^{1/2}$. Making these identifications, we can immediately use the results (6.2.11) to (6.2.13). The eigenfrequencies are therefore

$$\omega_{\pm}(N) = \pm\tfrac{1}{2}(\Delta^2 + 4\lambda^2 N)^{1/2} = \pm\tfrac{1}{2}\Omega_{\mathrm{R}}(N), \tag{6.2.31}$$

where $\Omega_{\mathrm{R}}(N)$ is the Rabi frequency for N quanta, and the coefficients $\alpha_{\pm}(N)$ and $\beta_{\pm}(N)$ in (6.2.27) and (6.2.28) are given by

$$\alpha_+(N) = \beta_-(N) = \left(\frac{\Omega_{\mathrm{R}}(N) - \Delta}{2\Omega_{\mathrm{R}}(N)}\right)^{1/2}, \tag{6.2.32}$$

$$\alpha_-(N) = \beta_+(N) = \mathrm{i}\left(\frac{2\lambda^2 N}{\Omega_{\mathrm{R}}(N)[\Omega_{\mathrm{R}}(N) - \Delta]}\right)^{1/2}. \tag{6.2.33}$$

The complete set of dressed states for this system comprises a pair of states $|+, N\rangle$ and $|-, N\rangle$ for each $N > 0$ and the single state $|1\rangle\,|0\rangle$. In solving for the dynamics, the initial atom–field state is expanded in terms of this set. The collapse and revival phenomena found in Section 2.4 arise from the beating between the quantum Rabi frequencies corresponding to different N.

The strength of the dressed state method developed in this section is that it may be applied to solve for the dynamics of systems which can be decomposed into pairs of coherently coupled states. It may be useful in situations where there are more than two bare states involved, as we have shown.

6.3 Three-state systems

In the previous section, we derived the dressed states for a two-level atom driven by a single classical electromagnetic field. We now extend this to the case of

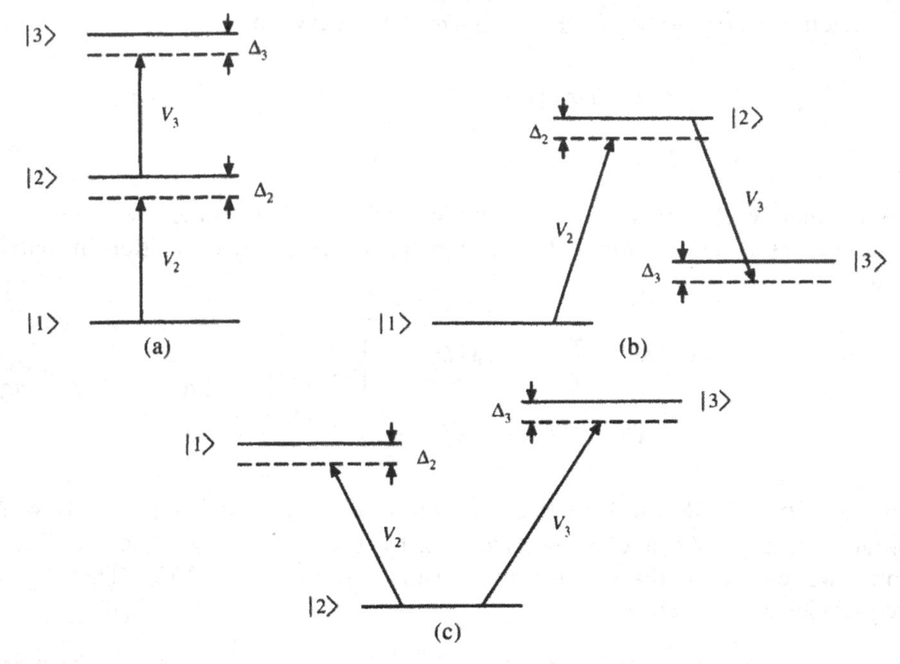

Fig. 6.3 Level schemes for the (a) ladder, (b) lambda, and (c) V configurations.

three bare states $|1\rangle$, $|2\rangle$, and $|3\rangle$ driven by two classical monochromatic fields. The field causing transitions between bare states $|1\rangle$ and $|2\rangle$ is detuned from resonance by frequency Δ_2, and has constant dipole matrix element V_2 and phase φ_2. The field causing transitions between bare states $|2\rangle$ and $|3\rangle$ is detuned by frequency Δ_3, and has dipole matrix element V_3 and phase φ_3. There are three possible configurations of the bare states, referred to as the 'ladder', 'lambda', and 'V' configurations (see Fig. 6.3). We define the positive detunings as shown in this figure in order that the formalism should cover all three cases. The time-independent Hamiltonian in the rotating-wave approximation for this system can be written in terms of projection operators as

$$\hat{H}_{\rm I} = \hbar\Delta_2\,|2\rangle\langle 2| + \hbar(\Delta_2+\Delta_3)\,|3\rangle\langle 3| - \frac{\hbar V_2}{2}[\exp(-{\rm i}\varphi_2)\,|2\rangle\langle 1| + \exp({\rm i}\varphi_2)\,|1\rangle\langle 2|]$$
$$- \frac{\hbar V_3}{2}[\exp({\rm i}\varphi_3)\,|2\rangle\langle 3| + \exp(-{\rm i}\varphi_3)\,|3\rangle\langle 2|], \tag{6.3.1}$$

where we have chosen an interaction picture in which the energy of the bare state $|1\rangle$ is zero.

The three dressed states, which we denote by $|a\rangle$, $|b\rangle$, and $|c\rangle$, are the eigenstates of $\hat{H}_{\rm I}$. We express these dressed states as superpositions of the bare states, so that

$$|i\rangle = \alpha_i\,|1\rangle + \beta_i\,|2\rangle + \gamma_i\,|3\rangle, \tag{6.3.2}$$

where $i = a, b$, or c. Using (6.3.1) and (6.3.2), the eigenvalue equation

$$\hat{H}_{\mathrm{I}} \, |i\rangle = \hbar \, \omega_i \, |i\rangle \tag{6.3.3}$$

becomes

$$\begin{aligned}
&\Big(\Delta_2 \, |2\rangle \langle 2| + (\Delta_2 + \Delta_3) \, |3\rangle \langle 3| - \frac{V_2}{2} [\exp(-\mathrm{i}\varphi_2) \, |2\rangle \langle 1| + \exp(\mathrm{i}\varphi_2) \, |1\rangle \langle 2|] \\
&\quad - \frac{V_3}{2} [\exp(\mathrm{i}\varphi_3) \, |2\rangle \langle 3| + \exp(-\mathrm{i}\varphi_3) \, |3\rangle \langle 2|]\Big) (\alpha_i \, |1\rangle + \beta_i \, |2\rangle + \gamma_i \, |3\rangle) \\
&= \omega_i (\alpha_i \, |1\rangle + \beta_i \, |2\rangle + \gamma_i \, |3\rangle).
\end{aligned} \tag{6.3.4}$$

Evaluating the inner products gives

$$\begin{aligned}
&\Big[-\frac{V_2}{2} \exp(\mathrm{i}\varphi_2) \beta_i \, |1\rangle + \Big(-\frac{V_2}{2} \exp(-\mathrm{i}\varphi_2) \alpha_i + \Delta_2 \, \beta_i - \frac{V_3}{2} \exp(\mathrm{i}\varphi_3) \gamma_i\Big) |2\rangle \\
&\quad + \Big(-\frac{V_3}{2} \exp(-\mathrm{i}\varphi_3) \beta_i + (\Delta_2 + \Delta_3) \gamma_i\Big) |3\rangle\Big] = \omega_i (\alpha_i \, |1\rangle + \beta_i \, |2\rangle + \gamma_i \, |3\rangle),
\end{aligned} \tag{6.3.5}$$

and comparing the coefficients of the bare states, we find three simultaneous equations for the complex constants α_i, β_i, and γ_i which can be written in matrix form as

$$\begin{pmatrix}
-\omega_i & -\frac{V_2}{2} \exp(\mathrm{i}\varphi_2) & 0 \\
-\frac{V_2}{2} \exp(-\mathrm{i}\varphi_2) & \Delta_2 - \omega_i & -\frac{V_3}{2} \exp(\mathrm{i}\varphi_3) \\
0 & -\frac{V_3}{2} \exp(-\mathrm{i}\varphi_3) & \Delta_2 + \Delta_3 - \omega_i
\end{pmatrix}
\begin{pmatrix} \alpha_i \\ \beta_i \\ \gamma_i \end{pmatrix} = 0. \tag{6.3.6}$$

The determinant of the matrix must be zero for a non-trivial solution. This condition leads to

$$\omega_i^3 - \omega_i^2 (2\Delta_2 + \Delta_3) + \omega_i \Big(\Delta_2 (\Delta_2 + \Delta_3) - \frac{V_2^2}{4} - \frac{V_3^2}{4}\Big) + \frac{V_2^2}{4} (\Delta_2 + \Delta_3) = 0, \tag{6.3.7}$$

which is a cubic equation for the three eigenfrequencies ω_i. This equation must, of course, have three real distinct roots. These can be found analytically using a

cubic equation (6.3.7) as $\omega_i^3 + A\omega_i^2 + B\omega_i + C =$
$_b$, and ω_c are then given by the three distinct

$$^{1/2}\cos\left[\tfrac{1}{3}\cos^{-1}\left(\mp\frac{27C+2A^3-9AB}{2(A^2-3B)^{3/2}}\right)\right], \quad (6.3.8)$$

(6.3.7),

$$(2\Delta_2+\Delta_3), \quad (6.3.9)$$

$$(\Delta_2+\Delta_3)-\tfrac{1}{4}(V_2^2+V_3^2), \quad (6.3.10)$$

$$^2_2(\Delta_2+\Delta_3). \quad (6.3.11)$$

of the matrix in (6.3.6) can be written in terms of

$$\frac{1}{{}_i^2+V_2^2)(\omega_i-\Delta_2-\Delta_3)^2+V_3^2\omega_i^2]}$$

$$\begin{pmatrix} V_2\exp(i\varphi_2)(\omega_i-\Delta_2-\Delta_3) \\ 2\omega_i(\omega_i-\Delta_2-\Delta_3) \\ -V_3\exp(-i\varphi_3)\omega_i \end{pmatrix}, \quad (6.3.12)$$

the dressed states in (6.3.2) are determined.
the case where the bare states $|1\rangle$ and $|3\rangle$ are
other, so that $\Delta_2+\Delta_3=0$. The two fields cause
$|1\rangle$ and $|3\rangle$, although the intermediate state $|2\rangle$
iting $\Delta_2=-\Delta_3=\Delta$, the cubic equation (6.3.7) for

$$\Delta\omega_i^2-\tfrac{1}{4}(V_2^2+V_3^2)\omega_i=0, \quad (6.3.13)$$

therefore given by

$$\tfrac{1}{2}\left[\Delta-\surd(\Delta^2+V_2^2+V_3^2)\right], \quad (6.3.14)$$

$$), \quad (6.3.15)$$

$$\tfrac{1}{2}\left[\Delta+\surd(\Delta^2+V_2^2+V_3^2)\right]. \quad (6.3.16)$$

zero eigenfrequency in this two-photon resonant
choosing an interaction picture in which the bare
f the resonant coupling between $|1\rangle$ and $|3\rangle$. From

state. For the c
Poissonian with

and mean phot
into (4.2.1), we

$M($

Alternatively, w
together with t
annihilation op

$M(\mu)$

We can now use
a coherent state

In the second e
Bose–Einstein

(see (3.5.10)). I
function

The correspondi

the general expression (6.3.12) for the normalized eigenvectors, we see immediately that

$$\begin{pmatrix} \alpha_a \\ \beta_a \\ \gamma_a \end{pmatrix} = \frac{1}{\sqrt{(4\omega_a^2 + V_2^2 + V_3^2)}} \begin{pmatrix} -V_2 \exp(\mathrm{i}\varphi_2) \\ 2\omega_a \\ -V_3 \exp(-\mathrm{i}\varphi_3) \end{pmatrix} \tag{6.3.17}$$

and

$$\begin{pmatrix} \alpha_c \\ \beta_c \\ \gamma_c \end{pmatrix} = \frac{1}{\sqrt{(4\omega_c^2 + V_2^2 + V_3^2)}} \begin{pmatrix} -V_2 \exp(\mathrm{i}\varphi_2) \\ 2\omega_c \\ -V_3 \exp(-\mathrm{i}\varphi_3) \end{pmatrix}. \tag{6.3.18}$$

The eigenvector corresponding to the zero eigenfrequency $\omega_b = 0$ is most easily found from the Hamiltonian matrix in (6.3.6) with $\Delta_2 + \Delta_3 = 0$, $\Delta_2 = \Delta$, and $\omega_i = \omega_b = 0$. From the resulting matrix we see that the normalized eigenvector is

$$\begin{pmatrix} \alpha_b \\ \beta_b \\ \gamma_b \end{pmatrix} = \frac{1}{\sqrt{(V_2^2 + V_3^2)}} \begin{pmatrix} -V_3 \exp(\mathrm{i}\varphi_3) \\ 0 \\ V_2 \exp(-\mathrm{i}\varphi_2) \end{pmatrix}. \tag{6.3.19}$$

The dressed state $|b\rangle$ formed from this eigenvector therefore is a superposition of the bare states $|1\rangle$ and $|3\rangle$ alone and contains no contribution from the bare state $|2\rangle$. This will be important later in this section. The evolved state of the atom, under the condition of two-photon resonance, is given by a sum over the dressed states so that

$$|\psi_1(t)\rangle = k_a \exp(-\mathrm{i}\omega_a t)|a\rangle + k_b|b\rangle + k_c \exp(-\mathrm{i}\omega_c t)|c\rangle, \tag{6.3.20}$$

where the constant amplitudes k_i are determined by the initial state of the atom. Taking, for example, $|\psi_1(0)\rangle = |\psi(0)\rangle = |1\rangle$, we have

$$|1\rangle = k_a|a\rangle + k_b|b\rangle + k_c|c\rangle \tag{6.3.21}$$

and hence

$$k_a = \langle a|1\rangle = \frac{-1}{\sqrt{(4\omega_a^2 + V_2^2 + V_3^2)}} V_2 \exp(-\mathrm{i}\varphi_2), \tag{6.3.22}$$

$$k_b = \langle b|1\rangle = \frac{-1}{\sqrt{(V_2^2 + V_3^2)}} V_3 \exp(-\mathrm{i}\varphi_3), \tag{6.3.23}$$

$$k_c = \langle c|1\rangle = \frac{-1}{\sqrt{(4\omega_c^2 + V_2^2 + V_3^2)}} V_2 \exp(-\mathrm{i}\varphi_2). \tag{6.3.24}$$

An interesting limiting case may be considered in which, although we retain the two-photon resonance condition, Δ (taken as positive for definiteness) is much larger than either V_2 or V_3. The bare state $|2\rangle$ is then far from resonance with either transition and we would expect it to remain essentially unpopulated and for population to cycle between bare states $|1\rangle$ and $|3\rangle$. The eigenfrequencies in this limit are found by expansion of (6.2.14) and (6.2.16) in powers of $(V_2^2+V_3^2)/\Delta^2$ to be

$$\omega_a \simeq -\frac{(V_2^2+V_3^2)}{4\Delta}, \quad \omega_b = 0, \quad \omega_c \simeq \Delta, \tag{6.3.25}$$

showing that ω_a is of a lower order of magnitude than both V_2 and V_3, whilst ω_c is of a higher order. Further, from (6.3.22) to (6.3.24),

$$k_a \simeq -\frac{V_2 \exp(-\mathrm{i}\varphi_2)}{\sqrt{(V_2^2+V_3^2)}}, \quad k_b = -\frac{V_3 \exp(-\mathrm{i}\varphi_3)}{\sqrt{(V_2^2+V_3^2)}}, \tag{6.3.26}$$

whilst k_c is of order $1/\Delta$ and may be neglected. Using the above results, the state (6.3.20) is, to this order of approximation, a superposition of only two dressed states given by

$$|\psi_1(t)\rangle \simeq \frac{-1}{\sqrt{(V_2^2+V_3^2)}}[V_2 \exp(-\mathrm{i}\varphi_2)\exp(-\mathrm{i}\omega_a t)|a\rangle + V_3 \exp(-\mathrm{i}\varphi_3)|b\rangle]. \tag{6.3.27}$$

The probability for being in the bare state $|1\rangle$ at time t is, using (6.3.26),

$$\begin{aligned} P_1(t) &= |\langle 1 | \psi_1(t)\rangle|^2 \\ &= \frac{1}{(V_2^2+V_3^2)}|V_2 \exp(-\mathrm{i}\varphi_2)\exp(-\mathrm{i}\omega_a t)\langle 1|a\rangle + V_3 \exp(-\mathrm{i}\varphi_3)\langle 1|b\rangle|^2 \\ &= \frac{1}{(V_2^2+V_3^2)}|V_2 \exp(-\mathrm{i}\varphi_2)\exp(-\mathrm{i}\omega_a t)k_a^* + V_3 \exp(-\mathrm{i}\varphi_3)k_b^*|^2 \\ &= \frac{1}{(V_2^2+V_3^2)^2}(V_2^4 + V_3^4 + 2V_2^2V_3^2 \cos \omega_a t) \\ &= 1 - \frac{V_2^2V_3^2}{8\Delta^2\omega_a^2}(1-\cos \omega_a t). \end{aligned} \tag{6.3.28}$$

This is of the same form as the two-state Rabi solution (2.3.12) with $|\omega_a| = (V_2^2+V_3^2)/4\Delta$ replacing the Rabi frequency Ω_{R} and an effective coupling between $|1\rangle$ and $|3\rangle$ given by $V = V_2V_3/2\Delta$. The evolution consists of coherent transitions between $|1\rangle$ and $|3\rangle$ at the two-photon Rabi frequency $|\omega_a|$. The bare state probability $P_1(t)$ has a minimum value of $(V_2^2-V_3^2)^2/(V_2^2+V_3^2)^2$, which is zero only if $V_2 = V_3$. Complete inversion only occurs, therefore, if the coupling strengths are equal. It is also worth noting that, to this order of approximation, $|c\rangle \simeq |2\rangle$, from (6.3.18), and that $|2\rangle$ remains unoccupied.

Returning now to the general solution for the two-photon resonant case, we discuss the effect of including an extra coupling from the bare state $|2\rangle$ to other states outside the three-state system. Here we are principally concerned with either ionization from state $|2\rangle$ in the 'lambda' system, or spontaneous emission from $|2\rangle$ to a state of lower energy in the 'V' system. The states $|a\rangle$, $|b\rangle$, and $|c\rangle$ are now only partly dressed states but can still account for the principal feature of the dynamics, that is population trapping.

In Section 5.3, we showed that irreversible losses from the system may, in part, be included by replacing the real frequency of the bare state which is coupled to states outside the system by an imaginary frequency which incorporates both the width and the shift of the level. The effect of this on the three-level system is that the frequency ω_2 of the bare state $|2\rangle$ gains a negative imaginary part $-\mathrm{i}\Gamma$. Consequently, $\Delta_2 = \Delta$ has a negative imaginary part $-\mathrm{i}\Gamma$ whilst the imaginary part of Δ_3 is $+\mathrm{i}\Gamma$ in order that the energy of bare state $|3\rangle$ remains real. The two-photon resonance condition $\Delta_2 + \Delta_3 = 0$ therefore continues to hold. The dressed state $|b\rangle$ still has a zero eigenfrequency and, from (6.3.19), is given by

$$|b\rangle = \frac{1}{\sqrt{(V_2^2 + V_3^2)}}[-V_3 \exp(\mathrm{i}\varphi_3)|1\rangle + V_2 \exp(-\mathrm{i}\varphi_2)|3\rangle]. \quad (6.3.29)$$

However, the eigenfrequencies ω_a and ω_c each have a negative imaginary part. This is straightforward, if tedious, to demonstrate algebraically from the expressions (6.3.14) and (6.3.16). Physically, we can see that the signs of their imaginary parts must be such as to cause exponential *decay* of the factors $\exp(-\mathrm{i}\omega_a t)$ and $\exp(-\mathrm{i}\omega_c t)$ as t increases, rather than exponential growth. If the atom is initially in bare state $|1\rangle$ then, for long times, the state of the three-level system tends to

$$|\psi_1(t \to \infty)\rangle = k_b|b\rangle = \frac{1}{(V_2^2 + V_3^2)}\{V_3^2|1\rangle - V_2 V_3 \exp[-\mathrm{i}(\varphi_2 + \varphi_3)]|3\rangle\}. \quad (6.3.30)$$

Of course, this does not include the states other than those in the three-level system to which state $|2\rangle$ is coupled. The populations in the bare states $|1\rangle$ and $|3\rangle$ therefore tend to

$$P_1(\infty) = \frac{V_3^4}{(V_2^2 + V_3^2)^2} \quad (6.3.31)$$

and

$$P_3(\infty) = \frac{V_2^2 V_3^2}{(V_2^2 + V_3^2)^2} \quad (6.3.32)$$

which are both non-zero. Contrary to what might have been expected, not all of the population flows out of the three-level system. Even though there is irreversible loss, the two-photon resonance condition has led to a dressed state $|b\rangle$ with a zero eigenfrequency, the population of which therefore does not decay. We then say that population is *trapped* in the system, and $|b\rangle$ is often referred to as the trapping state. If $V_2 = V_3$, we see that $P_1(\infty) = P_3(\infty) = 1/4$, and half of the population remains trapped in the system. More generally, by varying V_2 and V_3 it is possible to modify the total trapped population to lie anywhere between 0 and 1. We emphasize that the above calculation holds for an initial state of the system in which all the population occupies the bare state $|1\rangle$. Similar results hold if $|3\rangle$ is the initially occupied state. However, if the initial state is $|2\rangle$, then the trapping state is unpopulated and this leads to no population being trapped in the system at long times.

6.4 Dressing the Bixon–Jortner model

The method used in the previous two sections can be applied to obtain the dressed states for an arbitrary number of coupled discrete states. However, the complexity of the resulting eigenvalue equation usually means that analytic solution is not possible. Nevertheless there exist special cases for which it is possible to treat at least part of the problem analytically. As an extreme example, we consider the dressed states for the Bixon–Jortner model (see Section 2.5) in which a single state $|g\rangle$ is coupled to an unbounded quasi-continuum of states equally spaced by energy $\hbar\Delta$. The Hamiltonian for this model is

$$\hat{H}_{\mathrm{I}} = \hbar\Delta_g |g\rangle\langle g| + \hbar \sum_{n=-\infty}^{\infty} n\Delta |n\rangle\langle n| + \hbar W \sum_{n=-\infty}^{\infty} (|n\rangle\langle g| + |g\rangle\langle n|), \qquad (6.4.1)$$

where we have chosen an interaction picture in which the energy of the quasi-continuum state $|0\rangle$ is zero and $\hbar\Delta_g$ is the energy of $|g\rangle$. The solution presented in Section 2.5 corresponds to the special case $\Delta_g = 0$.

We seek dressed states of the form

$$|\omega\rangle = \alpha(\omega)|g\rangle + \sum_n \beta_n(\omega)|n\rangle, \qquad (6.4.2)$$

where here and subsequently summations run over all values of the index. We require that $|\omega\rangle$ satisfies the eigenvalue equation

$$\hat{H}_{\mathrm{I}}\,|\omega\rangle = \hbar\,\omega\,|\omega\rangle \qquad (6.4.3)$$

and using (6.4.1) and (6.4.2), we obtain the coupled equations

$$(\Delta_g - \omega)\alpha(\omega) + W\sum_n \beta_n(\omega) = 0, \qquad (6.4.4)$$

$$(n\Delta - \omega)\beta_n(\omega) + W\alpha(\omega) = 0 \qquad (6.4.5)$$

for the coefficients $\alpha(\omega)$ and $\beta_n(\omega)$. Solving (6.4.5) for $\beta_n(\omega)$ and summing over n gives

$$\sum_n \beta_n(\omega) = W\alpha(\omega)\sum_n \frac{1}{\omega - n\Delta} = \alpha(\omega)\frac{\pi W}{\Delta}\cot\left(\frac{\pi\omega}{\Delta}\right), \tag{6.4.6}$$

where we have used (A8.18) with $\omega_0 = \Delta$. Substituting (6.4.6) into (6.4.4) and cancelling $\alpha(\omega)$ gives the eigenvalue equation

$$\Delta_g - \omega + \frac{\pi W^2}{\Delta}\cot\left(\frac{\pi\omega}{\Delta}\right) = 0, \tag{6.4.7}$$

the solutions ω_p of which are the eigenfrequencies of the dressed states. These eigenfrequencies can only be found numerically but we note that between the energies of each pair of adjacent quasi-continuum states there is precisely one dressed state energy. Eliminating $\beta_n(\omega)$ in (6.4.2) using (6.4.5) gives the form

$$|\omega_p\rangle = \alpha(\omega_p)\left(|g\rangle + W\sum_n \frac{1}{\omega_p - n\Delta}|n\rangle\right) \tag{6.4.8}$$

of the dressed states and $\alpha(\omega_p)$ is determined by the requirement that $|\omega_p\rangle$ is normalized. We find

$$\langle\omega_p|\omega_p\rangle = 1 = |\alpha(\omega_p)|^2\left(1 + W^2\sum_n \frac{1}{(\omega_p - n\Delta)^2}\right). \tag{6.4.9}$$

The summation in (6.4.9) can be evaluated by noting that

$$\sum_n \frac{1}{(\omega_p - n\Delta)^2} = -\frac{\mathrm{d}}{\mathrm{d}\omega_p}\sum_n \frac{1}{(\omega_p - n\Delta)} = \frac{\pi^2}{\Delta^2}\operatorname{cosec}^2\left(\frac{\pi\omega_p}{\Delta}\right), \tag{6.4.10}$$

where we have again used (A8.18) with $\omega_0 = \Delta$. It then follows from (6.4.9) that

$$\alpha(\omega_p) = \left[1 + \frac{\pi^2 W^2}{\Delta^2}\operatorname{cosec}^2\left(\frac{\pi\omega_p}{\Delta}\right)\right]^{-1/2}, \tag{6.4.11}$$

where we have chosen $\alpha(\omega_p)$ to be real. Since the dressed states $|\omega_p\rangle$ are non-degenerate eigenstates of an Hermitian operator $\hat{H}_I$, they must, by (1.2.17), be mutually orthogonal. We verify this by calculating the overlap $\langle\omega_p|\omega_q\rangle$ for $p \neq q$ which, using (6.4.8), is given by

$$\langle\omega_p|\omega_q\rangle = \alpha(\omega_p)\alpha(\omega_q)\left(1 + W^2\sum_n \frac{1}{(\omega_p - n\Delta)(\omega_q - n\Delta)}\right). \tag{6.4.12}$$

The summation in (6.4.12) can be evaluated using partial fractions since

$$\sum_n \frac{1}{(\omega_p - n\Delta)(\omega_q - n\Delta)} = \frac{1}{(\omega_q - \omega_p)} \sum_n \left(\frac{1}{(\omega_p - n\Delta)} - \frac{1}{(\omega_q - n\Delta)} \right)$$
$$= \frac{1}{(\omega_q - \omega_p)} \left[\frac{\pi}{\Delta} \cot\left(\frac{\pi \omega_p}{\Delta} \right) - \frac{\pi}{\Delta} \cot\left(\frac{\pi \omega_q}{\Delta} \right) \right], \tag{6.4.13}$$

again using (A8.18). The cotangents can be replaced using the eigenvalue equation (6.4.7) and we then find that (6.4.13) is simply $-W^{-2}$ so that $\langle \omega_p | \omega_q \rangle$ in (6.4.12) is identically zero.

It remains to express the bare states $|g\rangle$ and $|n\rangle$ and the Hamiltonian in terms of the dressed states $|\omega_p\rangle$. We write

$$|g\rangle = \sum_p c_p |\omega_p\rangle \tag{6.4.14}$$

and take the overlap with $|\omega_q\rangle$ so that

$$\langle \omega_q | g \rangle = c_q = \alpha(\omega_q), \tag{6.4.15}$$

using (6.4.8). Hence we can write the ground state as

$$|g\rangle = \sum_p \alpha(\omega_p) |\omega_p\rangle \tag{6.4.16}$$

and similarly the quasi-continuum state $|n\rangle$ may be expressed as

$$|n\rangle = W \sum_p \frac{\alpha(\omega_p)}{(\omega_p - n\Delta)} |\omega_p\rangle. \tag{6.4.17}$$

If the state $|g\rangle$ is initially occupied then the time-evolved state is simply

$$|\psi_{\mathrm{I}}(t)\rangle = \sum_p \alpha(\omega_p) \exp(-\mathrm{i}\omega_p t) |\omega_p\rangle \tag{6.4.18}$$

which, once the eigenfrequencies ω_p are known, is completely determined. Finally, using (6.4.16) and (6.4.17), the Hamiltonian (6.4.1) can be written

$$\hat{H}_{\mathrm{I}} = \hbar \sum_p \sum_q \alpha(\omega_p)\alpha(\omega_q) |\omega_p\rangle\langle\omega_q| \left[\Delta_g + W^2 \sum_n \right.$$
$$\left. \times \left(\frac{n\Delta}{(\omega_p - n\Delta)(\omega_q - n\Delta)} + \frac{1}{(\omega_p - n\Delta)} + \frac{1}{(\omega_q - n\Delta)} \right) \right]. \tag{6.4.19}$$

For those terms in the double summation having $p = q$, the summation over n in (6.4.19) is

$$\sum_n \left(\frac{n\Delta}{(\omega_p - n\Delta)^2} + \frac{2}{(\omega_p - n\Delta)} \right) = \frac{\pi^2 \omega_p}{\Delta^2} \operatorname{cosec}^2\left(\frac{\pi\omega_p}{\Delta}\right) + \frac{\pi}{\Delta}\cot\left(\frac{\pi\omega_p}{\Delta}\right), \tag{6.4.20}$$

where we have used (A8.18) and (A8.21). The expression in square brackets in (6.4.19), with $p = q$, then becomes

$$\omega_p\left[1 + \frac{\pi^2 W^2}{\Delta^2}\operatorname{cosec}^2\left(\frac{\pi\omega_p}{\Delta}\right)\right] = \frac{\omega_p}{\alpha^2(\omega_p)}, \tag{6.4.21}$$

where we have used the eigenvalue equation (6.4.7) to substitute for the cotangent. For $p \neq q$, the summation over n in (6.4.19) is

$$\frac{1}{(\omega_q - \omega_p)} \sum_n \left(\frac{n\Delta}{(\omega_p - n\Delta)} - \frac{n\Delta}{(\omega_q - n\Delta)} + \frac{(\omega_q - \omega_p)}{(\omega_p - n\Delta)} + \frac{(\omega_q - \omega_p)}{(\omega_q - n\Delta)} \right)$$

$$= \frac{1}{(\omega_q - \omega_p)} \sum_n \left(\frac{\omega_q}{(\omega_p - n\Delta)} - \frac{\omega_p}{(\omega_q - n\Delta)} \right)$$

$$= \frac{1}{(\omega_q - \omega_p)}\left[\omega_q \frac{\pi}{\Delta}\cot\left(\frac{\pi\omega_p}{\Delta}\right) - \omega_p \frac{\pi}{\Delta}\cot\left(\frac{\pi\omega_q}{\Delta}\right)\right]. \tag{6.4.22}$$

Substituting for the cotangents using (6.4.7), we find that (6.4.22) becomes $-\Delta_g$ so that in (6.4.19) only those terms for which $p = q$ are non-zero and consequently, using (6.4.21), we obtain

$$\hat{H}_{\text{I}} = \hbar \sum_p \omega_p |\omega_p\rangle\langle\omega_p|. \tag{6.4.23}$$

The ability to write the Hamiltonian in this form, together with the orthonormality of the dressed states, constitutes a demonstration of the completeness of the dressed states $|\omega_p\rangle$. If these states are complete, we can resolve the identity as

$$\sum_p |\omega_p\rangle\langle\omega_p| = 1. \tag{6.4.24}$$

The form of the Hamiltonian (6.4.23) then follows on writing

$$\hat{H}_{\text{I}} = \hat{H}_{\text{I}} \sum_p |\omega_p\rangle\langle\omega_p| \tag{6.4.25}$$

and using the eigenvalue equation $\hat{H}_{\text{I}}|\omega_p\rangle = \hbar\,\omega_p|\omega_p\rangle$.

In the next section, we develop this analysis to dress the interaction between one or more discrete states and a true continuum of states. In such cases, the dressed states themselves form a continuum so that all possible energies are dressed state energies and a fully analytical treatment is possible.

6.5 Dressing dissipative systems: Fano theory

Until now, we have calculated dressed states for systems consisting only of discrete levels. Dissipative processes, such as ionization and spontaneous emission, involve couplings to continua. In Section 6.3, such couplings were treated as perturbations which cause the amplitudes of the partly dressed states to evolve in time. One can, however, employ a technique developed by Fano to dress coupled systems consisting of both discrete and continuum states. Using this method it is sometimes possible to diagonalize the Hamiltonian completely and so obtain the fully dressed states.

In its simplest form the Fano technique diagonalizes the system consisting of a single discrete state $|1\rangle$ coupled to a continuum of states $|\Delta\rangle$, satisfying the orthogonality relation $\langle\Delta\,|\,\Delta'\rangle = \delta(\Delta - \Delta')$, as in Section 5.2. The time-independent Hamiltonian in the rotating-wave approximation for this system is

$$\hat{H}_\mathrm{I} = \hbar \int \Delta\,|\Delta\rangle\langle\Delta|\,\mathrm{d}\Delta + \hbar \int W(\Delta)\{|\Delta\rangle\langle 1|\exp[-\mathrm{i}\varphi(\Delta)] + |1\rangle\langle\Delta|\exp[\mathrm{i}\varphi(\Delta)]\}\,\mathrm{d}\Delta, \tag{6.5.1}$$

where we have again chosen an interaction picture in which the energy of state $|1\rangle$ is zero. We seek dressed states $|\omega\rangle$ satisfying the eigenvalue equation

$$\hat{H}_\mathrm{I}\,|\omega\rangle = \hbar\,\omega\,|\omega\rangle. \tag{6.5.2}$$

Expressing $|\omega\rangle$ in terms of the bare states as

$$|\omega\rangle = \alpha(\omega)\,|1\rangle + \int \beta(\omega,\Delta)\,|\Delta\rangle\,\mathrm{d}\Delta, \tag{6.5.3}$$

we find from (6.5.1) and (6.5.2) that

$$\left\{\int \Delta\,|\Delta\rangle\langle\Delta|\,\mathrm{d}\Delta + \int W(\Delta)(|\Delta\rangle\langle 1|\exp[-\mathrm{i}\varphi(\Delta)] + |1\rangle\langle\Delta|\exp[\mathrm{i}\varphi(\Delta)])\,\mathrm{d}\Delta\right\}$$
$$\times\left(\alpha(\omega)\,|1\rangle + \int \beta(\omega,\Delta)\,|\Delta\rangle\,\mathrm{d}\Delta\right) = \omega\left(\alpha(\omega)\,|1\rangle + \int \beta(\omega,\Delta)\,|\Delta\rangle\,\mathrm{d}\Delta\right). \tag{6.5.4}$$

Evaluating the inner products gives

$$\int \Delta\beta(\omega,\Delta)|\Delta\rangle\,\mathrm{d}\Delta + \alpha(\omega)\int W(\Delta)|\Delta\rangle\exp[-\mathrm{i}\varphi(\Delta)]\,\mathrm{d}\Delta + \int W(\Delta)\beta(\omega,\Delta)$$

$$\times\exp[\mathrm{i}\varphi(\Delta)]\,|1\rangle\,\mathrm{d}\Delta = \omega\alpha(\omega)\,|1\rangle + \omega\int\beta(\omega,\Delta)\,|\Delta\rangle\,\mathrm{d}\Delta. \quad (6.5.5)$$

Comparing coefficients of the bare states in (6.5.5), we find

$$\Delta\beta(\omega,\Delta) + \alpha(\omega)W(\Delta)\exp[-\mathrm{i}\varphi(\Delta)] = \omega\beta(\omega,\Delta) \quad (6.5.6)$$

and

$$\int W(\Delta)\beta(\omega,\Delta)\exp[\mathrm{i}\varphi(\Delta)]\,\mathrm{d}\Delta = \omega\alpha(\omega). \quad (6.5.7)$$

We proceed, as in the previous section, by formally solving (6.5.6) to find $\beta(\omega,\Delta)$ in terms of $\alpha(\omega)$. When $\beta(\omega,\Delta)$ is then substituted into (6.4.7), $\alpha(\omega)$ cancels and the remaining equation must be self-consistent. However, $\beta(\omega,\Delta)$ found from (6.5.6) contains the factor $(\omega-\Delta)^{-1}$ which diverges at $\omega=\Delta$. This divergence is treated by following a procedure used by Dirac in which

$$\beta(\omega,\Delta) = \left(\frac{\mathbb{P}}{\omega-\Delta} + z(\omega)\delta(\omega-\Delta)\right)\alpha(\omega)W(\Delta)\exp[-\mathrm{i}\varphi(\Delta)], \quad (6.5.8)$$

where $\mathbb{P}$ denotes that the principal part (see Appendix 7) is taken whenever $(\omega-\Delta)^{-1}$ is integrated, and $z(\omega)$ is to be determined from consistency of the equations (6.5.6) and (6.5.7). Substituting (6.5.8) into (6.5.7) and cancelling $\alpha(\omega)$, we obtain

$$\omega = \int W^2(\Delta)\left(\frac{\mathbb{P}}{\omega-\Delta} + z(\omega)\delta(\omega-\Delta)\right)\mathrm{d}\Delta = F(\omega) + z(\omega)W^2(\omega), \quad (6.5.9)$$

where

$$F(\omega) = \mathbb{P}\int\frac{W^2(\Delta)}{\omega-\Delta}\,\mathrm{d}\Delta, \quad (6.5.10)$$

as in (5.4.9). An implicit assumption here is that $W^2(\Delta)$ is such that this principal part integral is finite, as it will be in many cases of practical importance. We will see that $F(\omega)$ can sometimes be interpreted as a frequency shift induced by the coupling. From (6.5.9), consistency of the equations requires that

$$z(\omega) = \frac{\omega - F(\omega)}{W^2(\omega)}. \quad (6.5.11)$$

Now that $z(\omega)$ is known, $\beta(\omega,\Delta)$ may be expressed in terms of $\alpha(\omega)$ using (6.5.8). In order to determine $\alpha(\omega)$, we require that the dressed states $|\omega\rangle$ in (6.5.3) should satisfy the orthonormality condition

$$\langle\omega'\,|\,\omega\rangle = \delta(\omega'-\omega). \quad (6.5.12)$$

Using (6.5.3) and the orthogonality relationships for the bare states, $\langle 1 \mid 1 \rangle = 1$, $\langle 1 \mid \Delta \rangle = 0$, $\langle \Delta' \mid \Delta \rangle = \delta(\Delta' - \Delta)$, we find that (6.5.12) becomes

$$\alpha^*(\omega')\alpha(\omega) + \int \beta^*(\omega', \Delta)\beta(\omega, \Delta)\,\mathrm{d}\Delta = \delta(\omega' - \omega). \qquad (6.5.13)$$

Substitution from (6.5.8) of $\beta(\omega, \Delta)$ in terms of $\alpha(\omega)$ and $\beta^*(\omega', \Delta)$ in terms of $\alpha^*(\omega')$ in (6.5.13) gives

$$\alpha^*(\omega')\alpha(\omega)\left[1 + \int W^2(\Delta)\left(\frac{\mathbb{P}}{\omega' - \Delta} + z(\omega')\delta(\omega' - \Delta)\right)\right.$$
$$\left. \times\left(\frac{\mathbb{P}}{\omega - \Delta} + z(\omega)\delta(\omega - \Delta)\right)\mathrm{d}\Delta\right] = \delta(\omega' - \omega). \quad (6.5.14)$$

The integrand in (6.5.14) contains the product of two principal parts. In Appendix 7 we show that this product may be expressed in terms of partial fractions and a singular term as

$$\frac{\mathbb{P}}{(\omega' - \Delta)} \cdot \frac{\mathbb{P}}{(\omega - \Delta)} = \frac{\mathbb{P}}{(\omega' - \omega)}\left(\frac{\mathbb{P}}{\omega - \Delta} - \frac{\mathbb{P}}{\omega' - \Delta}\right) + \pi^2\delta(\omega' - \Delta)\delta(\omega - \Delta). \quad (6.5.15)$$

Evaluating the integral in (6.5.14) using this result yields

$$\alpha^*(\omega')\alpha(\omega)\left(1 + \frac{\mathbb{P}}{(\omega' - \omega)}[F(\omega) - F(\omega') + z(\omega)W^2(\omega) - z(\omega')W^2(\omega')]\right.$$
$$\left. + [\pi^2 + z^2(\omega)]W^2(\omega)\delta(\omega' - \omega)\right) = \delta(\omega' - \omega). \quad (6.5.16)$$

The expression (6.5.11) for $z(\omega)$ in terms of $F(\omega)$ may now be used to show that all terms on the left-hand side of (6.5.16) which are not multiplied by $\delta(\omega' - \omega)$ cancel, leaving

$$|\alpha(\omega)|^2 = \frac{1}{W^2(\omega)[\pi^2 + z^2(\omega)]} = \frac{W^2(\omega)}{[\omega - F(\omega)]^2 + \pi^2 W^4(\omega)}. \quad (6.5.17)$$

The arbitrary phase in $\alpha(\omega)$ (and hence in the dressed state $|\omega\rangle$) may be chosen so that, for example,

$$\alpha(\omega) = \frac{W(\omega)}{\omega - F(\omega) - \mathrm{i}\pi W^2(\omega)}. \quad (6.5.18)$$

With this choice of phase, $\alpha(\omega)$ and, therefore, $\beta(\omega, \Delta)$ are completely determined and hence so are the dressed states $|\omega\rangle$.

As we will see, the set of dressed states $|\omega\rangle$ is complete so that any of the bare states may be expressed as an integral over the set $|\omega\rangle$. For example, considering the bare state $|1\rangle$ and writing

$$|1\rangle = \int g(\omega)|\omega\rangle \mathrm{d}\omega, \tag{6.5.19}$$

we see that $\langle\omega'|1\rangle = g(\omega') = \alpha^*(\omega')$ from (6.5.3) and hence that

$$|1\rangle = \int \alpha^*(\omega)|\omega\rangle \mathrm{d}\omega. \tag{6.5.20}$$

It may now be verified that normalization of the bare state $|1\rangle$ is maintained by calculating

$$\langle 1|1\rangle = \int \alpha(\omega')\langle\omega'|\mathrm{d}\omega' \int \alpha^*(\omega)|\omega\rangle \mathrm{d}\omega = \int |\alpha(\omega)|^2 \mathrm{d}\omega = 1. \tag{6.5.21}$$

In Appendix 8 we show that the last integral in (6.5.21) is indeed unity independently of the form of $W(\Delta)$ (provided, of course, that all the integrals exist). The bare continuum states may also be expressed in terms of the dressed states as

$$|\Delta\rangle = \int \beta^*(\omega,\Delta)|\omega\rangle \mathrm{d}\omega, \tag{6.5.22}$$

and it is straightforward to show from this that the delta-function normalization of the bare continuum states is maintained. The Hamiltonian $\hat{H}_\mathrm{I}$ can also be written in terms of the set $|\omega\rangle$. Using (6.5.20) and (6.5.22) we have

$$\begin{aligned}
\hat{H}_\mathrm{I} = \hbar \int \Delta\left(\int \beta^*(\omega',\Delta)|\omega'\rangle \mathrm{d}\omega'\right)\left(\int \beta(\omega,\Delta)\langle\omega|\mathrm{d}\omega\right)\mathrm{d}\Delta \\
+ \hbar \int\Big(W(\Delta)\exp[-\mathrm{i}\varphi(\Delta)]\int \beta^*(\omega',\Delta)|\omega'\rangle \mathrm{d}\omega' \int \alpha(\omega)\langle\omega|\mathrm{d}\omega \\
+ W(\Delta)\exp[\mathrm{i}\varphi(\Delta)]\int \alpha^*(\omega')|\omega'\rangle \mathrm{d}\omega' \int \beta(\omega,\Delta)\langle\omega|\mathrm{d}\omega\Big)\mathrm{d}\Delta \\
= \hbar \int \mathrm{d}\omega' \int \mathrm{d}\omega |\omega'\rangle\langle\omega| I(\omega,\omega'),
\end{aligned} \tag{6.5.23}$$

where

$$\begin{aligned}
I(\omega,\omega') = \int \Delta\beta^*(\omega',\Delta)\beta(\omega,\Delta)\mathrm{d}\Delta \\
+ \alpha(\omega)\int W(\Delta)\exp[-\mathrm{i}\varphi(\Delta)]\beta^*(\omega',\Delta)\mathrm{d}\Delta \\
+ \alpha^*(\omega')\int W(\Delta)\exp[\mathrm{i}\varphi(\Delta)]\beta(\omega,\Delta)\mathrm{d}\Delta.
\end{aligned} \tag{6.5.24}$$

Now from (6.5.8), the second integral in $I(\omega, \omega')$ is

$$\int W(\Delta)\exp[-\mathrm{i}\varphi(\Delta)]\,\beta^*(\omega',\Delta)\,\mathrm{d}\Delta = \alpha^*(\omega')\int W^2(\Delta) \times\left(\frac{\mathbb{P}}{\omega'-\Delta}+z(\omega')\delta(\omega'-\Delta)\right)\mathrm{d}\Delta$$
$$= \alpha^*(\omega')\{F(\omega')+z(\omega')W^2(\omega')\}$$
$$= \alpha^*(\omega')\omega', \qquad (6.5.25)$$

using (6.5.11). Similarly, the third integral in (6.5.24) is

$$\int W(\Delta)\exp[\mathrm{i}\varphi(\Delta)]\,\beta(\omega,\Delta)\,\mathrm{d}\Delta = \alpha(\omega)\omega. \qquad (6.5.26)$$

It remains to evaluate the first integral in (6.5.24). Again using (6.5.8), we have

$$\int \Delta\beta^*(\omega',\Delta)\beta(\omega,\Delta)\,\mathrm{d}\Delta = \alpha^*(\omega')\alpha(\omega)\int \Delta W^2(\Delta) \times\left(\frac{\mathbb{P}}{\omega'-\Delta}+z(\omega')\delta(\omega'-\Delta)\right)\left(\frac{\mathbb{P}}{\omega-\Delta}+z(\omega)\delta(\omega-\Delta)\right)\mathrm{d}\Delta. \qquad (6.5.27)$$

The same procedure as before is used to resolve the product of the principal part terms. Using (6.5.15) and the fact that

$$\frac{\Delta}{\omega-\Delta}-\frac{\Delta}{\omega'-\Delta}=\frac{\omega-(\omega-\Delta)}{\omega-\Delta}-\frac{\omega'-(\omega'-\Delta)}{\omega'-\Delta}=\frac{\omega}{\omega-\Delta}-\frac{\omega'}{\omega'-\Delta}, \qquad (6.5.28)$$

we find

$$I(\omega,\omega') = \alpha^*(\omega')\alpha(\omega)\left(\frac{\mathbb{P}}{\omega'-\omega}[\omega F(\omega)-\omega' F(\omega') + \omega z(\omega)W^2(\omega)-\omega' z(\omega')W^2(\omega')]+\omega+\omega' + \omega[\pi^2+z^2(\omega)]W^2(\omega)\delta(\omega'-\omega)\right). \qquad (6.5.29)$$

Eliminating $z(\omega)$ and $z(\omega')$ by means of (6.5.11), all the terms except for that containing the delta function cancel, and from (6.5.17) we see that the remaining term is

$$I(\omega,\omega') = \omega\delta(\omega'-\omega). \qquad (6.5.30)$$

Finally, the Hamiltonian in (6.5.23) becomes

$$\hat{H}_\mathrm{I} = \hbar\int \omega|\omega\rangle\langle\omega|\,\mathrm{d}\omega. \qquad (6.5.31)$$

The fact that the Hamiltonian has this form and that the normalization of the bare states $|1\rangle$ and $|\Delta\rangle$, when expressed in terms of the dressed states, is maintained constitutes a demonstration of the completeness of the set of dressed states $|\omega\rangle$.

Now that we have obtained the complete set of dressed states $|\omega\rangle$, we can expand the state $|\psi_{\mathrm{I}}(t)\rangle$ in terms of these dressed states. The time dependence of the coefficient of the dressed state $|\omega\rangle$ in this expansion is simply $\exp(-\mathrm{i}\,\omega t)$. In particular, if $|\psi_{\mathrm{I}}(0)\rangle = |\psi(0)\rangle = |1\rangle$, then the time-evolved state is, using (6.5.20),

$$|\psi_{\mathrm{I}}(t)\rangle = \int \alpha^*(\omega)\exp(-\mathrm{i}\,\omega t)\,|\omega\rangle\,\mathrm{d}\omega. \tag{6.5.32}$$

The evolution of any observable quantity can now be calculated, the specific dynamics depending on the form of the coupling $W(\Delta)$. Before considering special cases of $W(\Delta)$, we derive general expressions for the probability $P_1(t)$ of remaining in the bare state $|1\rangle$, and the probability density $P_\Delta(\infty)$ for the excitation of the bare continuum state $|\Delta\rangle$ at long times. From (6.5.32),

$$P_1(t) = |\langle 1\,|\,\psi_{\mathrm{I}}(t)\rangle|^2 = \left|\int |\alpha(\omega)|^2 \exp(-\mathrm{i}\,\omega t)\,\mathrm{d}\omega\right|^2. \tag{6.5.33}$$

As t increases, $P_1(t)$ will eventually decay to zero owing to the dephasing of contributions arising from the spectrum of dressed states. Similarly, the time-dependent bare continuum state probability density is, using (6.5.22),

$$P_\Delta(t) = |\langle \Delta\,|\,\psi_{\mathrm{I}}(t)\rangle|^2 = \left|\int \alpha^*(\omega)\beta(\omega,\Delta)\exp(-\mathrm{i}\,\omega t)\,\mathrm{d}\omega\right|^2. \tag{6.5.34}$$

Substituting the expression (6.5.8) for $\beta(\omega,\Delta)$ in terms of $\alpha(\omega)$ gives

$$P_\Delta(t) = W^2(\Delta)\left|\mathrm{P}\int \frac{|\alpha(\omega)|^2}{\omega-\Delta}\exp(-\mathrm{i}\,\omega t)\,\mathrm{d}\omega + z(\Delta)\,|\alpha(\Delta)|^2\exp(-\,\mathrm{i}\Delta t)\right|^2. \tag{6.5.35}$$

The integral in (6.5.35) will contain exponentially decaying contributions arising from poles of the analytic continuation of $|\alpha(\omega)|^2$ in the lower half plane (see Appendix 8). These contributions will tend to zero as t tends to infinity. There is, however, an additional term $-\pi\mathrm{i}\,|\alpha(\Delta)|^2\exp(-\mathrm{i}\,\omega t)$ arising from the residue at the pole at $\omega = \Delta$ on the real axis. Hence, in the limit of large t,

$$\begin{aligned} P_\Delta(\infty) &= W^2(\Delta)|-\pi\mathrm{i}\,|\alpha(\Delta)|^2\exp(-\mathrm{i}\Delta t) + z(\Delta)\,|\alpha(\Delta)|^2\exp(-\mathrm{i}\Delta t)|^2 \\ &= W^2(\Delta)\,|\alpha(\Delta)|^4\,[\pi^2 + z^2(\Delta)]. \end{aligned} \tag{6.5.36}$$

Using the general expression for $|\alpha(\omega)|^2$ in (6.5.17), we find

$$P_\Delta(\infty) = |\alpha(\Delta)|^2 = \frac{W^2(\Delta)}{[\Delta - F(\Delta)]^2 + \pi^2 W^4(\Delta)}. \tag{6.5.37}$$

This is precisely the result (5.4.8) found using the final value theorem. The probability for excitation to state $|\Delta\rangle$ is simply the probability that the initial state corresponded to a dressed state with energy $\omega=\Delta$. This is a natural consequence of the conservation of energy in the system. We note that if $W^2(\Delta)$ is a slowly varying function of Δ, then a maximum of $P_\Delta(\infty)$ will occur at a frequency shifted, due to $F(\Delta)$, from the bare energy $\Delta=0$ of state $|1\rangle$. The width of this spectral peak is determined by $W^2(\Delta)$. For other couplings, it may be more difficult to assign a simple interpretation to $F(\Delta)$.

Choosing the form of $W^2(\Delta)$ determines the principal part integral $F(\omega)$ in (6.5.10) and hence $\alpha(\omega)$ in (6.5.18). Once these quantities are found the time-evolved state is known from (6.5.32). Two important cases amenable to analytic solution are the flat continuum (for which $W(\omega)$ is a constant, W), and the Lorentzian continuum. In both of these, the continuum is unbounded, with the frequency ω extending from $-\infty$ to $+\infty$.

For the flat continuum the principal part integral $F(\omega)$ is identically zero giving

$$\alpha(\omega)=\frac{W}{\omega-\mathrm{i}\pi W^2}. \tag{6.5.38}$$

From (6.5.33), the probability of remaining in the bare state $|1\rangle$ is

$$P_1(t)=\left|\int_{-\infty}^{\infty}\frac{W^2}{\omega^2+\pi^2W^4}\exp(-\mathrm{i}\,\omega t)\,\mathrm{d}\omega\right|^2. \tag{6.5.39}$$

In Appendix 8, we evaluate the integral in (6.5.39) by finding the residue of the integrand at the simple pole at $\omega=-\mathrm{i}\pi W^2$ in the lower half plane. This leads to

$$P_1(t)=\exp(-2\pi W^2 t), \tag{6.5.40}$$

which we recognize as the Weisskopf–Wigner formula (5.3.9). The final state spectrum $P_\Delta(\infty)$ in (6.5.37) is the Lorentzian

$$P_\Delta(\infty)=\frac{W^2}{\Delta^2+\pi^2W^4}. \tag{6.5.41}$$

Both this and (6.5.40) are characteristic of the decay of a single excited state coupled to a broad continuum. Features of this type are found in models of spontaneous emission and of the photonionization of bound states.

If the coupling is not constant then some regions of the continuum are more strongly coupled to the discrete state than others. In the simplest case, the coupling has the Lorentzian form

$$W^2(\omega)=\frac{W^2\gamma}{\pi(\omega^2+\gamma^2)}, \tag{6.5.42}$$

where W and γ are constants characterizing the strength of the coupling and the width of the effective continuum. We note that the dimension of W in (6.5.42) is frequency, whereas in the flat continuum case, W^2 had the dimension of frequency. This model is exactly soluble for any choice of peak frequency in the Lorentzian $W^2(\omega)$, but for brevity we have chosen it to coincide with the energy of the bare state $|1\rangle$ (zero in this case). As we have seen above, a single discrete state coupled to a flat continuum is diagonalized as a Lorentzian continuum of dressed states. Coupling a discrete state to a Lorentzian continuum is therefore equivalent to excitation from this state to a flat continuum via an intermediate discrete state, where the interaction between the continuum and the intermediate state has already been diagonalized. There are many situations in which this type of coupling occurs, in particular two-photon ionization, as described in Section 5.3. The principal part integral $F(\omega)$ in (6.5.10) is

$$F(\omega) = \frac{W^2\gamma}{\pi} \mathbb{P} \int_{-\infty}^{\infty} \frac{d\Delta}{(\Delta^2 + \gamma^2)(\omega - \Delta)}. \tag{6.5.43}$$

We show in Appendix 8 that this integral can be evaluated by contour integration, closing the contour in the upper half plane, to give

$$F(\omega) = \frac{W^2\omega}{\omega^2 + \gamma^2}. \tag{6.5.44}$$

Hence, from (6.5.17),

$$|\alpha(\omega)|^2 = \frac{W^2\gamma}{\pi} \cdot \frac{1}{(\omega^2 - W^2)^2 + \gamma^2\omega^2}, \tag{6.5.45}$$

and from (6.5.33), the probability of remaining in the bare state $|1\rangle$ is

$$P_1(t) = \frac{W^4\gamma^2}{\pi^2} \left| \int_{-\infty}^{\infty} \frac{\exp(-i\omega t)}{(\omega^2 - W^2)^2 + \gamma^2\omega^2} d\omega \right|^2. \tag{6.5.46}$$

Using contour integration and closing the contour in the lower half plane gives

$$P_1(t) = W^4\gamma^2 \left| \frac{\exp(-i\omega_+ t)}{\omega_+[2(\omega_+^2 - W^2) + \gamma^2]} + \frac{\exp(-i\omega_- t)}{\omega_-[2(\omega_-^2 - W^2) + \gamma^2]} \right|^2, \tag{6.5.47}$$

where ω_+ and ω_- are the positions of the poles of (6.5.45) in the lower half plane given by

$$\omega_\pm = \tfrac{1}{2}[-i\gamma \pm \sqrt{(4W^2 - \gamma^2)}]. \tag{6.5.48}$$

Expressions (6.5.47) and (6.5.48) are general and agree with (5.3.24) if we identify γ with Γ and $2W$ with V. It is instructive to consider two limiting regimes of behaviour corresponding to large or small values of the ratio W/γ. If $W \gg \gamma$, then $\omega_{\pm} \simeq -i\gamma/2 \pm W$, and neglecting terms of order γ/W in (6.5.47) gives

$$P_1(t) \simeq \exp(-\gamma t)\cos^2 Wt, \tag{6.5.49}$$

which is the weak damping limit of (5.3.24) obtained for a discrete level coupled to a flat continuum via an intermediate level. In this case, the sharply peaked coupling leads to partially reversible transitions between the discrete state and the continuum, giving rise to damped Rabi oscillations. The coupling strength W is much greater than the effective width γ of the continuum and transitions to and from the continuum can therefore occur in a time which is short compared with the dephasing time (proportional to γ^{-1}). All the population will eventually decay into the continuum, as noted previously, but only after many Rabi periods. The final state spectrum $P_\Delta(\infty)$ is $|\alpha(\Delta)|^2$, the form of which is given in (6.5.45). In this strong coupling limit, $P_\Delta(\infty)$ can be approximated by

$$P_\Delta(\infty) \simeq \frac{W^2\gamma}{\pi}\,\frac{1}{\left[(\Delta+W)^2+\gamma^2/4\right]\left[(\Delta-W)^2+\gamma^2/4\right]}. \tag{6.5.50}$$

This Autler–Townes spectrum has two narrow peaks displaced from the bare discrete state energy (zero) by $\pm W$. These peaks correspond to the dressed states of the coupled two-level system described in Section 6.2 and the final state spectrum has the form characteristic of two discrete states embedded in a flat continuum.

In the weak coupling limit where $W \ll \gamma$, ω_+ and ω_- are purely imaginary leading, in (6.5.47), to irreversible decay of the discrete state population. Expanding (6.5.48) in powers of W/γ gives the approximate forms

$$\omega_+ \simeq -iW^2/\gamma, \quad \omega_- \simeq -i\gamma. \tag{6.5.51}$$

Inserting these into (6.5.47) and neglecting terms of order W/γ, we find

$$P_1(t) \simeq \exp(-2W^2 t/\gamma). \tag{6.5.52}$$

In this case, the coupling strength W is much less than the effective width γ of the continuum. Therefore, transitions to the continuum occur in a time which is long compared with the dephasing time. These are precisely the conditions under which the Weisskopf–Wigner approximation holds, leading to an exponential decay of the discrete state population. In this weak coupling limit, the final state spectrum in (6.5.37) can be approximated by

$$P_\Delta(\infty) \simeq \frac{W^2\gamma}{\pi}\,\frac{1}{(\Delta^2+\gamma^2)(\Delta^2+W^4/\gamma^2)}. \tag{6.5.53}$$

This spectrum has a single peak, centred at $\Delta = 0$, of width W^2/γ arising from the second Lorentzian factor in (6.5.53). The first Lorentzian factor provides a broad, slowly varying background. The essentially Lorentzian character of this spectrum is once again indicative of the decay of a single discrete state into a broad continuum.

In the above analysis, we considered a single discrete state $|1\rangle$ coupled to a continuum of states $|\Delta\rangle$. The Fano technique can also be applied to other, more complicated problems, and we illustrate this by diagonalizing a system in which a finite number of non-degenerate states $|n\rangle$ with energies $\hbar\omega_n$ are coupled to a single continuum but are not directly coupled to each other. This system can be diagonalized exactly for any choice of couplings. However, for simplicity, we choose the coupling $W_n \exp(-i\varphi_n)$ of state $|n\rangle$ to continuum state $|\Delta\rangle$ to be independent of Δ. The interaction picture Hamiltonian for this system is

$$\hat{H}_I = \hbar \sum_n \omega_n |n\rangle\langle n| + \hbar \int \Delta |\Delta\rangle\langle\Delta| \mathrm{d}\Delta$$

$$+ \hbar \sum_n \int W_n [|\Delta\rangle\langle n| \exp(-i\varphi_n) + |n\rangle\langle\Delta| \exp(i\varphi_n)] \mathrm{d}\Delta. \quad (6.5.54)$$

As before, we seek dressed states $|\omega\rangle$ expressed as superpositions of the bare states in the form

$$|\omega\rangle = \sum_n \alpha_n(\omega)|n\rangle + \int \beta(\omega,\Delta)|\Delta\rangle \mathrm{d}\Delta. \quad (6.5.55)$$

The equations for the coefficients $\alpha_n(\omega)$ and $\beta(\omega,\Delta)$ are found as before by comparing coefficients of the bare states in the eigenvalue equation (6.5.2). We obtain the following generalizations of (6.5.6) and (6.5.7):

$$\Delta\beta(\omega,\Delta) + \sum_n \alpha_n(\omega) W_n \exp(-i\varphi_n) = \omega\beta(\omega,\Delta), \quad (6.5.56)$$

$$\int W_n \exp(i\varphi_n)\beta(\omega,\Delta)\mathrm{d}\Delta = (\omega - \omega_n)\alpha_n(\omega). \quad (6.5.57)$$

The solution of (6.5.56) is, as in (6.5.8),

$$\beta(\omega,\Delta) = \left(\frac{\mathbb{P}}{\omega - \Delta} + z(\omega)\delta(\omega - \Delta)\right) \sum_m \alpha_m(\omega) W_m \exp(-i\varphi_m). \quad (6.5.58)$$

On substituting (6.5.58) into (6.5.57), we note that the principal part integral is zero because the couplings W_n have been chosen to be independent of frequency. Hence

$$z(\omega) W_n \exp(i\varphi_n) \sum_m \alpha_m(\omega) W_m \exp(-i\varphi_m) = (\omega - \omega_n)\alpha_n(\omega). \quad (6.5.59)$$

The analogous equation in the treatment of a single discrete state coupled to a continuum gave the form of $z(\omega)$ directly. Here, however, in order to find an expression for $z(\omega)$ which is independent of the $\alpha_n(\omega)$, we multiply (6.5.59) by $W_n \exp(-i\varphi_n)/(\omega - \omega_n)$ and then sum over n. The summations containing the α_n cancel, leaving

$$z(\omega) = \left(\sum_n \frac{W_n^2}{\omega - \omega_n} \right)^{-1}. \tag{6.5.60}$$

It only remains to find $\alpha_n(\omega)$. We do this by applying the normalization condition (6.5.12) to the dressed states, giving

$$\sum_n \alpha_n^*(\omega')\alpha_n(\omega) + \int \beta^*(\omega', \Delta)\beta(\omega, \Delta)\,\mathrm{d}\Delta = \delta(\omega' - \omega). \tag{6.5.61}$$

Evaluating the integral in (6.5.61), treating the product of the principal parts as before, we find

$$\begin{aligned}
&\sum_n \alpha_n^*(\omega')\alpha_n(\omega) + \sum_n \alpha_n^*(\omega')W_n \exp(i\varphi_n) \sum_m \alpha_m(\omega)W_m \exp(-i\varphi_n) \\
&\quad \times \left(\frac{z(\omega') - z(\omega)}{\omega - \omega'} \right) + \left| \sum_n \alpha_n(\omega)W_n \exp(-i\varphi_n) \right|^2 \\
&\quad \times [\pi^2 + z^2(\omega)]\delta(\omega' - \omega) = \delta(\omega' - \omega).
\end{aligned} \tag{6.5.62}$$

Using (6.5.59) to express $\alpha_n(\omega)$ in terms of $z(\omega)$, the first term of (6.5.62) becomes

$$\begin{aligned}
\sum_n \alpha_n^*(\omega')\alpha_n(\omega) &= z(\omega')z(\omega) \sum_n \frac{W_n^2}{(\omega' - \omega_n)(\omega - \omega_n)} \\
&\quad \times \sum_m \alpha_m^*(\omega')W_m \exp(i\varphi_m) \sum_n \alpha_n(\omega)W_n \exp(-i\varphi_n) \\
&= \frac{z(\omega')z(\omega)}{\omega - \omega'} \sum_n \left(\frac{W_n^2}{\omega' - \omega_n} - \frac{W_n^2}{\omega - \omega_n} \right) \\
&\quad \times \sum_m \alpha_m^*(\omega')W_m \exp(i\varphi_m) \sum_n \alpha_n(\omega)W_n \exp(-i\varphi_n) \\
&= \frac{z(\omega')z(\omega)}{\omega - \omega'} \left(\frac{1}{z(\omega')} - \frac{1}{z(\omega)} \right) \sum_m \alpha_m^*(\omega')W_m \exp(i\varphi_m) \\
&\quad \times \sum_n \alpha_n(\omega)W_n \exp(-i\varphi_n).
\end{aligned} \tag{6.5.63}$$

We see that this term cancels the second term of (6.5.62). The normalization condition is therefore satisfied if

$$\left|\sum_n \alpha_n(\omega) W_n \exp(-\mathrm{i}\varphi_n)\right|^2 = \frac{1}{\pi^2 + z^2(\omega)}. \tag{6.5.64}$$

There is again an arbitrary phase which we choose so that

$$\sum_n \alpha_n(\omega) W_n \exp(-\mathrm{i}\varphi_n) = \frac{1}{\pi + \mathrm{i}z(\omega)}. \tag{6.5.65}$$

Now that we have an explicit expression for this summation and for $z(\omega)$, (6.5.59) determines the form of the $\alpha_n(\omega)$ as

$$\alpha_n(\omega) = \frac{W_n \exp(\mathrm{i}\varphi_n) z(\omega)}{(\omega - \omega_n)[\pi + \mathrm{i}z(\omega)]}, \tag{6.5.66}$$

and (6.5.58) then determines the form of $\beta(\omega, \Delta)$.

Solving for the dynamics of the system simply requires expanding the initial state in terms of the complete set of dressed states. Consider for example the initial condition where $|\psi(0)\rangle$ is equal to one of the discrete states, $|m\rangle$, say. Then the time-evolved state is

$$|\psi_1(t)\rangle = \int \alpha_m^*(\omega) \exp(-\mathrm{i}\omega t) |\omega\rangle \,\mathrm{d}\omega. \tag{6.5.67}$$

The probability for remaining in the initial state at time t is

$$P_m(t) = \left|\int |\alpha_m(\omega)|^2 \exp(-\mathrm{i}\omega t) \,\mathrm{d}\omega\right|^2. \tag{6.5.68}$$

As before, this probability will ultimately decay to zero owing to the dephasing. Similarly, the probability that the system is found in one of the other discrete states $|n\rangle$ is

$$P_n(t) = \left|\int \alpha_n(\omega) \alpha_m^*(\omega) \exp(-\mathrm{i}\omega t) \,\mathrm{d}\omega\right|^2. \tag{6.5.69}$$

This probability is, of course, initially zero. However, it will subsequently be non-zero owing to transitions between the discrete states via the continuum associated with off-diagonal damping as discussed in Section 5.3. It will ultimately tend to zero owing to dephasing. It is worth emphasizing that this dephasing leads to complete transfer of population to the bare continuum states because we have assumed that the discrete states are non-degenerate. If the initial state $|m\rangle$ is degenerate with one or more of the other discrete states, then population will be trapped in a superposition of these discrete states.

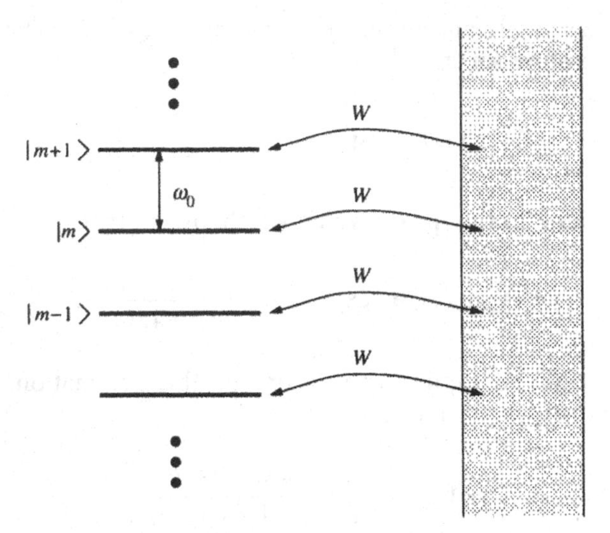

Fig. 6.4 A quasi-continuum coupled to a true continuum.

The final state spectrum is found by the method used to obtain (6.5.37), or by simply invoking energy conservation, to be

$$P_{\Delta}(\infty) = |\alpha_m(\Delta)|^2. \tag{6.5.70}$$

As an illustration, we explicitly calculate this spectrum when there are a large number of equally spaced discrete states with energy separation between adjacent states equal to $\hbar\omega_0$. We take the initially occupied state $|m\rangle$ of energy $m\hbar\omega_0$ to lie far from the discrete states with the lowest and highest energies and, for simplicity, take the couplings $W_n \exp(-\mathrm{i}\varphi_n)$ to be equal to a real constant W (see Fig. 6.4). We first find $z(\omega)$ which, from (6.5.60), is given by

$$z(\omega) = \frac{1}{W^2}\left(\sum_n \frac{1}{(\omega - n\omega_0)}\right)^{-1}. \tag{6.5.71}$$

In calculating the spectrum, we will be interested in energies close to the energy of the initially occupied discrete state $|m\rangle$. Since this has been chosen to lie far from the extreme states of the discrete set, we can approximate $z(\omega)$ by extending the limits of the summation over n in (6.5.71) from $-\infty$ to $+\infty$. In Appendix 8 we evaluate this summation and, on substituting the result into (6.5.71), obtain

$$z(\omega) = \frac{\omega_0}{\pi W^2}\tan\left(\frac{\pi\omega}{\omega_0}\right). \tag{6.5.72}$$

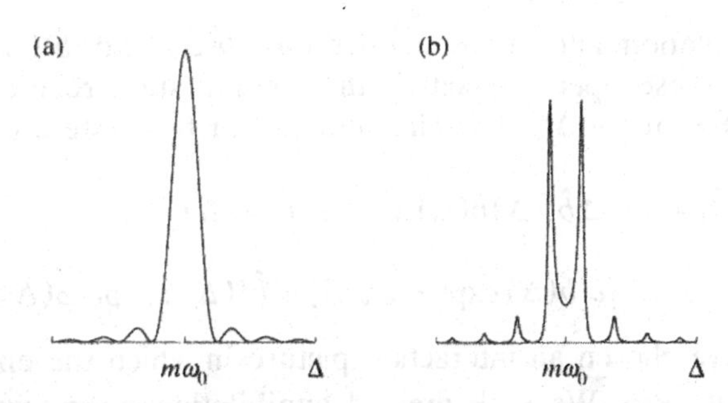

Fig. 6.5 The final state spectrum for a quasi-continuum coupled to a true continuum with state $|m\rangle$ initially occupied for (a) $\pi^2W^2/\omega_0 = 1$ and (b) $\pi^2W^2/\omega_0 = 4$.

From (6.5.66) and (6.5.70), the final state spectrum is

$$P_\Delta(\infty) = \frac{W^2}{(\Delta - m\omega_0)^2\left[(\pi^4W^4/\omega_0^2)\cot^2(\pi\Delta/\omega_0) + 1\right]}. \tag{6.5.73}$$

This spcctrum is plotted in Fig. 6.5 for $\pi^2W^2/\omega_0 = 1$ and for $\pi^2W^2/\omega_0 = 4$. If π^2W^2/ω_0 is less than $\sqrt{(3/2)}$, the spectrum consists of a principal peak at $\Delta = m\omega_0$ and subsidiary peaks separated from each other by zeros corresponding to the positions of the initially unoccupied discrete states. If π^2W^2/ω_0 is greater than $\sqrt{(3/2)}$, then the principal peak splits. This feature is similar to the Autler–Townes doublet shown in Fig. 6.1 in that it is a signature of damped Rabi oscillations between $|m\rangle$ and the other discrete states.

6.6 Dressed operators

So far in this chapter we have been concerned with dressed states, that is eigenstates of the full Hamiltonian. We have used these to study the dynamics of systems in the Schrödinger picture. A related technique suited to the Heisenberg picture involves dressed operators obeying eigenoperator equations. We can then express operators representing observables of interest in terms of the complete set of dressed operators having simple dynamics. Often it will be difficult to find the dressed operators but where this is possible, it can be a useful technique. In this section, we consider two examples of physical interest, the first describing an imperfect cavity and the second being the Jaynes–Cummings model of quantum optical resonance.

Optical and microwave cavities are often modelled as isolated systems, the mirrors separating the field inside the cavity from the environment outside. When the cavity is imperfect but still of high quality, the field in the cavity may be accurately described in terms of weakly damped quasi-modes. These quasi-modes correspond to the modes of a perfect cavity of the same dimensions. Consider first a single quantized cavity quasi-mode having annihilation and creation operators $\hat{a}$ and $\hat{a}^\dagger$, coupled to a continuum of external modes, having annihilation and creation operators $\hat{b}(\Delta)$ and $\hat{b}^{\dagger}(\Delta)$. In Section 3.2, we saw that these operators satisfy the commutation relations $[\hat{a}, \hat{a}^{\dagger}] = 1$ and $[\hat{b}(\Delta), \hat{b}^{\dagger}(\Delta')] = \delta(\Delta - \Delta')$. The Hamiltonian for this system is (see (5.5.1))

$$\hat{H}_{\mathrm{I}} = \hbar \int \Delta \hat{b}^{\dagger}(\Delta)\hat{b}(\Delta)\,\mathrm{d}\Delta + \hbar \int W(\Delta) \times \left\{\hat{a}^{\dagger}\hat{b}(\Delta)\exp[-\mathrm{i}\varphi(\Delta)] + \hat{b}^{\dagger}(\Delta)\hat{a}\exp[\mathrm{i}\varphi(\Delta)]\right\}\mathrm{d}\Delta, \tag{6.6.1}$$

where we have chosen an interaction picture in which the energy of the single quasi-mode is zero. We seek dressed annihilation operators $\hat{A}(\omega)$ which we define by the requirement that they satisfy the eigenoperator equation

$$\left[\hat{A}(\omega), \hat{H}_{\mathrm{I}}\right] = \hbar\,\omega \hat{A}(\omega). \tag{6.6.2}$$

As in the case of dressed states, this requirement ensures that the time dependence of the dressed operator $\hat{A}(\omega)$ is simply $\exp(-\mathrm{i}\,\omega t)$. Expressing $\hat{A}(\omega)$ in terms of the bare annihilation operators as

$$\hat{A}(\omega) = \alpha(\omega)\hat{a} + \int \beta(\omega, \Delta)\hat{b}(\Delta)\,\mathrm{d}\Delta, \tag{6.6.3}$$

we find from (6.6.1) and (6.6.2) that

$$\int \Delta\beta(\omega, \Delta)\hat{b}(\Delta)\,\mathrm{d}\Delta + \alpha(\omega)\int W(\Delta)\exp[-\mathrm{i}\varphi(\Delta)]\hat{b}(\Delta)\,\mathrm{d}\Delta + \hat{a}\int W(\Delta) \times \exp[\mathrm{i}\varphi(\Delta)]\,\beta(\omega, \Delta)\,\mathrm{d}\Delta = \omega\alpha(\omega)\hat{a} + \omega\int \beta(\omega, \Delta)\hat{b}(\Delta)\,\mathrm{d}\Delta. \tag{6.6.4}$$

Comparing coefficients of the bare annihilation operators in (6.6.4) gives

$$\Delta\beta(\omega, \Delta) + \alpha(\omega)W(\Delta)\exp[-\mathrm{i}\varphi(\Delta)] = \omega\beta(\omega, \Delta) \tag{6.6.5}$$

and

$$\int W(\Delta)\beta(\omega, \Delta)\exp[\mathrm{i}\varphi(\Delta)]\,\mathrm{d}\Delta = \omega\alpha(\omega). \tag{6.6.6}$$

These equations are precisely those derived in the previous section for a single discrete state coupled to a continuum. The analysis follows that of Section 6.5 exactly, except that the orthonormality condition (6.5.12) is replaced by the dressed operator commutator

$$[\hat{A}(\omega), \hat{A}^{\dagger}(\omega')] = \delta(\omega - \omega'). \tag{6.6.7}$$

The bare quasi-mode and continuum annihilation operators can then be expressed as

$$\hat{a} = \int \alpha^{*}(\omega)\hat{A}(\omega)\,\mathrm{d}\omega \tag{6.6.8}$$

and

$$\hat{b}(\Delta) = \int \beta^{*}(\omega, \Delta)\hat{A}(\omega)\,\mathrm{d}\omega, \tag{6.6.9}$$

analogous to (6.5.20) and (6.5.22), with $\alpha(\omega)$ and $\beta(\omega,\Delta)$ given by (6.5.8) and (6.5.18). The time-evolved form of these operators has a factor $\exp(-\mathrm{i}\omega t)$ multiplying $\hat{A}(\omega)$ within the integrals in (6.6.8) and (6.6.9). It is straightforward to verify that the commutators of $\hat{a}$ and $\hat{a}^\dagger$ and of $\hat{b}(\Delta)$ and $\hat{b}^\dagger(\Delta')$ are maintained, and that the Hamiltonian can be written as

$$\hat{H}_1 = \hbar \int \omega \hat{A}^\dagger(\omega)\hat{A}(\omega)\,\mathrm{d}\omega. \tag{6.6.10}$$

As in the dressed state analysis, these constitute a demonstration of the completeness of the set of dressed annihilation operators $\hat{A}(\omega)$.

We now turn our attention to treating a number of quasi-modes, with annihilation and creation operators $\hat{a}_n$ and $\hat{a}_n^\dagger$ and energies $n\hbar\omega_0$, such as the plane-wave modes of an imperfect Fabry–Perot cavity with fundamental frequency $\omega_0 = \pi c/L$. The Hamiltonian for this system is

$$\begin{aligned}\hat{H} = \hbar \sum_n n\omega_0 \hat{a}_n^\dagger \hat{a}_n + \hbar \int \Delta \hat{b}^\dagger(\Delta)\hat{b}(\Delta)\,\mathrm{d}\Delta \\ + \hbar \sum_n \int W_n(\Delta)\{\hat{a}_n^\dagger \hat{b}(\Delta)\exp[-\mathrm{i}\varphi_n(\Delta)] + \hat{b}^\dagger(\Delta)\hat{a}_n \exp[\mathrm{i}\varphi(\Delta)]\}\,\mathrm{d}\Delta,\end{aligned} \tag{6.6.11}$$

which is analogous to (6.5.54). The summations in (6.6.11) are over a large but finite number of modes. Above some high frequency ω_c, say, the mirrors become transparent and consequently we omit frequencies above ω_c from the analysis. It is possible to find a complete set of dressed annihilation operators for this system with arbitrary couplings, but for brevity we restrict our attention to the special case where $W_n(\Delta)\exp[-\mathrm{i}\varphi_n(\Delta)]$ is a real constant W. Physically this choice corresponds to a mirror reflectivity which is independent of frequency in the range of interest. As before, we seek dressed annihilation operators $\hat{A}(\omega)$ now expressed in the form

$$\hat{A}(\omega) = \sum_n \alpha_n(\omega)\hat{a}_n + \int \beta(\omega,\Delta)\hat{b}(\Delta)\,\mathrm{d}\Delta. \tag{6.6.12}$$

The eigenoperator equation $[\hat{A}(\omega), \hat{H}] = \hbar\omega\hat{A}(\omega)$ leads to

$$\Delta\beta(\omega,\Delta) + W\sum_n \alpha_n(\omega) = \omega\beta(\omega,\Delta), \tag{6.6.13}$$

$$W\int \beta(\omega,\Delta)\,\mathrm{d}\Delta = (\omega - n\omega_0)\alpha_n(\omega). \tag{6.6.14}$$

These equations are a special case of (6.5.56) and (6.5.57) and are solved in the same way. We find

$$\beta(\omega,\Delta) = \left(\frac{\mathbb{P}}{\omega-\Delta} + z(\omega)\delta(\omega-\Delta)\right) W \sum_m \alpha_m(\omega) \tag{6.6.15}$$

and

$$\alpha_n(\omega) = \frac{Wz(\omega)}{(\omega - n\omega_0)[\pi + \mathrm{i}z(\omega)]}, \tag{6.6.16}$$

where $z(\omega)$ is given by (6.5.71). If we restrict our attention to frequencies much greater than the fundamental cavity frequency ω_0 but much less than ω_c then we can use the result (6.5.72) for $z(\omega)$. In terms of the dressed operators, the quasi-mode annihilation operators can be expressed as

$$\hat{a}_n = \int \alpha_n^*(\omega)\hat{A}(\omega)\,\mathrm{d}\omega = \int \frac{Wz(\omega)\hat{A}(\omega)}{(\omega - n\omega_0)[\pi - \mathrm{i}z(\omega)]}\,\mathrm{d}\omega. \tag{6.6.17}$$

In the quasi-mode approach the ideal cavity mode functions and the operators $\hat{a}_n$ are used to expand the positive frequency component of the quantized intracavity field as

$$\begin{aligned}\hat{E}^{(+)}(x,t) &= \mathrm{i}\sum_n \left(\frac{\hbar n\omega_0}{\epsilon_0 \mathscr{V}}\right)^{1/2} \sin(n\omega_0 x/c)\hat{a}_n(t)\\ &= \mathrm{i}\sum_n \left(\frac{\hbar n\omega_0}{\epsilon_0 \mathscr{V}}\right)^{1/2} \sin(n\omega_0 x/c)\int \frac{Wz(\omega)\hat{A}(\omega)\exp(-\mathrm{i}\omega t)}{(\omega - n\omega_0)[\pi - \mathrm{i}z(\omega)]}\,\mathrm{d}\omega,\end{aligned} \tag{6.6.18}$$

where $\mathscr{V}$ is the effective mode volume. We use this expression to calculate the spectrum $S(\Omega)$ of the vacuum electric field fluctuations averaged over the cavity length given by

$$S(\Omega) = \frac{1}{2\pi L}\int_0^L \mathrm{d}x \int_{-\infty}^{\infty} \mathrm{d}\tau \,\langle 0|\,\hat{E}^{(-)}(x,t+\tau)\hat{E}^{(+)}(x,t)\,|0\rangle \exp(-\mathrm{i}\Omega\tau). \tag{6.6.19}$$

Using the fact that the mode functions are orthogonal and that the vacuum expectation value $\langle 0|\,\hat{A}(\omega)\hat{A}^\dagger(\omega')\,|0\rangle$ is $\delta(\omega - \omega')$, this spectrum becomes

$$S(\Omega) = \frac{\hbar\omega_0}{2\epsilon_0\mathscr{V}} \frac{W^2 z^2(\Omega)}{[\pi^2 + z^2(\Omega)]} \sum_n \frac{n}{(\Omega - n\omega_0)^2}. \tag{6.6.20}$$

Extending the summation in (6.5.20) from $n = -\infty$ to $n = +\infty$, as we did in calculating $z(\omega)$, allows us to evaluate $S(\Omega)$ (see Appendix 8). The spectrum is

$$\begin{aligned}S(\Omega) &= \frac{\pi\hbar\omega_0}{2\epsilon_0\mathscr{V}} \frac{W^2}{[\pi^4 W^4 \cot^2(\pi\Omega/\omega_0) + \omega_0^2]}\left[\frac{\pi\Omega}{\omega_0}\operatorname{cosec}^2\left(\frac{\pi\Omega}{\omega_0}\right) - \cot\left(\frac{\pi\Omega}{\omega_0}\right)\right]\\ &= \frac{\pi\hbar\omega_0}{2\epsilon_0\mathscr{V}} \frac{W^2}{[\pi^4 W^4 \cos^2(\pi\Omega/\omega_0) + \omega_0^2 \sin^2(\pi\Omega/\omega_0)]}\\ &\quad\times\left[\frac{\pi\Omega}{\omega_0} - \frac{1}{2}\sin\left(\frac{2\pi\Omega}{\omega_0}\right)\right].\end{aligned} \tag{6.6.21}$$

Since we have assumed that $\Omega \gg \omega_0$, consistency requires that the term $\frac{1}{2}\sin(2\pi\Omega/\omega_0)$ in (6.6.21) is neglected compared with $\pi\Omega/\omega_0$. The final form of the spectrum is

$$S(\Omega) = \frac{\hbar\Omega}{2\epsilon_0\mathscr{V}} \frac{\pi^2 W^2}{\left[\pi^4 W^4 + (\omega_0^2 - \pi^4 W^4)\sin^2(\pi\Omega/\omega_0)\right]}$$

$$= \frac{\hbar\Omega}{2\epsilon_0\mathscr{V}} \frac{W^2}{[\pi^2 W^4 + (c^2/L^2)\sin^2(\Omega L/c)]}, \tag{6.6.22}$$

where we have substituted for ω_0 and used the fact that, for a high-quality cavity, $\omega_0 \gg \pi^2 \mathrm{W}^2$. We recognize (6.6.22) as the free-space vacuum field strength for a mode of frequency Ω, multiplied by the familiar Fabry–Perot mode pattern.

In Section 2.4, we studied the Jaynes–Cummings model of quantum optical resonance in which a two-level atom interacts with a single cavity mode. We found that the atom exhibits Rabi oscillations with a Rabi frequency dependent on the number of quanta in the coupled atom–field system. For fields with a distribution of photon numbers, the oscillations beat against one another leading to collapses and revivals of the Rabi oscillations. In Section 6.2, we found the dressed states for the model which comprise a pair of eigenstates for each excitation number $N > 0$, together with the single state $|1\rangle|0\rangle$. We conclude this section by deriving dressed operators for the Jaynes–Cummings model following an approach developed by Ackerhalt. We anticipate that finding these operators will be more difficult than previously because the eigenfrequencies of the dressed states depend on the total number of excitations N. We expect, therefore, that the eigenfrequencies will themselves be operators which depend on the excitation number operator

$$\hat{N} = \hat{a}^\dagger \hat{a} + \hat{\sigma}_+ \hat{\sigma}_-, \tag{6.6.23}$$

where $\hat{a}^\dagger \hat{a}$ is the photon number operator and $\hat{\sigma}_+ \hat{\sigma}_-$ is the projector onto the excited atomic state.

We recall that the Hamiltonian for the Jaynes–Cummings model is

$$\hat{H}_\mathrm{I} = \frac{\hbar\Delta}{2}\hat{\sigma}_3 - \mathrm{i}\hbar\lambda\left(\hat{\sigma}_+ \hat{a} - \hat{a}^\dagger \hat{\sigma}_-\right). \tag{6.6.24}$$

The commutators of the annihilation operator $\hat{a}$ and the lowering operator $\hat{\sigma}_-$ with $\hat{H}_\mathrm{I}$ are, using (3.2.6) and (3.2.7),

$$\left[\hat{a}, \hat{H}_\mathrm{I}\right] = \mathrm{i}\hbar\lambda\hat{\sigma}_-, \tag{6.6.25}$$

$$\left[\hat{\sigma}_-, \hat{H}_\mathrm{I}\right] = \hbar\Delta\hat{\sigma}_- + \mathrm{i}\hbar\lambda\hat{\sigma}_3 \hat{a}$$

$$= -\mathrm{i}\hbar\lambda\hat{a} - 2\hat{H}_\mathrm{I}\hat{\sigma}_-. \tag{6.6.26}$$

It is not possible to construct a dressed operator satisfying an eigenoperator equation of the form of (6.6.2) as a linear combination of $\hat{a}$ and $\hat{\sigma}_-$. This is because the commutator in (6.6.26) is not simply a linear combination of $\hat{a}$ and $\hat{\sigma}_-$. In order to derive dressed operators, we must allow the eigenfrequencies and the expansion coefficients in these operators to be operators themselves. This produces the additional complication that the ordering of these is important. We seek the two dressed operators of the form

$$\hat{A}_\pm = \hat{\alpha}_\pm \hat{a} + \hat{\beta}_\pm \hat{\sigma}_- \tag{6.6.27}$$

obeying the eigenoperator equation

$$\left[\hat{A}_\pm, \hat{H}_\mathrm{I}\right] = \hbar\hat{\omega}_\pm \hat{A}_\pm. \tag{6.6.28}$$

Evaluating the commutator in this equation requires knowledge of the commutators of the operators $\hat{\alpha}_\pm$ and $\hat{\beta}_\pm$ with $\hat{H}_\mathrm{I}$. In order to proceed, we will assume that $\hat{\alpha}_\pm$, $\hat{\beta}_\pm$, and $\hat{\omega}_\pm$ commute with $\hat{H}_\mathrm{I}$. This assumption will have to be checked to ensure the self-consistency of the solution.

We first find the eigenfrequency operators $\hat{\omega}_\pm$ by noticing that

$$\begin{aligned}\hat{H}_\mathrm{I}^2 &= \hbar^2\left(\frac{\Delta^2}{4}\hat{\sigma}_3^2 + \lambda^2\left(\hat{a}^\dagger\hat{a}\hat{\sigma}_-\hat{\sigma}_+ + \hat{a}\hat{a}^\dagger\hat{\sigma}_+\hat{\sigma}_-\right)\right)\\ &= \hbar^2\left(\frac{\Delta^2}{4} + \lambda^2\hat{N}\right),\end{aligned} \tag{6.6.29}$$

using the properties of the Pauli operators and of $\hat{a}$ and $\hat{a}^\dagger$. Hence the commutators of $\hat{a}$ and $\hat{\sigma}_-$ with $\hat{H}_\mathrm{I}^2$ are

$$\left[\hat{a}, \hat{H}_\mathrm{I}^2\right] = \hbar^2\lambda^2\hat{a}, \tag{6.6.30}$$

$$\left[\hat{\sigma}_-, \hat{H}_\mathrm{I}^2\right] = \hbar^2\lambda^2\hat{\sigma}_-, \tag{6.6.31}$$

and consequently the dressed operators $\hat{A}_\pm$ satisfy the eigenoperator equation $[\hat{A}_\pm, \hat{H}_\mathrm{I}^2] = \hbar^2\lambda^2\hat{A}_\pm$. Using the fact that

$$\begin{aligned}\left[\hat{A}_\pm, \hat{H}_\mathrm{I}^2\right] &= \hat{A}_\pm\hat{H}_\mathrm{I}^2 - \hat{H}_\mathrm{I}^2\hat{A}_\pm\\ &= \hat{A}_\pm\hat{H}_\mathrm{I}^2 - \hat{H}_\mathrm{I}\hat{A}_\pm\hat{H}_\mathrm{I} + \hat{H}_\mathrm{I}\hat{A}_\pm\hat{H}_\mathrm{I} - \hat{H}_\mathrm{I}^2\hat{A}_\pm\\ &= \left[\left[\hat{A}_\pm, \hat{H}_\mathrm{I}\right], \hat{H}_\mathrm{I}\right] + 2\hat{H}_\mathrm{I}\left[\hat{A}_\pm, \hat{H}_\mathrm{I}\right],\end{aligned} \tag{6.6.32}$$

and imposing the condition (6.6.28), gives

$$\hbar^2\hat{\omega}_\pm^2\hat{A}_\pm + 2\hat{H}_\mathrm{I}\hbar\hat{\omega}_\pm\hat{A}_\pm - \hbar^2\lambda^2\hat{A}_\pm = 0. \tag{6.6.33}$$

This equation will be satisfied if $\hat{\omega}_\pm$ obey the operator equation

$$(\hbar\hat{\omega}_\pm)^2 + 2\hat{H}_\mathrm{I}\hbar\hat{\omega}_\pm - \hbar^2\lambda^2 = 0. \tag{6.6.34}$$

Solving this as a quadratic, we find

$$\hbar\hat{\omega}_{\pm} = -\hat{H}_{\mathrm{I}} \pm \sqrt{\left(\hat{H}_{\mathrm{I}}^2 + \hbar^2\lambda^2\right)}. \tag{6.6.35}$$

The operators $\hat{\omega}_+$ and $\hat{\omega}_-$ clearly commute with $\hat{H}_{\mathrm{I}}$ as assumed earlier.

Returning to the eigenoperator equation (6.6.28) and inserting the form (6.6.27) gives

$$\begin{aligned}\left[\hat{\alpha}_{\pm}\hat{a} + \hat{\beta}_{\pm}\hat{\sigma}_-, \hat{H}_{\mathrm{I}}\right] &= \hat{\alpha}_{\pm}\mathrm{i}\hbar\lambda\hat{\sigma}_- + \hat{\beta}_{\pm}\left(-\mathrm{i}\hbar\lambda\hat{a} - 2\hat{H}_{\mathrm{I}}\hat{\sigma}_-\right)\\ &= \hbar\hat{\omega}_{\pm}\left(\hat{\alpha}_{\pm}\hat{a} + \hat{\beta}_{\pm}\hat{\sigma}_-\right),\end{aligned} \tag{6.6.36}$$

where we have used (6.6.25) and (6.6.26) together with the assumption that $\hat{\alpha}_{\pm}$ and $\hat{\beta}_{\pm}$ commute with $\hat{H}_{\mathrm{I}}$. If we compare coefficients of $\hat{a}$ and $\hat{\sigma}_-$ in (6.6.36) and use (6.6.35), we obtain the two equations

$$\mathrm{i}\hat{\alpha}_{\pm}\lambda = -\hat{\omega}_{\mp}\hat{\beta}_{\pm}, \tag{6.6.37}$$

$$\hat{\omega}_{\pm}\hat{\alpha}_{\pm} = -\mathrm{i}\lambda\hat{\beta}_{\pm}. \tag{6.6.38}$$

Consistency requires that $\hat{\omega}_+\hat{\omega}_- = -\lambda^2$, which is indeed satisfied by the form of $\hat{\omega}_{\pm}$ given in (6.6.35). From (6.6.38), the forms of the dressed operators are

$$\hat{A}_{\pm} = \hat{\alpha}_{\pm}\left(\hat{a} + \frac{\mathrm{i}}{\lambda}\hat{\omega}_{\pm}\hat{\sigma}_-\right), \tag{6.6.39}$$

where $\hat{\alpha}_{\pm}$ is any operator which commutes with $\hat{H}_{\mathrm{I}}$. It may be readily verified, using (6.6.34) and the fact that $\hat{\alpha}_{\pm}$ and $\hat{\omega}_{\pm}$ commute with $\hat{H}_{\mathrm{I}}$, that (6.6.28) is satisfied by (6.6.39) and consequently that $\hat{A}_{\pm}$ are the required dressed operators. The fact that $\hat{\beta}_{\pm}$ commutes with $\hat{H}_{\mathrm{I}}$ follows from the fact that both $\hat{\alpha}_{\pm}$ and $\hat{\omega}_{\pm}$ commute with $\hat{H}_{\mathrm{I}}$. It is interesting to ask whether there is a connection between these operators and the dressed states for the Jaynes–Cummings model found at the end of Section 6.2. We shall not explore this in detail, but we note here that there is a sense in which the operators $\hat{A}_{\pm}$ and their Hermitian conjugates $\hat{A}^{\dagger}_{\pm}$ are associated with the dressed states $|\pm, N\rangle$: $\hat{A}_{\pm}$ act as lowering operators on the states $|\pm, N\rangle$, respectively, reducing the number of excitations by one, while $\hat{A}_{\pm}|\mp, N\rangle = 0$. Further, the products $\hat{A}^{\dagger}_+\hat{A}_+$ and $\hat{A}^{\dagger}_-\hat{A}_-$ are analogous to the number operators for single field modes. In particular

$$\langle +, N|\,\hat{A}^{\dagger}_+\hat{A}_+\,|+, N\rangle = \langle +, N|\,\hat{\alpha}^{\dagger}_+\hat{\alpha}_+\,|+, N\rangle\frac{N}{\alpha_+^2(N)}, \tag{6.6.40}$$

$$\langle -, N|\,\hat{A}^{\dagger}_-\hat{A}_-\,|-, N\rangle = \langle -, N|\,\hat{\alpha}^{\dagger}_-\hat{\alpha}_-\,|-, N\rangle\frac{N}{\alpha_-^2(N)}, \tag{6.6.41}$$

where $\alpha_+(N)$ and $\alpha_-(N)$ are given by (6.2.32) and (6.2.33). If we choose the operators $\hat{\alpha}_{\pm}$ so that

$$\langle \pm, N|\,\hat{\alpha}^{\dagger}_{\pm}\hat{\alpha}_{\pm}\,|\pm, N\rangle = \alpha_{\pm}^2(N), \tag{6.6.42}$$

then $\langle +, N| \hat{A}^{\dagger}_{+}\hat{A}_{+} |+, N\rangle = \langle -, N |\hat{A}^{\dagger}_{-}\hat{A}_{-} |-, N\rangle = N$. The expectation value of $\hat{A}^{\dagger}_{+}\hat{A}_{+}$ in any state is then the mean number of excitations associated with that part of the state which may be expanded in terms of the states $\{|+, N\rangle\}$; $\hat{A}^{\dagger}_{-}\hat{A}_{-}$ is similarly associated with the states $\{|-, N\rangle\}$.

The operators $\hat{a}$ and $\hat{\sigma}_{-}$ may now be expressed in terms of the dressed operators $\hat{A}_{\pm}$. At time $t = 0$ we write

$$\hat{a} = \hat{k}_{+}\hat{A}_{+} + \hat{k}_{-}\hat{A}_{-}, \tag{6.6.43}$$

where we assume that $\hat{k}_{\pm}$ commute with $\hat{H}_{\mathrm{I}}$. Substituting for $\hat{A}_{\pm}$ from (6.6.39) gives

$$\hat{a} = \hat{k}_{+}\hat{\alpha}_{+}\left(\hat{a} + \frac{\mathrm{i}}{\lambda}\,\hat{\omega}_{+}\hat{\sigma}_{-}\right) + \hat{k}_{-}\hat{\alpha}_{-}\left(\hat{a} + \frac{\mathrm{i}}{\lambda}\,\hat{\omega}_{-}\hat{\sigma}_{-}\right). \tag{6.6.44}$$

If we compare the coefficients of $\hat{a}$ and $\hat{\sigma}_{-}$ in (6.6.44) we obtain

$$\hat{k}_{+}\hat{\alpha}_{+} + \hat{k}_{-}\hat{\alpha}_{-} = 1, \tag{6.6.45}$$

$$\hat{k}_{+}\hat{\alpha}_{+}\hat{\omega}_{+} + \hat{k}_{-}\hat{\alpha}_{-}\hat{\omega}_{-} = 0. \tag{6.6.46}$$

Solving for $\hat{k}_{\pm}\hat{\alpha}_{\pm}$, we find

$$\hat{k}_{+}\hat{\alpha}_{+} = \frac{\hat{\omega}_{-}}{(\hat{\omega}_{-} - \hat{\omega}_{+})}, \quad \hat{k}_{-}\hat{\alpha}_{-} = \frac{\hat{\omega}_{+}}{(\hat{\omega}_{+} - \hat{\omega}_{-})}, \tag{6.6.47}$$

where division by an operator represents the inverse of that operator. The time-evolved annihilation operator is therefore

$$\hat{a}(t) = \frac{1}{(\hat{\omega}_{-} - \hat{\omega}_{+})}\left[\exp(-\mathrm{i}\,\hat{\omega}_{+}t)\left(\hat{\omega}_{-}\hat{a} - \mathrm{i}\lambda\hat{\sigma}_{-}\right) - \exp(-\mathrm{i}\,\hat{\omega}_{-}t)\left(\hat{\omega}_{+}\hat{a} - \mathrm{i}\lambda\hat{\sigma}_{-}\right)\right], \tag{6.6.48}$$

where we have used the fact that $\hat{\omega}_{+}\hat{\omega}_{-} = -\lambda^2$. We can obtain the evolution of the lowering operator $\hat{\sigma}_{-}$ directly from (6.6.48) using (6.6.25) and the eigenoperator equation (6.6.28). We find

$$\begin{aligned}\hat{\sigma}_{-}(t) &= \frac{1}{\mathrm{i}\hbar\lambda}\left[\hat{a}(t), \hat{H}_{\mathrm{I}}\right]\\ &= \frac{1}{(\hat{\omega}_{-} - \hat{\omega}_{+})}\left[\exp(-\mathrm{i}\,\hat{\omega}_{+}t)\left(\mathrm{i}\lambda\hat{a} - \hat{\omega}_{+}\hat{\sigma}_{-}\right) - \exp(-\mathrm{i}\,\hat{\omega}_{-}t)\left(\mathrm{i}\lambda\hat{a} - \hat{\omega}_{-}\hat{\sigma}_{-}\right)\right].\end{aligned} \tag{6.6.49}$$

Now that we have explicit expressions for the operators $\hat{a}(t)$ and $\hat{\sigma}_{-}(t)$ in terms of the initial operators, we only need to specify the initial state in order for the dynamics to be completely determined. Any quantity which may be calculated in the Schrödinger picture can also be found using the time-dependent operators given in (6.6.48) and (6.6.49). Some properties are easier to calculate using the Heisenberg picture, in particular expressions involving more than one time, such as correlation functions. As an example we consider the correlation function $\langle \hat{a}^\dagger(t)\hat{a}(t')\rangle$ for the initial state $|2\rangle\,|0\rangle$ corresponding to an excited atom coupled to the vacuum field. Using (6.6.48) and its Hermitian conjugate we have

$$\langle \hat{a}^\dagger(t)\hat{a}(t')\rangle = \langle 0|\langle 2|\exp(\mathrm{i}\omega t)\Big[\big(\hat{a}^\dagger\hat{\omega}_- + \mathrm{i}\lambda\hat{\sigma}_+\big)\exp(\mathrm{i}\hat{\omega}_+ t) - \big(\hat{a}^\dagger\hat{\omega}_+ + \mathrm{i}\lambda\hat{\sigma}_+\big) \times\exp(\mathrm{i}\hat{\omega}_- t)\Big]\frac{1}{(\hat{\omega}_- - \hat{\omega}_+)^2}\Big[\exp(-\mathrm{i}\hat{\omega}_+ t')\big(\hat{\omega}_-\hat{a} - \mathrm{i}\lambda\hat{\sigma}_-\big) - \exp(-\mathrm{i}\hat{\omega}_- t')\big(\hat{\omega}_+\hat{a} - \mathrm{i}\lambda\hat{\sigma}_-\big)\Big]\exp(-\mathrm{i}\omega t')\,|2\rangle\,|0\rangle, \tag{6.6.50}$$

where ω is the natural frequency of the mode and the factors $\exp(\mathrm{i}\omega t)$ and $\exp(-\mathrm{i}\omega t')$ arise from the free evolution of the state $|2\rangle\,|0\rangle$ in the Heisenberg interaction picture. Since $\hat{a}\,|2\rangle\,|0\rangle = 0$ and $\hat{\sigma}_-\,|2\rangle\,|0\rangle = |1\rangle\,|0\rangle$, we can simplify (6.6.50) to

$$\langle \hat{a}^\dagger(t)\hat{a}(t')\rangle = \lambda^2\exp[\mathrm{i}\omega(t-t')]\langle 0|\langle 1|[\exp(\mathrm{i}\hat{\omega}_+ t) - \exp(\mathrm{i}\hat{\omega}_- t)] \times\frac{1}{(\hat{\omega}_- - \hat{\omega}_+)^2}[\exp(-\mathrm{i}\hat{\omega}_+ t') - \exp(-\mathrm{i}\hat{\omega}_- t')]\,|1\rangle\,|0\rangle. \tag{6.6.51}$$

The state $|1\rangle\,|0\rangle$ is an eigenstate of the Hamiltonian $\hat{H}_\mathrm{I}$ with eigenvalue $-\hbar\Delta/2$ and hence, from (6.6.35), it is also an eigenstate of $\hat{\omega}_\pm$ with eigenvalues $\Delta/2 \pm \sqrt{(\Delta^2/4 + \lambda^2)} = [\Delta \pm \Omega_\mathrm{R}(1)]/2$, where $\Omega_\mathrm{R}(1)$ is the one-photon Rabi frequency defined by (6.2.31). Inserting these eigenvalues into (6.6.51) and simplifying gives

$$\langle \hat{a}^\dagger(t)\hat{a}(t')\rangle = \frac{4\lambda^2}{\Omega_\mathrm{R}^2(1)}\exp[\mathrm{i}(\omega+\Delta/2)(t-t')]\sin\left(\frac{\Omega_\mathrm{R}(1)t}{2}\right)\sin\left(\frac{\Omega_\mathrm{R}(1)t'}{2}\right). \tag{6.6.52}$$

Setting $t' = t$ we find that the mean photon number oscillates at the one-quantum Rabi frequency:

$$\langle \hat{a}^\dagger(t)\hat{a}(t)\rangle = \frac{2\lambda^2}{\Omega_\mathrm{R}^2(1)}\{1 - \cos[\Omega_\mathrm{R}(1)t]\}. \tag{6.6.53}$$

We have seen that the dressed operators have simple time dependence and that obtaining them allows us to solve for the dynamics in the Heisenberg picture. The problem reduces to expressing the initial forms of the operators of interest in terms of the dressed operators. This procedure is analogous to expressing the initial state in terms of the dressed states for calculations in the Schrödinger picture. As with the dressed states, obtaining the dressed operators can simplify some calculations and add insight in interpreting results.

Appendix 1

KRONECKER DELTA AND THE PERMUTATION SYMBOL

If i and j are integers, then the Kronecker delta δ_{ij} is defined to be

$$\delta_{ij} = \begin{cases} 1 & \text{if } i = j, \\ 0 & \text{if } i \neq j. \end{cases} \tag{A1.1}$$

In this appendix, we shall assume that i and j range from 1 to N, although they can take *any* integer values, positive, negative, or zero. If we consider δ_{ij} to be the element in the ith row and jth column of an $N \times N$ matrix (δ_{ij}) then this matrix is clearly the $N \times N$ identity matrix. Since the product of two $N \times N$ identity matrices is again the identity matrix, it follows that

$$\sum_{j=1}^{N} \delta_{ij}\delta_{jk} = \delta_{ik}, \tag{A1.2}$$

the left-hand side being the element in the ith row and kth column of the product of the two identity matrices (δ_{ij}) and (δ_{jk}). Similarly if an $N \times 1$ column vector with elements A_j is multiplied by the identity matrix, it remains unchanged and hence

$$\sum_{j=1}^{N} \delta_{ij}A_j = A_i. \tag{A1.3}$$

More generally, in any summation over the index j involving the Kronecker delta δ_{ij}, only the value of the term for which $i = j$ can be non-zero. This 'sifting property' arises frequently in calculations involving the Kronecker delta. For example,

$$\sum_{q=1}^{N} B_{klq}\delta_{pq} = B_{klp}. \tag{A1.4}$$

Note here that k, l, and p are free indices that can take any values from 1 to N, whereas q is a dummy index over which we are summing (and could be written using any letter other than k, l, or p).

If $N = 3$ and we have three unit vectors $\mathbf{e}_1$, $\mathbf{e}_2$, and $\mathbf{e}_3$ which form a mutually orthogonal, right-handed set, then we can write

$$\delta_{ij} = \mathbf{e}_i \cdot \mathbf{e}_j. \tag{A1.5}$$

In terms of these unit vectors, the permutation (or alternating) symbol ϵ_{ijk} can be defined as

$$\epsilon_{ijk} = \mathbf{e}_i \cdot (\mathbf{e}_j \times \mathbf{e}_k). \tag{A1.6}$$

From this definition, it is easy to see that

$$\epsilon_{ijk} = \begin{cases} +1 & \text{if } i, j, k \text{ are a cyclic permutation of } 1, 2, 3, \\ -1 & \text{if } i, j, k \text{ are an anticyclic permutation of } 1, 2, 3, \\ 0 & \text{if any two (or all) indices are equal.} \end{cases} \tag{A1.7}$$

Hence cyclically permuting the indices i, j, k leaves ϵ_{ijk} unaffected whereas interchanging any two indices changes its sign so that

$$\epsilon_{ijk} = \epsilon_{kij} = \epsilon_{jki}, \quad \epsilon_{jik} = -\epsilon_{ijk}. \tag{A1.8}$$

It may be readily verified that the cross-product of any two unit vectors referred to above can be written in terms of ϵ_{ijk} as

$$\mathbf{e}_i \times \mathbf{e}_j = \sum_{k=1}^{3} \epsilon_{ijk} \mathbf{e}_k. \tag{A1.9}$$

A relationship between the permutation symbol and the Kronecker delta can be derived using the following standard result from vector algebra: for any vectors **a**, **b**, **c**, and **d**,

$$(\mathbf{a} \times \mathbf{b}) \cdot (\mathbf{c} \times \mathbf{d}) = (\mathbf{a} \cdot \mathbf{c})(\mathbf{b} \cdot \mathbf{d}) - (\mathbf{a} \cdot \mathbf{d})(\mathbf{b} \cdot \mathbf{c}). \tag{A1.10}$$

Putting $\mathbf{a} = \mathbf{e}_i$, $\mathbf{b} = \mathbf{e}_j$, $\mathbf{c} = \mathbf{e}_k$, and $\mathbf{d} = \mathbf{e}_l$, we obtain

$$\begin{aligned} (\mathbf{e}_i \times \mathbf{e}_j) \cdot (\mathbf{e}_k \times \mathbf{e}_l) &= (\mathbf{e}_i \cdot \mathbf{e}_k)(\mathbf{e}_j \cdot \mathbf{e}_l) - (\mathbf{e}_i \cdot \mathbf{e}_l)(\mathbf{e}_j \cdot \mathbf{e}_k) \\ &= \delta_{ik} \delta_{jl} - \delta_{il} \delta_{jk}, \end{aligned} \tag{A1.11}$$

using (A1.5). The left-hand side of (A1.11) can be rewritten using (A1.9) as

$$\begin{aligned} (\mathbf{e}_i \times \mathbf{e}_j) \cdot (\mathbf{e}_k \times \mathbf{e}_l) &= \left(\sum_{p=1}^{3} \epsilon_{ijp} \mathbf{e}_p \right) \cdot \left(\sum_{q=1}^{3} \epsilon_{klq} \mathbf{e}_q \right) \\ &= \sum_{p=1}^{3} \sum_{q=1}^{3} \epsilon_{ijp} \epsilon_{klq} \mathbf{e}_p \cdot \mathbf{e}_q \\ &= \sum_{p=1}^{3} \sum_{q=1}^{3} \epsilon_{ijp} \epsilon_{klq} \delta_{pq} \\ &= \sum_{p=1}^{3} \epsilon_{ijp} \epsilon_{klp}, \end{aligned} \tag{A1.12}$$

where we have again used (A1.5), and (A1.4) with $B_{klq} = \epsilon_{klq}$. Combining (A1.11) and (A1.12), we find

$$\sum_{p=1}^{3} \epsilon_{ijp} \epsilon_{klp} = \delta_{ik} \delta_{jl} - \delta_{il} \delta_{jk}. \tag{A1.13}$$

The Pauli operators $\hat{\sigma}_1$, $\hat{\sigma}_2$, and $\hat{\sigma}_3$ discussed in Chapters 2 and 3 have the commutators (3.2.8) to (3.2.10) given by $[\hat{\sigma}_1, \hat{\sigma}_2] = 2\mathrm{i}\hat{\sigma}_3$, $[\hat{\sigma}_2, \hat{\sigma}_3] = 2\mathrm{i}\hat{\sigma}_1$, and $[\hat{\sigma}_3, \hat{\sigma}_1] = 2\mathrm{i}\hat{\sigma}_2$. Note that the second and third of these commutators are cyclic permutations of the indices 1, 2, and 3 in the first commutator and that consequently

$$\left[\hat{\sigma}_j, \hat{\sigma}_k\right] = 2\mathrm{i} \sum_{l=1}^{3} \epsilon_{jkl} \hat{\sigma}_l. \tag{A1.14}$$

The property of ϵ_{jkl} that it is zero if any two indices are equal ensures that, if $j = k$, the commutator of a Pauli operator with itself is zero, whilst the property that $\epsilon_{kjl} = -\epsilon_{jkl}$ ensures that a sign change results if $\hat{\sigma}_j$ and $\hat{\sigma}_k$ are exchanged in (A1.14). Further, as mentioned in Chapter 3, the anticommutator of two Pauli operators can be expressed as

$$\left\{\hat{\sigma}_j, \hat{\sigma}_k\right\} = 2\delta_{jk}. \tag{A1.15}$$

From (A1.14) and (A1.15), we can write the product of any two Pauli operators in the compact form

$$\hat{\sigma}_j \hat{\sigma}_k = \delta_{jk} + \mathrm{i} \sum_{l=1}^{3} \epsilon_{jkl} \hat{\sigma}_l. \tag{A1.16}$$

Appendix 2
THE DIRAC DELTA FUNCTION

The Dirac delta function is not a function in the usual sense of being defined for values of its argument, but is a generalized function in that it is defined by a limiting procedure and only has meaning within an integral. It is often written and manipulated without reference to an integral but it should be remembered that results so obtained are strictly prescriptions for manipulating integrals containing delta functions.

The delta function $\delta(x)$ may be defined as the limit as $\epsilon \to 0$ of a normalized function $\phi(x, \epsilon)$ of x and ϵ which is symmetric about its main or only peak at $x = 0$. Such functions tend to zero in the limit $\epsilon \to 0$ for all *non-zero* values of x, and hence, since the function is normalized to unit area, the value at $x = 0$ must tend to infinity. Examples of suitable functions are $\epsilon/[\pi(x^2 + \epsilon^2)]$, $(\pi\epsilon)^{-1/2} \exp(-x^2/\epsilon)$, and $(\pi x)^{-1} \sin(x/\epsilon)$. A delta function $\delta(x - a)$ peaked at $x = a$ is simply obtained by replacing x by $x - a$ in the function and then taking the limit $\epsilon \to 0$. The integral of the delta function is, for $p < q$,

$$\int_p^q \delta(x-a)\,dx = \begin{cases} 1 & \text{for } p < a < q, \\ \frac{1}{2} & \text{for } p = a \text{ or } q = a, \\ 0 & \text{otherwise,} \end{cases} \tag{A2.1}$$

where the value unity arises from the normalization and the value $\frac{1}{2}$ is obtained because only one-half of the symmetric function lies within the range of integration. The Heaviside unit step function $H(x - a)$ can then be defined using (A2.1) as

$$H(x-a) = \int_{-\infty}^{x} \delta(x'-a)\,dx' = \begin{cases} 0 & \text{for } x < a, \\ \frac{1}{2} & \text{for } x = a, \\ 1 & \text{for } x > a. \end{cases} \tag{A2.2}$$

The important property of $\delta(x - a)$ is the so-called sifting property, which states that if $f(x)$ is a function for which $f(a)$ is defined then

$$\int_p^q \delta(x-a) f(x)\,dx = \begin{cases} f(a) & \text{for } p < a < q, \\ \frac{1}{2} f(a) & \text{for } p = a \text{ or } q = a, \\ 0 & \text{otherwise.} \end{cases} \tag{A2.3}$$

This result may be used more generally: if $f(x)$ is itself a generalized function then (A2.3) is meaningful if $f(a)$ is meaningful. For example,

$$\int_{-\infty}^{\infty} \delta(x-a)\delta(x-b)\,dx = \delta(a-b) \tag{A2.4}$$

within an integral over either a or b. Comparing (A2.1) with (A2.3), we see that

$$\int_p^q \delta(x-a)f(x)\,\mathrm{d}x = \int_p^q \delta(x-a)f(a)\,\mathrm{d}x \tag{A2.5}$$

and hence we can write

$$\delta(x-a)f(x) = \delta(x-a)f(a) \tag{A2.6}$$

in the sense that, as before, equality holds when either side of (A2.6) lies within an integral.

The Fourier transform of $\delta(x)$ can be calculated using the sifting property (A2.3) as

$$\mathscr{F}\{\delta(x)\} = \frac{1}{2\pi}\int_{-\infty}^{\infty} \delta(x)\exp(\pm ikx)\,\mathrm{d}x = \frac{1}{2\pi}. \tag{A2.7}$$

Hence a representation of $\delta(x)$ obtained from the inverse transform of $(2\pi)^{-1}$ is

$$\delta(x) = \frac{1}{2\pi}\int_{-\infty}^{\infty} \exp(\mp \mathrm{i}kx)\,\mathrm{d}k. \tag{A2.8}$$

A simple change of variable then shows that

$$\delta(bx) = \frac{1}{|b|}\,\delta(x). \tag{A2.9}$$

The extension of the definition of the delta function into two or more dimensions is achieved by considering the product of individual delta functions, one for each dimension. For example, if $\mathbf{k} = (k_x, k_y, k_z)$ and $\mathbf{k}' = (k'_x, k'_y, k'_z)$ are two vectors then we define the three-dimensional delta function $\delta^{(3)}(\mathbf{k}-\mathbf{k}')$ as

$$\delta^{(3)}(\mathbf{k}-\mathbf{k}') = \delta(k_x - k'_x)\delta(k_y - k'_y)\delta(k_z - k'_z). \tag{A2.10}$$

Two-dimensional delta functions are useful in problems involving functions of complex variables, so that for a complex variable $\xi = \xi_r + \mathrm{i}\xi_i$, we define

$$\delta^{(2)}(\xi) = \delta(\xi_r)\delta(\xi_i) = \delta^{(2)}(\xi^*). \tag{A2.11}$$

Considering the one-dimensional delta functions in (A2.11) to be the limit of Gaussians leads to the identification

$$\begin{aligned}\delta(\xi_r)\delta(\xi_i) &= \lim_{\epsilon\to 0} \frac{1}{\pi\epsilon}\exp[-(\xi_r^2 + \xi_i^2)/\epsilon] \\ &= \lim_{\epsilon\to 0} \frac{1}{\pi\epsilon}\exp(-|\xi|^2/\epsilon),\end{aligned} \tag{A2.12}$$

with $\delta^{(2)}(\xi)$ (see (4.4.18)). It follows from (A2.8) and (A2.11) that a representation of $\delta^{(2)}(\xi)$ is

$$\begin{aligned}\delta^{(2)}(\xi) &= \frac{1}{4\pi^2}\int_{-\infty}^{\infty}\int_{-\infty}^{\infty} \mathrm{d}(2\alpha_r)\,\mathrm{d}(2\alpha_i)\exp(-2\mathrm{i}\alpha_r\xi_i + 2\mathrm{i}\alpha_i\xi_r) \\ &= \frac{1}{\pi^2}\int_{-\infty}^{\infty} \mathrm{d}^2\alpha\,\exp(\alpha\xi^* - \alpha^*\xi),\end{aligned} \tag{A2.13}$$

where $d^2\alpha = d\alpha_r\, d\alpha_i$ and integration is implied over the whole of the complex α-plane.

Consider the integral

$$\int_{-\infty}^{\infty} \frac{d\phi(x,\epsilon)}{dx} f(x)\, dx = [\phi(x,\epsilon)f(x)]_{-\infty}^{\infty} - \int_{-\infty}^{\infty} \phi(x,\epsilon)f'(x)\, dx, \quad (A2.14)$$

where the limit as $\epsilon \to 0$ of $\phi(x,\epsilon)$ is $\delta(x)$. Assuming the integrated term in (A2.14) is zero, as it will be for all functions $f(x)$ for which the first integral exists, then in the limit $\epsilon \to 0$ the right-hand side of (A2.14) becomes $-f'(0)$, using (A2.3). This gives rise to the symbol $\delta'(x)$ which we call the first derivative of the delta function, given by the limit $\delta'(x) = \lim_{\epsilon \to 0} \phi'(x,\epsilon)$, with the property

$$\int_{-\infty}^{\infty} \delta'(x) f(x)\, dx = -f'(0), \quad (A2.15)$$

provided $f'(0)$ exists. We can extend this to higher-order derivatives of $\delta(x)$, written $d^m\delta(x)/dx^m$, with the properties

$$\int_{-\infty}^{\infty} \frac{d^m\delta(x)}{dx^m} f(x)\, dx = \left(-\frac{d}{dx}\right)^m f(x)\bigg|_{x=0}. \quad (A2.16)$$

An alternative representation of $\delta'(x)$ can be found by considering

$$\int_{-\infty}^{\infty} \delta'(x) x g(x)\, dx = -\frac{d}{dx}[xg(x)]\bigg|_{x=0} = -g(0), \quad (A2.17)$$

using (A2.15), provided $g(0)$ exists. Further, using the sifting property (A2.3) with $f(x) = -g(x)$, we have

$$-g(0) = \int_{-\infty}^{\infty} \delta(x)[-g(x)]\, dx = \int_{-\infty}^{\infty} \frac{-\delta(x)}{x} xg(x)\, dx. \quad (A2.18)$$

Comparing (A2.17) with (A2.18) leads to the identification

$$\delta'(x) = -\frac{\delta(x)}{x} \quad (A2.19)$$

with the usual understanding that this should be interpreted as a rule for manipulating integrals containing delta functions. The two-dimensional extension of (A2.16) for $\delta^{(2)}(\xi)$ and $f(\xi)$, where ξ is a complex variable, is

$$\int_{-\infty}^{\infty} d^2\xi\, f(\xi)\left(\frac{\partial}{\partial\xi}\right)^m \left(\frac{\partial}{\partial\xi^*}\right)^n \delta^{(2)}(\xi) = \left(-\frac{\partial}{\partial\xi}\right)^m \left(-\frac{\partial}{\partial\xi^*}\right)^n f(\xi)\bigg|_{\xi=0}. \quad (A2.20)$$

Appendix 3

SPECIAL FUNCTIONS

In this appendix, we summarize the properties of the special functions required in Chapters 2, 3, and 4.

The gamma function $\Gamma(x)$ is defined as

$$\Gamma(x) = \int_0^\infty t^{x-1} \exp(-t)\, dt, \tag{A3.1}$$

where we require $x > 0$ for the integral to converge. We have from (A3.1), for example, $\Gamma(1) = 1$ and $\Gamma(\frac{1}{2}) = \sqrt{\pi}$, whilst $\lim_{x \to 0} \Gamma(x) = \infty$. Now consider $\Gamma(x + 1)$; integrating by parts we obtain

$$\Gamma(x+1) = \int_0^\infty t^x \exp(-t)\, dt = [-t^x \exp(-t)]_0^\infty + x \int_0^\infty t^{x-1} \exp(-t)\, dt. \tag{A3.2}$$

The integrated term is zero and hence, using (A3.1),

$$\Gamma(x+1) = x\Gamma(x). \tag{A3.3}$$

This relationship can be used to *define* $\Gamma(x)$ for negative values of x. For example, choosing $x = -\frac{1}{2}$ we have $\Gamma(\frac{1}{2}) = -\Gamma(-\frac{1}{2})/2$ so that $\Gamma(-\frac{1}{2}) = -2\Gamma(\frac{1}{2}) = -2\sqrt{\pi}$. Since $\Gamma(0^+) = \infty$ it follows that $\Gamma(x)$ diverges at the negative integers using this definition. Suppose now that $x = n$, where $n \geq 1$ is an integer. Then, using (A3.3), we have

$$\Gamma(n+1) = n\Gamma(n) = n(n-1)\Gamma(n-1) = \cdots = n!\,\Gamma(1). \tag{A3.4}$$

Since $\Gamma(1) = 1$, we find

$$\Gamma(n+1) = n!. \tag{A3.5}$$

In addition to the above, $\Gamma(x)$ satisfies the reflection formula

$$\Gamma(x)\Gamma(1-x) = \frac{\pi}{\sin(\pi x)}. \tag{A3.6}$$

The beta function $B(p, q)$ is defined to be

$$B(p,q) = \int_0^1 x^{p-1}(1-x)^{q-1}\, dx, \tag{A3.7}$$

where $p > 0$, $q > 0$. An alternative form can be found from (A3.7) by making the substitution $x = \sin^2\varphi$. Then, since $dx = 2 \sin\varphi \cos\varphi\, d\varphi$, we have

$$B(p,q) = 2 \int_0^{\pi/2} (\sin\varphi)^{2p-1} (\cos\varphi)^{2q-1}\, d\varphi. \tag{A3.8}$$

The relationship between the beta function and the gamma function is

$$B(p,q) = \frac{\Gamma(p)\Gamma(q)}{\Gamma(p+q)}. \tag{A3.9}$$

In Chapter 3, we require the value of the integral

$$I_1 = \int_0^\pi (\sin \tfrac{1}{2}\theta)^{2(j+m)+1}(\cos \tfrac{1}{2}\theta)^{2(j-m)+1}\,\mathrm{d}\theta \tag{A3.10}$$

(see (3.8.12)). Putting $\varphi = \frac{1}{2}\theta$, we obtain

$$I_1 = 2\int_0^{\pi/2} (\sin \varphi)^{2(j+m)+1}(\cos \varphi)^{2(j-m)+1}\,\mathrm{d}\varphi = B(j+m+1, j-m+1), \tag{A3.11}$$

using (A3.8). Then, from (A3.9) and (A3.5), we find

$$I_1 = \frac{\Gamma(j+m+1)\Gamma(j-m+1)}{\Gamma(2j+2)}$$

$$= \frac{(j+m)!\,(j-m)!}{(2j+1)!} = \frac{1}{(2j+1)C_{j+m}^{2j}}, \tag{A3.12}$$

using $C_l^k = k!/[l!(k-l)!]$.

The Hermite polynomials $H_n(x)$ satisfy the differential equation

$$\frac{\mathrm{d}^2 H_n}{\mathrm{d}x^2} - 2x\frac{\mathrm{d}H_n}{\mathrm{d}x} + 2nH_n = 0, \tag{A3.13}$$

where $n = 0, 1, 2, \ldots$. In particular

$$H_0(x) = 1, \quad H_1(x) = 2x, \quad H_2(x) = 4x^2 - 2. \tag{A3.14}$$

The orthogonality property of the Hermite polynomials is expressed by the integral

$$\int_{-\infty}^{\infty} \exp(-x^2) H_n(x) H_m(x)\,\mathrm{d}x = n!\,2^n \pi^{1/2} \delta_{nm}. \tag{A3.15}$$

The generating function is

$$\exp(2tx - t^2) = \sum_{n=0}^{\infty} \frac{1}{n!} t^n H_n(x). \tag{A3.16}$$

The Legendre polynomials $P_n(x)$ satisfy the differential equation

$$(1-x^2)\frac{\mathrm{d}^2P_n}{\mathrm{d}x^2} - 2x\frac{\mathrm{d}P_n}{\mathrm{d}x} + n(n+1)P_n = 0, \tag{A3.17}$$

where $n = 0, 1, 2, \ldots$ and $|x| \leqslant 1$. In particular

$$P_0(x) = 1, \quad P_1(x) = x, \quad P_2(x) = (3x^2-1)/2, \quad P_3(x) = (5x^3-3x)/2. \tag{A3.18}$$

The orthogonality property of the Legendre polynomials is expressed by the integral

$$\int_{-1}^{1} P_n(x)P_m(x)\,\mathrm{d}x = \frac{2}{2n+1}\,\delta_{nm}. \tag{A3.19}$$

The generating function is

$$\frac{1}{\sqrt{(1-2xt+t^2)}} = \sum_{n=0}^{\infty} t^n P_n(x), \tag{A3.20}$$

where $|t| < 1$.

The Laguerre polynomials $L_n(x)$ satisfy the differential equation

$$x\frac{\mathrm{d}^2L_n}{\mathrm{d}x^2} + (1-x)\frac{\mathrm{d}L_n}{\mathrm{d}x} + nL_n = 0, \tag{A3.21}$$

where $n = 0, 1, 2, \ldots$ and $x \geqslant 0$. In particular

$$L_0(x) = 1, \quad L_1(x) = 1-x, \quad L_2(x) = (x^2-4x+2)/2. \tag{A3.22}$$

The orthogonality property of the Laguerre polynomials is expressed by the integral

$$\int_0^{\infty} \exp(-x)L_n(x)L_m(x)\,\mathrm{d}x = \delta_{nm}. \tag{A3.23}$$

The generating function is

$$\frac{1}{1-z}\exp[xz/(z-1)] = \sum_{n=0}^{\infty} z^n L_n(x), \tag{A3.24}$$

where $|z| < 1$. The Laguerre polynomial $L_n(x)$ can be obtained from the relation

$$L_n(x) = \frac{1}{n!}\exp(x)\frac{\mathrm{d}^n}{\mathrm{d}x^n}[x^n\exp(-x)] \tag{A3.25}$$

and expressed as the power series

$$L_n(x) = \sum_{m=0}^{n} \frac{(-x)^m}{(m!)^2}\frac{n!}{(n-m)!}. \tag{A3.26}$$

The associated Laguerre polynomials $L_n^{(k)}(x)$ are defined by

$$L_n^{(k)}(x) = \frac{\mathrm{d}^k}{\mathrm{d}x^k} L_n(x) \tag{A3.27}$$

and hence $L_n^{(k)}(x)$ is identically zero for all x if $k > n$. If $k \leqslant n$, it is straightforward to show from (A3.26) and (A3.27) that

$$L_n^{(k)}(0) = (-1)^k \frac{n!}{k!\,(n-k)!} = (-1)^k C_k^n. \tag{A3.28}$$

The Bessel functions $J_n(x)$ of order n satisfy the differential equation

$$x^2 \frac{\mathrm{d}^2 J_n}{\mathrm{d}x^2} + x \frac{\mathrm{d}J_n}{\mathrm{d}x} + (x^2 - n^2) J_n = 0, \tag{A3.29}$$

where $n = 0, 1, 2, \ldots,$ and can be expressed as the power series

$$J_n(x) = \sum_{k=0}^{\infty} \frac{(-1)^k}{k!\,(n+k)!} \left(\frac{x}{2}\right)^{n+2k}. \tag{A3.30}$$

The generating function is

$$\exp[x(t - 1/t)/2] = \sum_{n=-\infty}^{\infty} t^n J_n(x), \tag{A3.31}$$

where $J_{-n}(x) = (-1)^n J_n(x)$. Important recurrence relations are

$$2 \frac{\mathrm{d}J_n(x)}{\mathrm{d}x} = J_{n-1}(x) - J_{n+1}(x), \tag{A3.32}$$

$$\frac{2n}{x} J_n(x) = J_{n-1}(x) + J_{n+1}(x). \tag{A3.33}$$

Note that $\mathrm{d}J_0/\mathrm{d}x = -J_1(x)$. The integral form of $J_n(x)$ is

$$J_n(x) = \frac{1}{\pi} \int_0^{\pi} \cos(n\theta - x \sin\theta)\,\mathrm{d}\theta. \tag{A3.34}$$

This integral may be written in a number of equivalent forms. In particular

$$\begin{aligned} J_0(x) &= \frac{1}{\pi} \int_0^{\pi} \cos(x \sin\theta)\,\mathrm{d}\theta \\ &= \frac{1}{2\pi} \int_0^{2\pi} \cos(x \sin\theta)\,\mathrm{d}\theta \\ &= \frac{1}{2\pi} \int_0^{2\pi} \exp(\mathrm{i}x \sin\theta)\,\mathrm{d}\theta, \end{aligned} \tag{A3.35}$$

since $\int_0^{2\pi} \sin(x \sin\theta)\,d\theta \equiv 0$. Further, since $\sin\theta$ is periodic with period 2π, the final integral in (A3.35) can have any limits which differ by 2π. Alternatively, θ can be shifted by an arbitrary angle φ so that

$$2\pi J_0(x) = \int_0^{2\pi} \exp[\mathrm{i}x \sin(\theta - \varphi)]\,d\theta. \tag{A3.36}$$

A number of relationships exist between the Laguerre polynomials and the Bessel function J_0. Consider first the integral

$$I_2 = \int_0^{\infty} u^m \exp(x-u) J_0(2\surd(xu))\,du. \tag{A3.37}$$

Substituting the series expansion (A3.30) with $n = 0$ for $J_0(2\surd(xu))$ we obtain

$$\begin{aligned} I_2 &= \exp(x) \sum_{k=0}^{\infty} \frac{(-1)^k}{(k!)^2} x^k \int_0^{\infty} \exp(-u) u^{m+k}\,du \\ &= \exp(x) \sum_{k=0}^{\infty} \frac{(-1)^k}{(k!)^2} x^k (m+k)!. \end{aligned} \tag{A3.38}$$

Writing $(m+k)!/k! = (m+k)(m+k-1)\ldots(k+1)$, we see that (A3.38) can be expressed as

$$I_2 = \exp(x) \frac{d^m}{dx^m} \sum_{k=0}^{\infty} \frac{(-1)^k}{k!} x^{m+k} = \exp(x) \frac{d^m}{dx^m}[x^m \exp(-x)]. \tag{A3.39}$$

Finally, using (A3.25), we have the result

$$\int_0^{\infty} u^m \exp(x-u) J_0(2\surd(xu))\,du = m!\, L_m(x). \tag{A3.40}$$

This result can be used to evaluate the integral

$$I_3 = \frac{2}{\pi} \int_0^{\infty} x \exp[-(1-p)x^2/2] L_n(x^2) J_0(2|\alpha| x)\,dx, \tag{A3.41}$$

for $p < 1$, required in Chapter 4. Writing $(1-p)x^2/2 = u$ in (A3.41) we obtain

$$I_3 = \frac{2}{\pi(1-p)} \int_0^{\infty} \exp(-u) L_n\left(\frac{2u}{1-p}\right) J_0\left(2\sqrt{\frac{2|\alpha|^2 u}{1-p}}\right) du. \tag{A3.42}$$

Substituting the series expansion (A3.26) for L_n and performing the resulting integral using (A3.40) we obtain

$$I_3 = \frac{2}{\pi(1-p)} \sum_{m=0}^{n} \frac{(-1)^m n!}{m!(n-m)!} \left(\frac{2}{1-p}\right)^m L_m\left(\frac{2|\alpha|^2}{1-p}\right) \exp\left[-2|\alpha|^2/(1-p)\right]. \tag{A3.43}$$

Again substituting the series expansion for L_m and reversing the order of the two summations in (A3.43) we find

$$I_3 = \frac{2}{\pi(1-p)} \exp\left[-2|\alpha|^2/(1-p)\right] \sum_{k=0}^{n} \frac{(-1)^k}{(k!)^2} \left(\frac{2|\alpha|^2}{1-p}\right)^k$$

$$\times \sum_{m=k}^{n} \frac{n!(-1)^m}{(n-m)!(m-k)!} \left(\frac{2}{1-p}\right)^m$$

$$= \frac{2}{\pi(1-p)} \exp\left[-2|\alpha|^2/(1-p)\right] \sum_{k=0}^{n} \frac{n!}{(k!)^2(n-k)!} \left(\frac{4|\alpha|^2}{(1-p)^2}\right)^k$$

$$\times \sum_{j=0}^{n-k} \frac{(n-k)!}{j!(n-k-j)!} \left(-\frac{2}{1-p}\right)^j, \tag{A3.44}$$

where we have written $m-k=j$. The summation over j is simply the binomial expansion of $[1-2/(1-p)]^{n-k} = (-1)^{n-k}[(1+p)/(1-p)]^{n-k}$ so that

$$I_3 = \frac{2}{\pi(1-p)} (-1)^n \left(\frac{1+p}{1-p}\right)^n \exp\left[-2|\alpha|^2/(1-p)\right]$$

$$\times \sum_{k=0}^{n} \left(-\frac{4|\alpha|^2}{1-p^2}\right) \frac{n!}{(k!)^2(n-k)!}$$

$$= \frac{2}{\pi(1-p)} (-1)^n \left(\frac{1+p}{1-p}\right)^n \exp\left[-2|\alpha|^2/(1-p)\right] L_n\left(\frac{4|\alpha|^2}{1-p^2}\right), \tag{A3.45}$$

again using (A3.26). If $p = -1$, the result is found by taking the limit as $p \to -1$ of (A3.45). The argument of the Laguerre polynomial becomes large in this limit, whilst the factor $(1+p)^n$ becomes small. From (A3.26) we have

$$\lim_{p \to -1} (1+p)^n L_n\left(\frac{4|\alpha|^2}{1-p^2}\right) = \lim_{p \to -1} (1+p)^n \frac{1}{n!} \left(-\frac{4|\alpha|^2}{1-p^2}\right)^n$$

$$= \frac{(-2|\alpha|^2)^n}{n!}. \tag{A3.46}$$

Hence

$$I_3(p=-1) = \frac{|\alpha|^{2n}}{n!\,\pi} \exp(-|\alpha|^2), \tag{A3.47}$$

as in (4.5.32).

Appendix 4
QUADRATURE EIGENSTATES

In Section 3.2, we introduced the quadrature operator $\hat{x}_\lambda$ for a single-mode field (see (3.2.20)). Here we describe the properties of this operator and its eigenstates $|x_\lambda\rangle$ in more detail. We recall that the commutator of $\hat{x}_\lambda$ and $\hat{x}_{\lambda+\pi/2}$ is

$$[\hat{x}_\lambda, \hat{x}_{\lambda+\pi/2}] = \mathrm{i} \tag{A4.1}$$

and that

$$\hat{x}_\lambda |x_\lambda\rangle = x_\lambda |x_\lambda\rangle. \tag{A4.2}$$

Further, the quadrature representation $\psi(x_\lambda)$ of a single-mode state $|\psi\rangle$ is defined by the relation $\psi(x_\lambda) = \langle x_\lambda | \psi\rangle$. Here we show that these relations imply that the eigenstates of $\hat{x}_\lambda$ cannot satisfy *Kronecker*-delta orthogonality. For suppose that $\hat{x}_\lambda$ has a countable set of orthogonal eigenstates $|x_\lambda^{(m)}\rangle$ with eigenvalues $x_\lambda^{(m)}$, so that

$$\hat{x}_\lambda |x_\lambda^{(m)}\rangle = x_\lambda^{(m)} |x_\lambda^{(m)}\rangle \tag{A4.3}$$

and

$$\langle x_\lambda^{(m)} | x_\lambda^{(n)}\rangle = \delta_{mn}. \tag{A4.4}$$

The matrix element of the commutator on the left-hand side of (A4.1) would then be

$$\langle x_\lambda^{(m)}| [\hat{x}_\lambda, \hat{x}_{\lambda+\pi/2}] |x_\lambda^{(n)}\rangle = (x_\lambda^{(m)} - x_\lambda^{(n)}) \langle x_\lambda^{(m)}| \hat{x}_{\lambda+\pi/2} |x_\lambda^{(n)}\rangle, \tag{A4.5}$$

whereas the matrix element of the right-hand side of (A4.1) would be simply $\mathrm{i}\,\delta_{mn}$. These clearly cannot be equal since the former is zero when $m = n$, whereas the latter is only non-zero when $m = n$. Hence the orthonormality in (A4.4) is inconsistent with the commutator (A4.1). The resolution of this difficulty is to employ delta *function* orthogonality for the eigenstates $|x_\lambda\rangle$ having a continuum of eigenvalues x_λ, so that

$$\langle x_\lambda | x'_\lambda\rangle = \delta(x_\lambda - x'_\lambda). \tag{A4.6}$$

In doing this, we have extended the set of possible states beyond those of square-integrable functions which can be represented in Hilbert space. This is because $|\langle x_\lambda | x'_\lambda\rangle|^2$ is not defined. Now taking matrix elements of (A4.1) in the basis of eigenstates $|x_\lambda\rangle$ we find

$$\begin{aligned}\langle x_\lambda|[\hat{x}_\lambda, \hat{x}_{\lambda+\pi/2}]|x'_\lambda\rangle &= (x_\lambda - x'_\lambda)\langle x_\lambda| \hat{x}_{\lambda+\pi/2} |x'_\lambda\rangle \\ &= \mathrm{i}\,\delta(x_\lambda - x'_\lambda).\end{aligned} \tag{A4.7}$$

Hence

$$\langle x_\lambda|\,\hat{x}_{\lambda+\pi/2}\,|x'_\lambda\rangle = \mathrm{i}\,\frac{\delta(x_\lambda - x'_\lambda)}{(x_\lambda - x'_\lambda)} = -\mathrm{i}\,\frac{\partial}{\partial x_\lambda}\,\delta(x_\lambda - x'_\lambda), \tag{A4.8}$$

where we have used the form of the derivative of the delta function given in Appendix 2. This result together with the resolution of the identity in terms of the quadrature eigenstates, given in (3.3.27), leads to a representation for the action of $\hat{x}_{\lambda+\pi/2}$ on $\psi(x_\lambda)$ since

$$\begin{aligned}\langle x_\lambda|\,\hat{x}_{\lambda+\pi/2}\,|\psi\rangle &= \int_{-\infty}^{\infty} \mathrm{d}x'_\lambda\,\langle x_\lambda|\,\hat{x}_{\lambda+\pi/2}\,|x'_\lambda\rangle\langle x'_\lambda\,|\,\psi\rangle\\ &= -\mathrm{i}\,\frac{\mathrm{d}}{\mathrm{d}x_\lambda}\int_{-\infty}^{\infty} \mathrm{d}x'_\lambda\,\delta(x_\lambda - x'_\lambda)\langle x'_\lambda\,|\,\psi\rangle\\ &= -\mathrm{i}\,\frac{\mathrm{d}}{\mathrm{d}x_\lambda}\,\psi(x_\lambda).\end{aligned} \tag{A4.9}$$

This is just the familiar position representation of the action of the momentum operator on the state $|\psi\rangle$. The representation of the action of $\hat{x}_\lambda$ on $|\psi\rangle$ is simply $x_\lambda\psi(x_\lambda)$ since

$$\langle x_\lambda|\,\hat{x}_\lambda\,|\psi\rangle = x_\lambda\psi(x_\lambda). \tag{A4.10}$$

We can use (1.3.33), (1.3.36), (3.2.36) and (A3.16) to write the quadrature eigenstate in the form

$$|x_\lambda\rangle = \pi^{-1/4}\exp\left[-\tfrac{1}{2}x_\lambda^2 + \sqrt{2}\,\exp(\mathrm{i}\lambda)x_\lambda\hat{a}^\dagger - \tfrac{1}{2}\exp(2\mathrm{i}\lambda)\hat{a}^{\dagger 2}\right]|0\rangle. \tag{A4.11}$$

From this, we can obtain the quadrature representation of any state. For example, the quadrature representation of the coherent state is

$$\begin{aligned}\langle x_\lambda\,|\,\alpha\rangle &= \pi^{-1/4}\exp\left[-\tfrac{1}{2}x_\lambda^2 + \sqrt{2}\,\exp(-\mathrm{i}\lambda)x_\lambda\alpha - \tfrac{1}{2}\exp(-2\mathrm{i}\lambda)\alpha^2\right]\exp\left(-\tfrac{1}{2}|\alpha|^2\right)\\ &= \pi^{-1/4}\exp\left[\mathrm{i}\langle\hat{x}_{\lambda+\pi/2}\rangle x_\lambda\right]\exp\left[-\tfrac{1}{2}(x_\lambda - \langle\hat{x}_\lambda\rangle)^2\right]\\ &\quad\times\exp\left[-\tfrac{1}{2}\mathrm{i}\langle\hat{x}_\lambda\rangle\langle\hat{x}_{\lambda+\pi/2}\rangle\right],\end{aligned} \tag{A4.12}$$

in agreement with (3.6.45), where κ is the final factor in (A4.12).

Appendix 5

OPERATOR ORDERING THEOREMS

In Chapter 3, we derived a number of theorems relating different orderings of the exponential function of operators. Here we summarize these theorems and give some alternative forms and more general expressions.

For two operators $\hat{A}$ and $\hat{B}$ which commute with their commutator $[\hat{A}, \hat{B}]$ we have

$$\exp[\theta(\hat{A}+\hat{B})] = \exp(\theta\hat{A})\exp(\theta\hat{B})\exp\left(-\tfrac{1}{2}\theta^2[\hat{A},\hat{B}]\right). \qquad \text{(A5.1)}$$

Important examples of this are

$$\begin{aligned}\hat{D}(\alpha) &= \exp(\alpha\hat{a}^\dagger - \alpha^*\hat{a}) = \exp(\alpha\hat{a}^\dagger)\exp(-\alpha^*\hat{a})\exp\left(-|\alpha|^2/2\right)\\ &= \exp(-\alpha^*\hat{a})\exp(\alpha\hat{a}^\dagger)\exp\left(|\alpha|^2/2\right), \qquad \text{(A5.2)}\end{aligned}$$

$$\hat{D}(\alpha)\hat{D}(\beta) = \hat{D}(\alpha+\beta)\exp[\tfrac{1}{2}(\alpha\beta^* - \alpha^*\beta)], \qquad \text{(A5.3)}$$

and the continuum operator analogue of these given by

$$\begin{aligned}\hat{D}[\alpha(\omega)] &= \exp\left(\int \mathrm{d}\omega[\alpha(\omega)\hat{b}^\dagger(\omega) - \alpha^*(\omega)\hat{b}(\omega)]\right)\\ &= \exp\left(\int \mathrm{d}\omega\, \alpha(\omega)\hat{b}^\dagger(\omega)\right)\exp\left(-\int \mathrm{d}\omega\, \alpha^*(\omega)\hat{b}(\omega)\right)\\ &\quad\times \exp\left(-\tfrac{1}{2}\int \mathrm{d}\omega|\alpha(\omega)|^2\right)\\ &= \exp\left(-\int \mathrm{d}\omega\, \alpha^*(\omega)\hat{b}(\omega)\right)\exp\left(\int \mathrm{d}\omega\, \alpha(\omega)\hat{b}^\dagger(\omega)\right)\\ &\quad\times \exp\left(\tfrac{1}{2}\int \mathrm{d}\omega|\alpha(\omega)|^2\right) \qquad \text{(A5.4)}\end{aligned}$$

and

$$\begin{aligned}\hat{D}[\alpha(\omega)]\hat{D}[\beta(\omega)] &= \hat{D}[\alpha(\omega)+\beta(\omega)]\\ &\quad\times \exp\left(\tfrac{1}{2}\int \mathrm{d}\omega[\alpha(\omega)\beta^*(\omega) - \alpha^*(\omega)\beta(\omega)]\right). \qquad \text{(A5.5)}\end{aligned}$$

The exponential function of the number operator is related to its normal ordered form by

$$\exp(\theta\hat{a}^\dagger\hat{a}) = :\exp\{[\exp(\theta)-1]\hat{a}^\dagger\hat{a}\}: \qquad \text{(A5.6)}$$

with the continuum operator generalization

$$\exp\left(\int d\omega\, \theta(\omega)\hat{b}^\dagger(\omega)\hat{b}(\omega)\right) = :\exp\left(\int d\omega\{\exp[\theta(\omega)] - 1\}\hat{b}^\dagger(\omega)\hat{b}(\omega)\right):. \tag{A5.7}$$

There is an antinormal ordered analogue of (A5.6) given by

$$\exp(\theta\hat{a}\hat{a}^\dagger) = \vdots \exp\{[1 - \exp(-\theta)]\hat{a}^\dagger\hat{a}\} \vdots \tag{A5.8}$$

but there is no antinormal ordered continuum generalization.

For two operators $\hat{A}$ and $\hat{B}$ with commutator $[\hat{A}, \hat{B}] = -\hat{A}$, we have

$$\begin{aligned} \exp[\theta(\hat{A} + \hat{B})] &= \exp(\theta\hat{B})\exp\{[1 - \exp(-\theta)]\hat{A}\} \\ &= \exp\{[\exp(\theta) - 1]\hat{A}\}\exp(\theta\hat{B}). \end{aligned} \tag{A5.9}$$

This result may also be applied to superoperators as in Section 5.6.

The angular momentum operators $\hat{J}_3$, $\hat{J}_+$, and $\hat{J}_-$ satisfy the commutation relations $[\hat{J}_3, \hat{J}_\pm] = \pm\hat{J}_\pm$ and $[\hat{J}_+, \hat{J}_-] = 2\hat{J}_3$. We have

$$\exp\left[i\theta\left(\hat{J}_+ + \hat{J}_-\right)\right] = \exp\left[i(\tan\theta)\hat{J}_+\right]\exp\left[-\ln(\cos^2\theta)\hat{J}_3\right]\exp\left[(i\tan\theta)\hat{J}_-\right], \tag{A5.10}$$

which may be generalized to

$$\begin{aligned} \exp\left(\lambda_+\hat{J}_+ + \lambda_-\hat{J}_- + \lambda_3\hat{J}_3\right) &= \exp\left(\Lambda_+\hat{J}_+\right)\exp\left[(\ln\Lambda_3)\hat{J}_3\right]\exp\left(\Lambda_-\hat{J}_-\right) \\ &= \exp\left(\Lambda_-\hat{J}_-\right)\exp\left[-(\ln\Lambda_3)\hat{J}_3\right]\exp\left(\Lambda_+\hat{J}_+\right), \end{aligned} \tag{A5.11}$$

where

$$\Lambda_3 = \left(\cosh\alpha - \frac{\lambda_3}{2\alpha}\sinh\alpha\right)^{-2}, \tag{A5.12}$$

$$\Lambda_\pm = \frac{2\lambda_\pm \sinh\alpha}{2\alpha\cosh\alpha - \lambda_3\sinh\alpha}, \tag{A5.13}$$

and

$$\alpha^2 = \tfrac{1}{4}\lambda_3^2 + \lambda_+\lambda_-. \tag{A5.14}$$

These theorems can be applied to a pair of field modes with annihilation operators $\hat{a}$ and $\hat{b}$ by making the identifications

$$\hat{J}_3 = \tfrac{1}{2}(\hat{a}^\dagger\hat{a} - \hat{b}^\dagger\hat{b}), \tag{A5.15}$$

$$\hat{J}_+ = \hat{a}^\dagger\hat{b} = \hat{J}_-^\dagger. \tag{A5.16}$$

The operators $\hat{K}_3$, $\hat{K}_+$, and $\hat{K}_-$ satisfy the commutation relations $[\hat{K}_3, \hat{K}_\pm] = \pm\hat{K}_\pm$ and $[\hat{K}_+, \hat{K}_-] = -2\hat{K}_3$. We have

$$\exp\left[i\theta\left(\hat{K}_+ + \hat{K}_-\right)\right] = \exp\left[i(\tanh\theta)\hat{K}_+\right]\exp\left[-\ln(\cosh^2\theta)\hat{K}_3\right]\exp\left[i(\tanh\theta)\hat{K}_-\right], \tag{A5.17}$$

which may be generalized to

$$\begin{aligned}\exp\left(\gamma_+\hat{K}_+ + \gamma_-\hat{K}_- + \gamma_3\hat{K}_3\right) &= \exp\left(\Gamma_+\hat{K}_+\right)\exp\left[(\ln\Gamma_3)\hat{K}_3\right]\exp\left(\Gamma_-\hat{K}_-\right)\\ &= \exp\left(\Gamma_-\hat{K}_-\right)\exp\left[-(\ln\Gamma_3)\hat{K}_3\right]\exp\left(\Gamma_+\hat{K}_+\right),\end{aligned} \tag{A5.18}$$

where

$$\Gamma_3 = \left(\cosh\beta - \frac{\gamma_3}{2\beta}\sinh\beta\right)^{-2}, \tag{A5.19}$$

$$\Gamma_\pm = \frac{2\gamma_\pm\sinh\beta}{2\beta\cosh\beta - \gamma_3\sinh\beta}, \tag{A5.20}$$

and

$$\beta^2 = \tfrac{1}{4}\gamma_3^2 - \gamma_+\gamma_-. \tag{A5.21}$$

These theorems can be applied either to a single field mode by making the identifications

$$\hat{K}_3 = \tfrac{1}{4}(\hat{a}^\dagger\hat{a} + \hat{a}\hat{a}^\dagger), \tag{A5.22}$$

$$\hat{K}_+ = \tfrac{1}{2}\hat{a}^{\dagger 2} = \hat{K}_-^\dagger, \tag{A5.23}$$

or to a pair of modes by making the identifications

$$\hat{K}_3 = \tfrac{1}{2}(\hat{a}^\dagger\hat{a} + \hat{b}\hat{b}^\dagger), \tag{A5.24}$$

$$\hat{K}_+ = \hat{a}^\dagger\hat{b}^\dagger = \hat{K}_-^\dagger. \tag{A5.25}$$

Two operators are equivalent if their matrix elements between any two basis states are equal, for all possible pairs of basis states. This can be used to derive ordering theorems. We illustrate this by obtaining (A5.6) using the number state and coherent state bases. As in Chapter 3, let

$$\exp(\theta\hat{a}^\dagger\hat{a}) = :\exp\left[p(\theta)\hat{a}^\dagger\hat{a}\right]:. \tag{A5.26}$$

The number state matrix elements of each side of (A5.26) are

$$\langle n|\exp(\theta\hat{a}^{\dagger}\hat{a})\,|n'\rangle = \exp(\theta n)\,\delta_{nn'} \tag{A5.27}$$

and

$$\begin{aligned}\langle n|{:}\exp\left[p(\theta)\hat{a}^{\dagger}\hat{a}\right]{:}|n'\rangle &= \sum_{l=0}^{n} C_l^n[p(\theta)]^l\,\delta_{nn'}\\ &= [1+p(\theta)]^n\,\delta_{nn'},\end{aligned} \tag{A5.28}$$

using (3.4.1). It follows immediately that $p(\theta)=\exp(\theta)-1$. Alternatively we can use the coherent state basis to establish this result. It is sufficient to equate only the diagonal matrix elements in this basis since, apart from a normalization factor $\exp(-|\alpha|^2/2)$, $|\alpha\rangle$ depends only on α while $\langle\alpha|$ depends only on α^*, and these are treated as two independent variables. The diagonal coherent state matrix elements of each side of (A5.26) are

$$\begin{aligned}\langle\alpha|\exp(\theta\hat{a}^{\dagger}\hat{a})\,|\alpha\rangle &= \langle\alpha\,|\,\alpha\exp(\theta)\rangle\exp\left\{\tfrac{1}{2}|\alpha|^2[\exp(2\theta)-1]\right\}\\ &= \exp\left\{-|\alpha|^2[1-\exp(\theta)]\right\},\end{aligned} \tag{A5.29}$$

using (3.6.32) and (3.6.24), and

$$\langle\alpha|{:}\exp\left[p(\theta)\hat{a}^{\dagger}\hat{a}\right]{:}|\alpha\rangle = \exp\left[|\alpha|^2 p(\theta)\right], \tag{A5.30}$$

using $\hat{a}\,|\alpha\rangle=\alpha\,|\alpha\rangle$. Equating (A5.29) and (A5.30) again gives $p(\theta)=\exp(\theta)-1$.

Appendix 6

THE POLE APPROXIMATION

In Section 5.3 we studied the dissipation caused by the interaction between a discrete state and a continuum. We found an expression for the Laplace transform of the amplitude $c_1(t)$ of the discrete state in the form

$$\bar{c}_1(s) = [s + I(s)]^{-1}, \tag{A6.1}$$

where

$$I(s) = \int \frac{W^2(\Delta)}{s + \mathrm{i}\Delta}\,\mathrm{d}\Delta. \tag{A6.2}$$

For convenience in this appendix we have replaced Δ_f by Δ and written W_f^2 as $W^2(\Delta)$. As we have seen, coupling to a flat continuum for which $W(\Delta) = W_0$ leads to exponential decay of $c_1(t)$ at a rate $\Gamma = \pi W_0^2$. If $W(\Delta)$ is not a constant but is still a sufficiently slowly varying function of Δ then we can hope to find an approximate solution corresponding to an exponential decay. Accordingly we assume that $\bar{c}_1(s)$ has a single dominant pole, that is one pole with a residue much greater than any others (should they exist). Hence we approximate $\bar{c}_1(s)$ by

$$\bar{c}_1(s) \simeq \frac{1}{s + s_0} \tag{A6.3}$$

with $\mathrm{Re}(s_0) > 0$, corresponding to an exponential decay in the time domain given by

$$c_1(t) = \exp(-s_0 t). \tag{A6.4}$$

We therefore require a self-consistent solution to the integral equation

$$s_0 = I(s_0) = \int \frac{W^2(\Delta)}{s_0 + \mathrm{i}\Delta}\,\mathrm{d}\Delta. \tag{A6.5}$$

In general s_0 will be complex and we write $s_0 = x + \mathrm{i}y$ with x and y real. Since exponential growth of the amplitude $c_1(t)$ is unphysical, we must have $x \geqslant 0$ so that x will be a decay rate and y a frequency shift. Separating the real and imaginary parts of (A6.5) gives the following simultaneous equations for x and y:

$$x = \mathrm{Re}[I(s_0)] = \int \frac{xW^2(\Delta)}{x^2 + (\Delta + y)^2}\,\mathrm{d}\Delta, \tag{A6.6}$$

$$y = \mathrm{Im}[I(s_0)] = -\int \frac{(\Delta + y)W^2(\Delta)}{x^2 + (\Delta + y)^2}\,\mathrm{d}\Delta. \tag{A6.7}$$

We are interested here in broad, slowly varying coupling to the continuum. Hence we assume that the value of x is such that $W^2(\Delta)$ in (A6.6) is slowly varying compared with the Lorentzian $x/[x^2+(\Delta+y)^2]$. This assumption, which must be checked for consistency once x has been calculated, means that we can approximate $W^2(\Delta)$ by its value at the peak of the Lorentzian ($\Delta=-y$) and find

$$x \simeq W^2(-y)\int \frac{x}{x^2+(\Delta+y)^2}\,\mathrm{d}\Delta = \pi W^2(-y). \tag{A6.8}$$

The decay rate is given by the value of W^2 at a frequency shifted by $-y$ from resonance with the original discrete state. We must now check the self-consistency of our solution by verifying that $W^2(\Delta)$ varies only slowly over the width $2x$ of the Lorentzian.

In some situations it may be necessary or of interest to calculate the frequency shift y explicitly. This means finding a solution to the equation

$$y = -\int \frac{(\Delta+y)W^2(\Delta)}{x^2+(\Delta+y)^2}\,\mathrm{d}\Delta. \tag{A6.9}$$

In order to proceed we use the property, verified in the calculation of x, that W^2 is approximately constant as Δ changes by $2x$. Since $(\Delta+y)/[x^2+(\Delta+y)^2]$ in (A6.9) is an odd function of Δ about $\Delta=-y$ and W^2 is slowly varying, the contribution to the integral from the region $-y-x\leqslant\Delta\leqslant -y+x$ is approximately zero. Hence the non-zero contribution to y comes from the regions $|\Delta+y|>x$ and we can approximate (A6.9) by omitting the region $-x\leqslant\Delta+y\leqslant x$ giving

$$y = -\mathbb{P}\int \frac{W^2(\Delta)}{\Delta+y}\,\mathrm{d}\Delta. \tag{A6.10}$$

In general, this will be a transcendental equation for y. However, if the continuum is sufficiently flat so that $W(\Delta)\simeq W(\Delta\pm y)$ then we can approximate $W^2(\Delta)$ by $W^2(\Delta+y)$ in (A6.10) and $W^2(-y)$ by $W^2(0)$ in (A6.8). Combining these we find

$$s_0 = \pi W^2(0) - \mathrm{i}\mathbb{P}\int \frac{W^2(\Delta)}{\Delta}\,\mathrm{d}\Delta. \tag{A6.11}$$

We can obtain this directly from (A6.5) by applying the limiting procedure

$$s_0 = \lim_{s\to 0^+} I(s) \tag{A6.12}$$

using (A7.7).

We end this appendix with a few remarks concerning the assumptions leading to (A6.11) and the range of its validity. We have assumed that $W^2(\Delta)$ is slowly varying on two scales, firstly on the scale x so that $W^2(\Delta) \simeq W^2(\Delta \pm x)$ and secondly on the scale y so that $W^2(\Delta) \simeq W^2(\Delta \pm y)$. If the first of these assumptions is violated then the whole procedure outlined above is inconsistent and we must find another approach. In this case we will find a more complicated evolution than simple exponential decay. However, if $W^2(\Delta) \simeq W^2(\Delta \pm x)$ but y is sufficiently large that the second assumption is violated then we can still say something about the decay rate x but not about the frequency shift y. We recall the comment about the natural measured frequency or energy of the discrete state. If we include the now unknown frequency shift into a redefinition of the natural measured energy of the discrete state then $x = \pi W^2(0)$ where the condition $\Delta = 0$ now corresponds to the coupling at the *observed* resonance position. This procedure of redefining the natural frequency in terms of the observed frequency is analogous to the procedure of renormalization in quantum field theory.

Appendix 7
PRINCIPAL PART INTEGRALS

Consider an integral $\int_a^b f(x)\,\mathrm{d}x$ where $f(x)$ diverges at the point $x = x_0$ within the range of integration. The integral is then not defined in the usual sense because $f(x)$ becomes infinite at $x = x_0$. However, we may define the principal part integral by

$$\mathbb{P}\int_a^b f(x)\,\mathrm{d}x = \lim_{\delta\to 0}\left(\int_a^{x_0-\delta} f(x)\,\mathrm{d}x + \int_{x_0+\delta}^b f(x)\,\mathrm{d}x\right). \tag{A7.1}$$

This limit will lead to a finite value in many cases. Here we will be concerned only with functions $f(x)$ having a single simple pole at $x = x_0$; that is, $f(x)$ behaves as $(x - x_0)^{-1}$ near this point. We write $f(x) = g(x)/(x - x_0)$ where $g(x)$ is finite in the range of integration. An alternative way to express the principal part integral is

$$\mathbb{P}\int_a^b f(x)\,\mathrm{d}x = \lim_{\epsilon\to 0}\int_a^b \frac{(x-x_0)g(x)}{(x-x_0)^2+\epsilon^2}\,\mathrm{d}x. \tag{A7.2}$$

The integral must be evaluated *before* the limit is taken.

Now consider the integral I defined by the limit

$$I = \lim_{\epsilon\to 0^+}\int_a^b \frac{g(x)}{x - x_0 + \mathrm{i}\epsilon}\,\mathrm{d}x, \tag{A7.3}$$

where ϵ tends to zero from above. Separating the real and imaginary parts of the integrand leads to

$$I = \lim_{\epsilon\to 0^+}\left(\int_a^b \frac{(x-x_0)g(x)}{(x-x_0)^2+\epsilon^2}\,\mathrm{d}x - \mathrm{i}\pi\int_a^b \frac{(\epsilon/\pi)g(x)}{(x-x_0)^2+\epsilon^2}\,\mathrm{d}x\right). \tag{A7.4}$$

We recognize the real part of I as the principal part integral given in (A7.2). In the limit $\epsilon \to 0^+$, the Lorentzian part of the integrand of the imaginary part of I tends to the delta function $\delta(x - x_0)$. Hence

$$I = \mathbb{P}\int_a^b \frac{g(x)}{(x-x_0)}\,\mathrm{d}x - \mathrm{i}\pi g(x_0). \tag{A7.5}$$

It is usual to write

$$\lim_{\epsilon\to 0^+}\frac{1}{x - x_0 + \mathrm{i}\epsilon} = \frac{\mathbb{P}}{x - x_0} - \mathrm{i}\pi\delta(x - x_0), \tag{A7.6}$$

with the understanding that this equality holds when the quantities appear within an integral. Similarly

$$\lim_{\epsilon\to 0^+} \frac{1}{x-x_0-\mathrm{i}\epsilon} = \frac{\mathbb{P}}{x-x_0} + \mathrm{i}\pi\delta(x-x_0). \qquad \text{(A7.7)}$$

In Chapter 6 we used the identity

$$\frac{\mathbb{P}}{\omega'-\Delta}\cdot\frac{\mathbb{P}}{\omega-\Delta} = \frac{\mathbb{P}}{\omega'-\omega}\left(\frac{\mathbb{P}}{\omega-\Delta} - \frac{\mathbb{P}}{\omega'-\Delta}\right) + \pi^2\,\delta(\omega'-\Delta)\,\delta(\omega-\Delta). \qquad \text{(A7.8)}$$

Here we prove this identity using the properties of the principal part established above. Consider the double limit

$$L = \lim_{\epsilon,\epsilon'\to 0^+} \frac{1}{\omega'-\Delta-\mathrm{i}\epsilon}\cdot\frac{1}{\omega-\Delta+\mathrm{i}\epsilon'}, \qquad \text{(A7.9)}$$

where ϵ and ϵ' tend to zero independently. Using (A7.6) and (A7.7), this limit can be written

$$\begin{aligned} L &= \left(\frac{\mathbb{P}}{\omega'-\Delta} + \mathrm{i}\pi\delta(\omega'-\Delta)\right)\left(\frac{\mathbb{P}}{\omega-\Delta} - \mathrm{i}\pi\delta(\omega-\Delta)\right) \\ &= \frac{\mathbb{P}}{\omega'-\Delta}\cdot\frac{\mathbb{P}}{\omega-\Delta} + \mathrm{i}\pi\delta(\omega'-\Delta)\frac{\mathbb{P}}{\omega-\Delta} - \mathrm{i}\pi\delta(\omega-\Delta)\frac{\mathbb{P}}{\omega'-\Delta} \\ &\quad + \pi^2\,\delta(\omega'-\Delta)\,\delta(\omega-\Delta). \end{aligned} \qquad \text{(A7.10)}$$

Alternatively, we can separate L into partial fractions as follows:

$$L = \lim_{\epsilon,\epsilon'\to 0^+}\left[\frac{1}{\omega'-\omega-\mathrm{i}(\epsilon+\epsilon')}\left(\frac{1}{\omega-\Delta+\mathrm{i}\epsilon'} - \frac{1}{\omega'-\Delta-\mathrm{i}\epsilon}\right)\right]. \qquad \text{(A7.11)}$$

Again using (A7.6) and (A7.7), this becomes

$$\begin{aligned} L &= \frac{\mathbb{P}}{\omega'-\omega}\left(\frac{\mathbb{P}}{\omega-\Delta} - \frac{\mathbb{P}}{\omega'-\Delta}\right) - \mathrm{i}\frac{\mathbb{P}}{\omega'-\omega}[\pi\delta(\omega-\Delta) + \pi\delta(\omega'-\Delta)] \\ &\quad + \mathrm{i}\pi\delta(\omega'-\omega)\left(\frac{\mathbb{P}}{\omega-\Delta} - \frac{\mathbb{P}}{\omega'-\Delta}\right) \\ &\quad + \pi\delta(\omega'-\omega)[\pi\delta(\omega-\Delta) + \pi\delta(\omega'-\Delta)]. \end{aligned} \qquad \text{(A7.12)}$$

Using the property $\delta(x-y)f(y) = \delta(x-y)f(x)$, we see that the third term is zero and that

$$\begin{aligned} L &= \frac{\mathbb{P}}{\omega'-\omega}\left(\frac{\mathbb{P}}{\omega-\Delta} - \frac{\mathbb{P}}{\omega'-\Delta}\right) + \mathrm{i}\pi\delta(\omega'-\Delta)\frac{\mathbb{P}}{\omega-\Delta} \\ &\quad - \mathrm{i}\pi\delta(\omega-\Delta)\frac{\mathbb{P}}{\omega'-\Delta} + 2\pi^2\,\delta(\omega'-\Delta)\,\delta(\omega-\Delta). \end{aligned} \qquad \text{(A7.13)}$$

Comparing (A7.10) with (A7.13) gives the required result stated in (A7.8).

Appendix 8
CONTOUR INTEGRALS

The Cauchy residue theorem states that the integral anticlockwise around a closed contour in the complex plane of any function $f(z)$ which is analytic apart from poles within the contour is equal to $2\pi i$ multiplied by the sum of the residues of $f(z)$ at these poles. We use this theorem to evaluate some of the integrals required in Chapters 5 and 6.

First we consider the integral

$$I_1 = \int_{-\infty}^{\infty} \frac{W^2}{\omega^2 + \pi^2 W^4} \exp(-i\omega t)\, d\omega, \tag{A8.1}$$

required in deriving (5.5.11) and (6.4.40). As written, this is the form required in Chapter 6; for those integrals in (5.5.10), we identify ω with Δ, W with W_0, and note that t in (A8.1) can be set equal to zero. We choose a contour consisting of the part of the real axis from $-R$ to $+R$, closed by the semicircle $R\exp(i\theta)$ in the lower half plane, as shown in Fig. A8.1. The lower half plane is chosen to ensure that the integrand tends to zero on the semicircle as $R \to \infty$. The simple pole of the integrand in the lower half plane is at $\omega = -i\pi W^2$ and the residue there is $W^2 \exp(-\pi W^2 t)/(-2\pi i W^2)$. Letting $R \to \infty$ gives

$$I_1 = -2\pi i W^2 \exp(-\pi W^2 t)/(-2\pi i W^2) = \exp(-\pi W^2 t), \tag{A8.2}$$

where the minus sign arises because the contour is traversed in the *clockwise* direction. Note that if the integrand contained $\exp(i\omega t)$ rather than $\exp(-i\omega t)$, then we could close the contour in the upper half plane and obtain the same result (A8.2).

The second integral

$$I_2 = \mathbb{P} \int_{-\infty}^{\infty} \frac{d\Delta}{(\Delta^2 + \gamma^2)(\omega - \Delta)} \tag{A8.3}$$

is required in deriving (6.4.44). We choose a contour consisting of the part of the real axis from $-R$ to $+R$, indented by constructing a small semicircle of radius ρ

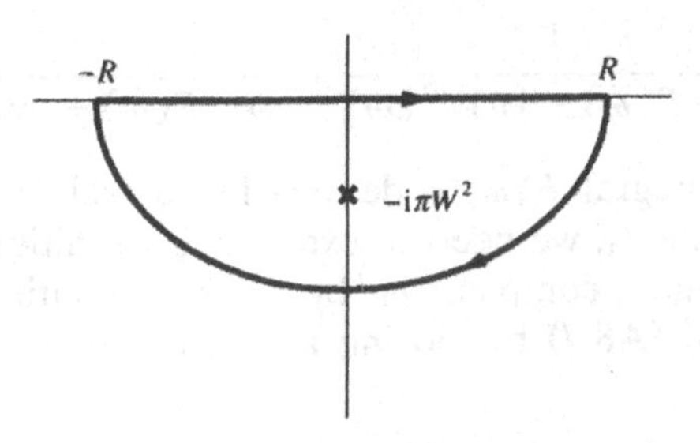

Fig. A8.1 A semicircular contour closed in the lower half plane.

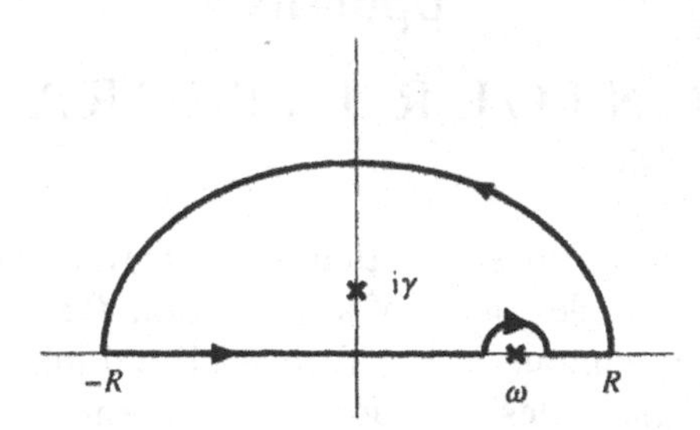

Fig. A8.2 An indented semicircular contour closed in the upper half plane.

in the upper half plane, closed by the semicircle $R\exp(\mathrm{i}\theta)$ also in the upper half plane, as shown in Fig. A8.2. The simple pole of the integrand in the upper half plane is at $\Delta = \mathrm{i}\gamma$ and the residue there is $[2\mathrm{i}\gamma(\omega - \mathrm{i}\gamma)]^{-1}$. The integral around the large semicircle tends to zero as $R \to \infty$ while the integral around the small semicircle as $\rho \to 0$ is

$$\lim_{\rho\to 0}\int_{\pi}^{0} \frac{\mathrm{i}\rho\exp(\mathrm{i}\theta)\,\mathrm{d}\theta}{-\rho\exp(\mathrm{i}\theta)(\omega^2+\gamma^2)} = \frac{\pi\mathrm{i}}{(\omega^2+\gamma^2)}. \tag{A8.4}$$

Hence

$$I_2 = \frac{2\pi\mathrm{i}}{2\mathrm{i}\gamma(\omega-\mathrm{i}\gamma)} - \frac{\pi\mathrm{i}}{(\omega^2+\gamma^2)} = \frac{\pi\omega}{\gamma(\omega^2+\gamma^2)}. \tag{A8.5}$$

The third integral we evaluate is that in (6.5.21), required to establish the completeness of the dressed states $|\omega\rangle$. Consider

$$I_3 = \int_{-\infty}^{\infty} |\alpha(\omega)|^2\,\mathrm{d}\omega = \int_{-\infty}^{\infty} \frac{W^2(\omega)}{[\omega - F(\omega)]^2 + \pi^2 W^4(\omega)}\,\mathrm{d}\omega, \tag{A8.6}$$

where

$$F(\omega) = \mathbb{P}\int_{-\infty}^{\infty} \frac{W^2(\Delta)}{\omega - \Delta}\,\mathrm{d}\Delta. \tag{A8.7}$$

We proceed by separating the integrand of I_3 into partial fractions, so that

$$I_3 = \frac{1}{2\pi\mathrm{i}}\int_{-\infty}^{\infty}\left(\frac{1}{\omega - F(\omega) - \mathrm{i}\pi W^2(\omega)} - \frac{1}{\omega - F(\omega) + \mathrm{i}\pi W^2(\omega)}\right)\mathrm{d}\omega. \tag{A8.8}$$

The principal part integral $F(\omega)$ is defined for ω real. In order to use contour integration to evaluate I_3, we need to extend the definition of $F(\omega)$, and hence of $|\alpha(\omega)|^2$, to include complex ω by analytic continuation. We formally evaluate the integral (A8.7) by closing a contour in the upper half plane, as for I_2, giving

$$F(\omega) = 2\pi\mathrm{i}S(\omega) - \pi\mathrm{i}W^2(\omega), \tag{A8.9}$$

where $S(\omega)$ is the sum of the residues of $W^2(\Delta)/(\omega-\Delta)$ at the poles of $W^2(\Delta)$ in the upper half plane. Consider the function $G(\omega)$ defined, for values of ω in the upper half plane, by

$$G(\omega)=\int_{-\infty}^{\infty}\frac{W^2(\Delta)}{\omega-\Delta}\,\mathrm{d}\Delta. \tag{A8.10}$$

Evaluating this by contour integration in the upper half plane gives

$$G(\omega)=2\pi\mathrm{i}S(\omega)-2\pi\mathrm{i}W^2(\omega). \tag{A8.11}$$

Comparing this with $F(\omega)$ given in (A8.9), we see that $F(\omega)$ becomes $G(\omega)+\pi\mathrm{i}W^2(\omega)$ in the upper half plane. Similarly in the lower half plane, $G(\omega)=2\pi\mathrm{i}S(\omega)$ and $F(\omega)$ becomes $G(\omega)-\pi\mathrm{i}W^2(\omega)$. The integrand of I_3 in (A8.8) analytically continued into the upper half plane is therefore given by

$$\frac{1}{2\pi\mathrm{i}}\left(\frac{1}{\omega-G(\omega)-2\pi\mathrm{i}W^2(\omega)}-\frac{1}{\omega-G(\omega)}\right). \tag{A8.12}$$

The second term in this expression has no poles in the upper half plane because $\omega-G(\omega)$ has no zeros there, as we now show. Writing $\omega=x+\mathrm{i}y$ and setting $\omega-G(\omega)=0$ leads to

$$\omega-\int_{-\infty}^{\infty}\frac{W^2(\Delta)}{\omega-\Delta}\,\mathrm{d}\Delta=x+\mathrm{i}y-\int_{-\infty}^{\infty}\frac{(x-\mathrm{i}y)W^2(\Delta)}{(x-\Delta)^2+y^2}\,\mathrm{d}\Delta=0, \tag{A8.13}$$

in which the real and imaginary parts must each be zero. However, the imaginary part is strictly positive if $y>0$, corresponding to the upper half plane, and hence there are no solutions. We now evaluate I_3 by contour integration closing the contour, as before, with a large semicircle of radius R in the upper half plane. We add to I_3 the integral of (A8.12) around the semicircle, this additional integral being identically zero in the limit $R\to\infty$. The second terms of I_3 and of the integral of (A8.12) combine to give a zero contribution because $[\omega-G(\omega)]^{-1}$ has no poles in the upper half plane. Hence I_3 is given by the remaining integrals, as follows:

$$\begin{aligned}I_3&=\frac{1}{2\pi\mathrm{i}}\lim_{R\to\infty}\left(\int_{-R}^{R}\frac{\mathrm{d}\omega}{\omega-F(\omega)-\mathrm{i}\pi W^2(\omega)}\right.\\&\quad\left.+\int_0^{\pi}\frac{\mathrm{i}R\exp(\mathrm{i}\theta)\,\mathrm{d}\theta}{R\exp(\mathrm{i}\theta)-G(R\exp(\mathrm{i}\theta))-2\pi\mathrm{i}W^2(R\exp(\mathrm{i}\theta))}\right)\\&=\frac{1}{2\pi\mathrm{i}}\left(\int_{-\infty}^{\infty}\frac{\mathrm{i}\pi W^2(\omega)\,\mathrm{d}\omega}{[\omega-F(\omega)]^2+\pi^2W^4(\omega)}\right.\\&\quad\left.+\int_{-\infty}^{\infty}\frac{[\omega-F(\omega)]\,\mathrm{d}\omega}{[\omega-F(\omega)]^2+\pi^2W^4(\omega)}+\pi\mathrm{i}\right),\end{aligned} \tag{A8.14}$$

where we note that $W^2(R\exp(\mathrm{i}\theta))$ must tend to a finite value and $G(R\exp(\mathrm{i}\theta))$ must tend to zero as $R\to\infty$ in order for $F(\omega)$ to be finite. Since I_3 is real, the second term of (A8.14) must be zero, as it is purely imaginary. The first term is just $I_3/2$ while the last is simply $1/2$. Hence $I_3 = 1$ as required.

The Cauchy residue theorem can also be used to sum particular infinite series. Suppose $f(z)$ is a function which is analytic at the integers $z = 0, \pm1, \pm2, \ldots$ and tends to zero at least as fast as $|z|^{-2}$ as $|z|\to\infty$. Then

$$\sum_{n=-\infty}^{+\infty} f(n) = -\{\text{sum of the residues of } \pi\cot(\pi z)\cdot f(z) \text{ at the poles of } f(z)\}. \tag{A8.15}$$

In Chapter 6 we require the sum

$$S_1 = \sum_{n=-\infty}^{+\infty} \frac{1}{\omega - n\omega_0}. \tag{A8.16}$$

It appears that the terms in this sum do not tend to zero sufficiently fast as n increases to allow use of (A8.15). However, this can be overcome by writing S_1 as

$$\begin{aligned} S_1 &= \frac{1}{\omega} + \sum_{n=1}^{\infty}\left(\frac{1}{\omega - n\omega_0} + \frac{1}{\omega + n\omega_0}\right) \\ &= \frac{1}{\omega} + \sum_{n=1}^{\infty} \frac{2\omega}{\omega^2 - n^2\omega_0^2} \\ &= \sum_{n=-\infty}^{+\infty} \frac{\omega}{\omega^2 - n^2\omega_0^2}, \end{aligned} \tag{A8.17}$$

the terms of which are of order n^{-2} for large n and therefore tend to zero sufficiently fast as $n\to\infty$ to allow the application of (A8.15). The simple poles of $f(z) = \omega/[\omega^2 - z^2\omega_0^2]$ are at $z = \pm\omega/\omega_0$. The residues of $\pi\cot(\pi z)\cdot f(z)$ at these points are both $-(\pi/2\omega_0)\cot(\pi\omega/\omega_0)$. Hence

$$S_1 = \frac{\pi}{\omega_0}\cot\left(\frac{\pi\omega}{\omega_0}\right). \tag{A8.18}$$

We also require the sum

$$\begin{aligned} S_2 &= \sum_{n=-\infty}^{\infty} \frac{n}{(\Omega - n\omega_0)^2} \\ &= \sum_{n=1}^{\infty} n\left(\frac{1}{(\Omega - n\omega_0)^2} - \frac{1}{(\Omega + n\omega_0)^2}\right) \\ &= \sum_{n=-\infty}^{\infty} \frac{2\Omega\omega_0 n^2}{(\Omega^2 - n^2\omega_0^2)^2}. \end{aligned} \tag{A8.19}$$

Again, the terms are of order n^{-2} for large n as required. The double poles of $f(z) = 2\Omega\omega_0 z^2/(\Omega^2 - z^2\omega_0^2)^2$ are at $z = \pm\Omega/\omega_0$. The residue of $\pi\cot(\pi z)\cdot f(z)$ at either pole may be calculated, using the formula for double poles, as

$$\frac{\mathrm{d}}{\mathrm{d}z}\left[(z \mp \Omega/\omega_0)^2 \pi\cot(\pi z)\cdot f(z)\right]\Big|_{z=\pm\Omega/\omega_0}$$

$$= \frac{\pi}{2\omega_0^2}\left[\cot\left(\frac{\pi\Omega}{\omega_0}\right) - \frac{\pi\Omega}{\omega_0}\operatorname{cosec}^2\left(\frac{\pi\Omega}{\omega_0}\right)\right]. \quad \text{(A8.20)}$$

From (A8.15), summing the residues, we have

$$S_2 = \frac{\pi}{\omega_0^2}\left[\frac{\pi\Omega}{\omega_0}\operatorname{cosec}^2\left(\frac{\pi\Omega}{\omega_0}\right) - \cot\left(\frac{\pi\Omega}{\omega_0}\right)\right]. \quad \text{(A8.21)}$$

Appendix 9

LAPLACE TRANSFORMS AND THE FINAL VALUE THEOREM

Suppose that $f(t)$ is a function defined for $t \geqslant 0$. The Laplace transform of $f(t)$ is given by

$$\mathscr{L}\{f(t)\} = \bar{f}(s) = \int_0^{\infty} f(t)\exp(-st)\,\mathrm{d}t, \tag{A9.1}$$

where s is, in general, complex. If $f(t)$ is of exponential order, that is if $|f(t)\exp(-\alpha t)| \leqslant M$ as $t \to \infty$, where α and M are suitable (real) constants, then the integral in (A9.1) defining the Laplace transform will exist for $\mathrm{Re}(s) > \alpha$. The general inversion formula, giving the function $f(t)$ in terms of its Laplace transform $\bar{f}(s)$, is

$$f(t) = \frac{1}{2\pi \mathrm{i}} \int_{\gamma - \mathrm{i}\infty}^{\gamma + \mathrm{i}\infty} \exp(st)\bar{f}(s)\,\mathrm{d}s, \tag{A9.2}$$

where the path of integration in the complex s-plane is along the Bromwich contour. This is a straight line with equation $s = \gamma + \mathrm{i}y$, $-\infty < y < \infty$, γ being chosen so that all the singularities of $\bar{f}(s)$ lie to the left of this line. We often evaluate the integral in (A9.2) by taking a finite straight line, closing the contour in a suitable way, and then letting the closed contour expand to infinity, as shown in Fig. A9.1. The required integral is obtained by using the Cauchy residue theorem. If $\bar{f}(s)$ is analytic apart from poles (which lie to the left of the Bromwich contour by construction), we obtain

$$f(t) = \{\text{sum of residues of } \exp(st)\bar{f}(s) \text{ at the poles of } \bar{f}(s)\}. \tag{A9.3}$$

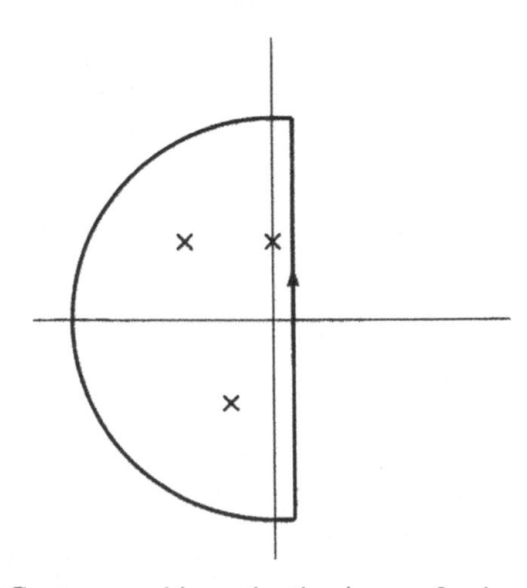

Fig. A9.1 Contour used in evaluating inverse Laplace transforms.

In order to calculate the residue r_0 of $\exp(st)\bar{f}(s)$ at a pole s_0 of order m of $\bar{f}(s)$, we can employ the general formula

$$r_0 = \lim_{s \to s_0} \frac{1}{(m-1)!} \frac{\mathrm{d}^{m-1}}{\mathrm{d}s^{m-1}} \left[(s - s_0)^m \exp(st)\bar{f}(s) \right]. \tag{A9.4}$$

In particular, for a simple pole (pole of order 1) we find

$$r_0 = \lim_{s \to s_0} (s - s_0) \exp(st)\bar{f}(s) = \left[\lim_{s \to s_0} (s - s_0)\bar{f}(s) \right] \exp(s_0 t). \tag{A9.5}$$

In Chapter 5, we obtain the Laplace transform $\bar{\hat{a}}(s)$ of the time-dependent annihilation operator $\hat{a}(t)$ given by

$$\bar{\hat{a}}(s) = \frac{\hat{a}(0)}{s + \Gamma} - \mathrm{i}W_0 \int \frac{\hat{b}(\Delta, 0)}{(s + \Gamma)(s + \mathrm{i}\Delta)} \mathrm{d}\Delta \tag{A9.6}$$

(see (5.5.7)), where $\Gamma = \pi W_0^2$. To invert this expression, we note that the first term has a simple pole at $s = -\Gamma$ of residue $\hat{a}(0)$. The residue of $\bar{\hat{a}}(s)\exp(st)$ arising from this term will therefore be $\hat{a}(0)\exp(-\Gamma t)$. Likewise, the second term of (A9.6) contains the expression $[(s + \Gamma)(s + \mathrm{i}\Delta)]^{-1}$ which has two poles. For the simple pole at $s = -\Gamma$, this term has a residue $(-\Gamma + \mathrm{i}\Delta)^{-1} = -(\Gamma - \mathrm{i}\Delta)^{-1}$, while for the simple pole at $s = -\mathrm{i}\Delta$, the term has a residue $(\Gamma - \mathrm{i}\Delta)^{-1}$. Summing the residues of $\bar{\hat{a}}(s)\exp(st)$ as required by (A9.3) and retaining, of course, the integration over Δ in (A9.6), we obtain

$$\hat{a}(t) = \hat{a}(0)\exp(-\Gamma t) - \mathrm{i}W_0 \int \frac{\hat{b}(\Delta, 0)}{(\Gamma - \mathrm{i}\Delta)} [\exp(-\mathrm{i}\Delta t) - \exp(-\Gamma t)] \,\mathrm{d}\Delta, \tag{A9.7}$$

as stated in (5.5.8).

It is readily seen from (A9.4) that the residue of $\exp(st)\bar{f}(s)$ at a pole s_0 of $\bar{f}(s)$ will contain the factor $\exp(s_0 t)$. In physical problems of the type considered in this book, we are usually concerned with quantities which oscillate or exponentially *decay*. This will mean that $\bar{f}(s)$ has poles which lie either on the imaginary axis, giving an oscillatory term, or in the left-hand half plane, giving a term which exponentially decays in magnitude. Consider such a function $\bar{f}(s)$ having poles s_i $(i = 1, 2, \ldots)$ in the left-hand half plane, as well as a single simple pole $s_0 = \mathrm{i}\theta$ lying on the imaginary axis. Using (A9.5), the residue at s_0 is $A\exp(\mathrm{i}\theta t)$ where

$$A = \lim_{s \to \mathrm{i}\theta} (s - \mathrm{i}\theta)\bar{f}(s). \tag{A9.8}$$

The other residues will each be of the form $B_i \exp(s_i t)$ $(i = 1, 2, \ldots)$ and hence, from (A9.3),

$$f(t) = A\exp(\mathrm{i}\theta t) + \sum_i B_i \exp(s_i t). \tag{A9.9}$$

Since each s_i has a negative real part, each term of the summation over i in (A9.9) exponentially decays in magnitude as t increases. If we let $t \to \infty$, only the first term, arising from the simple pole on the imaginary axis, remains and we obtain

$$|f(t \to \infty)|^2 = |A|^2 = \left| \lim_{s \to i\theta} (s - i\theta)\bar{f}(s) \right|^2, \tag{A9.10}$$

using (A9.8). This is the form of the final value theorem used in Chapter 5 (see (5.4.2)). It is important to realize that the final value theorem in the form (A9.10) *only* applies if there is precisely one simple pole on the imaginary axis. If there is more than one pole on the imaginary axis then $f(t)$ will contain more than one term with a purely imaginary exponent. All such terms will persist in the long-time limit so that $|f(t \to \infty)|^2$ is not constant but oscillates. Hence, in this case, there is no steady-state limit.

If $\bar{f}(s)$ and $\bar{g}(s)$ are the Laplace transforms of $f(t)$ and $g(t)$, respectively, then the Laplace transform of the convolution integral

$$f * g = \int_0^t f(t-\tau)g(\tau)\,d\tau \tag{A9.11}$$

is given by

$$\begin{aligned}\mathscr{L}\{f * g\} &= \int_0^\infty \exp(-st)\,dt \int_0^t f(t-\tau)g(\tau)\,d\tau \\ &= \int_0^\infty g(\tau)\,d\tau \int_\tau^\infty \exp(-st)f(t-\tau)\,dt,\end{aligned} \tag{A9.12}$$

on reversing the order of the integrations. Writing $t - \tau = u$ in (A9.12), we find

$$\begin{aligned}\mathscr{L}\{f * g\} &= \int_0^\infty g(\tau)\,d\tau \int_0^\infty \exp[-s(u+\tau)]f(u)\,du \\ &= \int_0^\infty \exp(-su)f(u)\,du \int_0^\infty \exp(-s\tau)g(\tau)\,d\tau \\ &= \bar{f}(s)\bar{g}(s).\end{aligned} \tag{A9.13}$$

We use this result in Section 2.3.

If the Laplace transform of $f(t)$ is $\bar{f}(s)$ and $H(t-a)$ is the Heaviside step function (see (A2.2)) with $a > 0$, then

$$\begin{aligned}\mathscr{L}\{H(t-a)f(t-a)\} &= \int_0^\infty \exp(-st)H(t-a)f(t-a)\,dt \\ &= \int_a^\infty \exp(-st)f(t-a)\,dt.\end{aligned} \tag{A9.14}$$

Writing $t - a = u$ in (A9.14), we find

$$\mathscr{L}\{H(t-a)f(t-a)\} = \int_0^\infty \exp[-s(a+u)]f(u)\,du$$
$$= \exp(-sa)\bar{f}(s). \tag{A9.15}$$

This result can be used to calculate the Laplace inversion of the function

$$\bar{b}(s) = \left[s + \frac{\pi W^2}{\Delta}\coth\left(\frac{\pi s}{\Delta}\right)\right]^{-1} \tag{A9.16}$$

required in Chapter 2 (see (2.5.7)). We first write $\coth(\pi s/\Delta)$ in terms of exponentials so that

$$\coth\left(\frac{\pi s}{\Delta}\right) = \frac{1 + \exp(-2\pi s/\Delta)}{1 - \exp(-2\pi s/\Delta)}$$
$$= 1 + \frac{2\exp(-2\pi s/\Delta)}{1 - \exp(-2\pi s/\Delta)}. \tag{A9.17}$$

Substituting (A9.17) into (A9.16) gives

$$\bar{b}(s) = \left(s + \frac{\pi W^2}{\Delta}\right)^{-1}\left(1 + \frac{(2\pi W^2/\Delta)\exp(-2\pi s/\Delta)}{(s + \pi W^2/\Delta)[1 - \exp(-2\pi s/\Delta)]}\right)^{-1}$$
$$= \sum_{m=0}^{\infty} \frac{(-2\pi W^2/\Delta)^m}{(s + \pi W^2/\Delta)^{m+1}} \frac{\exp(-2\pi m s/\Delta)}{[1 - \exp(-2\pi s/\Delta)]^m}. \tag{A9.18}$$

It is possible to choose $\mathrm{Re}(s) > 0$ large enough so that this expansion is convergent. We will find that $b(t)$ obtained by inverting (A9.18) is a convergent series for all finite t and is the required solution, as may be verified by back-substitution. Now

$$\frac{\theta^m}{(1-\theta)^m} = \begin{cases} 1 & m = 0, \\ \displaystyle\sum_{k=m}^{\infty} \frac{\theta^k (k-1)!}{(k-m)!(m-1)!} & m > 0, \end{cases} \tag{A9.19}$$

for $|\theta| < 1$, and using this result in (A9.18) with $\theta = \exp(-2\pi s/\Delta)$ gives

$$\bar{b}(s) = \frac{1}{s + \pi W^2/\Delta} + \sum_{m=1}^{\infty} \frac{(-2\pi W^2/\Delta)^m}{(s + \pi W^2/\Delta)^{m+1}} \sum_{k=m}^{\infty} \frac{(k-1)!\exp(-2\pi s k/\Delta)}{(k-m)!(m-1)!}. \tag{A9.20}$$

In (A9.20), we therefore require the inversion

$$\mathscr{L}^{-1}\left\{\frac{\exp(-2\pi sk/\Delta)}{(s+\pi W^2/\Delta)^{m+1}}\right\}=H(t-2\pi k/\Delta)\frac{(t-2\pi k/\Delta)^m}{m!}$$
$$\times\exp\left[-\frac{\pi W^2}{\Delta}\left(t-\frac{2\pi k}{\Delta}\right)\right], \qquad (A9.21)$$

using (A9.15). Writing

$$T=\frac{\Delta t}{2\pi},\quad \beta=\frac{2\pi^2W^2}{\Delta^2}, \qquad (A9.22)$$

we find that the inversion of (A9.20) is

$$b(T)=\exp(-\beta T)$$
$$\times\left(1+\sum_{m=1}^{\infty}\sum_{k=m}^{\infty}\frac{(k-1)!(-2\beta)^m}{(k-m)!\,m!\,(m-1)!}(T-k)^m H(T-k)\exp(\beta k)\right)$$
$$=\exp(-\beta T)\left(1+\sum_{k=1}^{\infty}\exp(\beta k)H(T-k)\sum_{m=1}^{k}\frac{[-2\beta(T-k)]^m(k-1)!}{(k-m)!\,m!\,(m-1)!}\right), \qquad (A9.23)$$

on reversing the order of the summations. Finally, using (A3.26) and (A3.27), (A9.23) can be written

$$b(T)=\exp(-\beta T)\left(1-2\beta\sum_{k=1}^{\infty}\exp(\beta k)\frac{(T-k)}{k}H(T-k)L_k^{(1)}[2\beta(T-k)]\right), \qquad (A9.24)$$

as in (2.5.8), where $L_k^{(1)}$ is the associated Laguerre polynomial.

Appendix 10

OPERATOR ORDERING IN THE HEISENBERG EQUATIONS

The Heisenberg operator equations will in general contain products of non-commuting operators and care must be taken to preserve the ordering of these. Interchanging pairs of commuting operators in the Hamiltonian leads to other forms of the Hamiltonian which are equivalent in the sense that the dynamics is unchanged. It is important, therefore, that any approximation methods used to solve the Heisenberg equations produce solutions which do not depend on the form of the Hamiltonian chosen.

In Chapter 5, we used a decorrelation approximation to derive the equations of motion (5.5.66) to (5.5.68) for a two-level atom coupled to a broad-band radiation field. In this appendix, we show that these equations are independent of the ordering of commuting operators in the Hamiltonian. The interaction term in the Hamiltonian (5.5.43) contains products of equal time and therefore commuting atom and field operators. An equivalent form of this Hamiltonian is therefore

$$\begin{aligned}\hat{H}_{\mathrm{I}} = \hbar \int \Delta \hat{b}^{\dagger}(\Delta,t)\hat{b}(\Delta,t)\,\mathrm{d}\Delta + \lambda\hbar \int W(\Delta)\big\{\hat{\sigma}_{+}(t)\hat{b}(\Delta,t)\exp[-\mathrm{i}\varphi(\Delta)]\\ +\hat{b}^{\dagger}(\Delta,t)\hat{\sigma}_{-}(t)\exp[\mathrm{i}\varphi(\Delta)]\big\}\,\mathrm{d}\Delta\\ +(1-\lambda)\hbar\int W(\Delta)\big\{\hat{b}(\Delta,t)\hat{\sigma}_{+}(t)\exp[-\mathrm{i}\varphi(\Delta)]\\ +\hat{\sigma}_{-}(t)\hat{b}^{\dagger}(\Delta,t)\exp[\mathrm{i}\varphi(\Delta)]\big\}\,\mathrm{d}\Delta,\end{aligned} \tag{A10.1}$$

where λ is any real constant. The dynamics generated by (A10.1) is independent of λ and an approximation method will not be valid unless it leads to a solution which is also independent of λ. The Heisenberg equations derived using (A10.1) are

$$\dot{\hat{b}}(\Delta,t) = -\mathrm{i}\Delta\hat{b}(\Delta,t) - \mathrm{i}W(\Delta)\exp[\mathrm{i}\varphi(\Delta)]\hat{\sigma}_{-}(t), \tag{A10.2}$$

$$\dot{\hat{b}}^{\dagger}(\Delta,t) = \mathrm{i}\Delta\hat{b}^{\dagger}(\Delta,t) + \mathrm{i}W(\Delta)\exp[-\mathrm{i}\varphi(\Delta)]\hat{\sigma}_{+}(t), \tag{A10.3}$$

$$\dot{\hat{\sigma}}_{-}(t) = \mathrm{i}\int W(\Delta)\exp[-\mathrm{i}\varphi(\Delta)]\big[\lambda\hat{\sigma}_{3}(t)\hat{b}(\Delta,t) + (1-\lambda)\hat{b}(\Delta,t)\hat{\sigma}_{3}(t)\big]\,\mathrm{d}\Delta, \tag{A10.4}$$

$$\dot{\hat{\sigma}}_{+}(t) = -\mathrm{i}\int W(\Delta)\exp[\mathrm{i}\varphi(\Delta)]\big[\lambda\hat{b}^{\dagger}(\Delta,t)\hat{\sigma}_{3}(t) + (1-\lambda)\hat{\sigma}_{3}(t)\hat{b}^{\dagger}(\Delta,t)\big]\,\mathrm{d}\Delta, \tag{A10.5}$$

$$\dot{\hat{\sigma}}_3(t) = -2\mathrm{i}\int W(\Delta)\exp[-\mathrm{i}\varphi(\Delta)]\left[\lambda\hat{\sigma}_+(t)\hat{b}(\Delta,t) + (1-\lambda)\hat{b}(\Delta,t)\hat{\sigma}_+(t)\right]\mathrm{d}\Delta$$
$$+ 2\mathrm{i}\int W(\Delta)\exp[\mathrm{i}\varphi(\Delta)]\left[\lambda\hat{b}^\dagger(\Delta,t)\hat{\sigma}_-(t) + (1-\lambda)\hat{\sigma}_-(t)\hat{b}^\dagger(\Delta,t)\right]\mathrm{d}\Delta. \tag{A10.6}$$

The equations of motion for the atomic operators, obtained by formally integrating (A10.2) and (A10.3), substituting into (A10.4) to (A10.6), making the Markov approximation, and ignoring the small frequency shift, as described in Section 5.5, are

$$\dot{\hat{\sigma}}_-(t) = -\Gamma(2\lambda-1)\hat{\sigma}_-(t) - \lambda\hat{\sigma}_3(t)\hat{F}(t) - (1-\lambda)\hat{F}(t)\hat{\sigma}_3(t), \tag{A10.7}$$
$$\dot{\hat{\sigma}}_+(t) = -\Gamma(2\lambda-1)\hat{\sigma}_+(t) - \lambda\hat{F}^\dagger(t)\hat{\sigma}_3(t) - (1-\lambda)\hat{\sigma}_3(t)\hat{F}^\dagger(t), \tag{A10.8}$$
$$\dot{\hat{\sigma}}_3(t) = -2\Gamma(2\lambda-1)\hat{\sigma}_3(t) - 2\Gamma + 2\lambda\left[\hat{\sigma}_+(t)\hat{F}(t) + \hat{F}^\dagger(t)\hat{\sigma}_-(t)\right]$$
$$+ 2(1-\lambda)\left[\hat{F}(t)\hat{\sigma}_+(t) + \hat{\sigma}_-(t)\hat{F}^\dagger(t)\right]. \tag{A10.9}$$

Note that since the Langevin operator $\hat{F}(t)$ depends on the initial field operators $\hat{b}(\Delta,0)$ it does not commute with the atomic operators $\hat{\sigma}_-(t)$, $\hat{\sigma}_+(t)$, and $\hat{\sigma}_3(t)$. It we take the expectation values of (A10.7) to (A10.9) and decorrelate the expectation values of products of atomic and Langevin operators, we obtain the equations

$$\langle\dot{\hat{\sigma}}_-(t)\rangle = -\Gamma(2\lambda-1)\langle\hat{\sigma}_-(t)\rangle, \tag{A10.10}$$
$$\langle\dot{\hat{\sigma}}_+(t)\rangle = -\Gamma(2\lambda-1)\langle\hat{\sigma}_+(t)\rangle, \tag{A10.11}$$
$$\langle\dot{\hat{\sigma}}_3(t)\rangle = -2\Gamma(2\lambda-1)\langle\hat{\sigma}_3(t)\rangle - 2\Gamma. \tag{A10.12}$$

These equations and their solutions depend on the value of λ and so the decorrelation which led to them cannot be a valid approximation. Instead, we follow the method described in Section 5.5 of formal integration of (A10.7) to (A10.9) and back-substitution, followed by decorrelation. The formal solutions of (A10.7) to (A10.9) are

$$\hat{\sigma}_-(t) = \hat{\sigma}_-(0)\exp[-\Gamma(2\lambda-1)t] - \int_0^t \exp[-\Gamma(2\lambda-1)(t-t')]$$
$$\times\left[\lambda\hat{\sigma}_3(t')\hat{F}(t') + (1-\lambda)\hat{F}(t')\hat{\sigma}_3(t')\right]\mathrm{d}t', \tag{A10.13}$$
$$\hat{\sigma}_+(t) = \hat{\sigma}_+(0)\exp[-\Gamma(2\lambda-1)t] - \int_0^t \exp[-\Gamma(2\lambda-1)(t-t')]$$
$$\times\left[\lambda\hat{F}^\dagger(t')\hat{\sigma}_3(t') + (1-\lambda)\hat{\sigma}_3(t')\hat{F}^\dagger(t')\right]\mathrm{d}t', \tag{A10.14}$$
$$\hat{\sigma}_3(t) + \frac{1}{2\lambda-1} = \left\{\hat{\sigma}_3(0) + \frac{1}{2\lambda-1}\right\}\exp[-2\Gamma(2\lambda-1)t]$$
$$+ 2\int_0^t \exp[-2\Gamma(2\lambda-1)(t-t')]$$
$$\times\left\{\lambda\left[\hat{\sigma}_+(t')\hat{F}(t') + \hat{F}^\dagger(t')\hat{\sigma}_-(t')\right]\right.$$
$$\left.+(1-\lambda)\left[\hat{F}(t')\hat{\sigma}_+(t') + \hat{\sigma}_-(t')\hat{F}^\dagger(t')\right]\right\}\mathrm{d}t'. \tag{A10.15}$$

We substitute (A10.13) to (A10.15) into those terms in (A10.17) to (A10.9) containing the product of an atom operator and a Langevin operator. Taking the expectation value of the resulting equations and decorrelating at this stage using

$$\langle \hat{F}(t)\rangle = \langle \hat{F}^\dagger(t)\rangle = 0, \tag{A10.16}$$

$$\langle \hat{F}(t)\hat{F}(t')\rangle = \langle \hat{F}^\dagger(t)\hat{F}^\dagger(t')\rangle = 0, \tag{A10.17}$$

$$\langle \hat{F}^\dagger(t)\hat{F}(t')\rangle = 2\Gamma\bar{n}(0)\delta(t-t'), \tag{A10.18}$$

$$\langle \hat{F}(t)\hat{F}^\dagger(t')\rangle = 2\Gamma[\bar{n}(0)+1]\delta(t-t') \tag{A10.19}$$

gives the equations

$$\left\langle \dot{\hat{\sigma}}_-(t)\right\rangle = -\Gamma[2\bar{n}(0)+1]\langle \hat{\sigma}_-(t)\rangle, \tag{A10.20}$$

$$\left\langle \dot{\hat{\sigma}}_+(t)\right\rangle = -\Gamma[2\bar{n}(0)+1]\langle \hat{\sigma}_+(t)\rangle, \tag{A10.21}$$

$$\left\langle \dot{\hat{\sigma}}_3(t)\right\rangle = -2\Gamma\{[2\bar{n}(0)+1]\langle \hat{\sigma}_3(t)\rangle + 1\}. \tag{A10.22}$$

As required, these equations are independent of λ and are identical to (5.5.66) to (5.5.68) derived for the case $\lambda = 1$.

Appendix 11

THE METHOD OF CHARACTERISTICS FOR PARTIAL DIFFERENTIAL EQUATIONS

We will discuss in this appendix the method of characteristics as applied to first-order partial differential equations. Consider first a function $u(x, y)$ of two independent variables x and y satisfying the equation

$$P(x,y)\frac{\partial u}{\partial x}+Q(x,y)\frac{\partial u}{\partial y}=R(x,y,u). \tag{A11.1}$$

The curves in the x, y-plane parametrized by t and defined by the relation

$$\frac{\mathrm{d}x}{P}=\frac{\mathrm{d}y}{Q}=\mathrm{d}t \tag{A11.2}$$

are called the characteristic curves (or, simply, the characteristics) of the partial differential equation (A11.1). Integration of the ordinary differential equation (A11.2) generates an arbitrary constant, each choice of this constant yielding one member of the family of characteristic curves. Along a characteristic, (A11.1) can be written in a form which involves the total derivative of $u(x, y)$, allowing integration to be achieved, as follows. Consider the element of arc length $\mathrm{d}s = \surd(\mathrm{d}x^2 + \mathrm{d}y^2)$ on a characteristic curve. Then, from (A11.2),

$$\mathrm{d}s^2=\frac{P^2}{Q^2}\,\mathrm{d}y^2+\mathrm{d}y^2=\mathrm{d}x^2+\frac{Q^2}{P^2}\,\mathrm{d}x^2. \tag{A11.3}$$

Hence

$$\frac{\mathrm{d}x}{P}=\frac{\mathrm{d}y}{Q}=\frac{\mathrm{d}s}{\surd(P^2+Q^2)}=\mathrm{d}t. \tag{A11.4}$$

Multiplying (A11.1) by $\mathrm{d}s$ gives

$$P\,\mathrm{d}s\,\frac{\partial u}{\partial x}+Q\,\mathrm{d}s\,\frac{\partial u}{\partial y}=R\,\mathrm{d}s \tag{A11.5}$$

and hence, using (A11.4), we have that

$$\surd(P^2+Q^2)\,\mathrm{d}x\,\frac{\partial u}{\partial x}+\surd(P^2+Q^2)\,\mathrm{d}y\,\frac{\partial u}{\partial y}=R\,\mathrm{d}s. \tag{A11.6}$$

Dividing (A11.6) by $\surd(P^2 + Q^2)$, we see that the left-hand side becomes the total derivative of u so that

$$\mathrm{d}u=\frac{\partial u}{\partial x}\,\mathrm{d}x+\frac{\partial u}{\partial y}\,\mathrm{d}y=\frac{R\,\mathrm{d}s}{\surd(P^2+Q^2)}=R\,\mathrm{d}t, \tag{A11.7}$$

using (A11.4). Finally, using (A11.7), (A11.2) may be written

$$\frac{\mathrm{d}x}{P} = \frac{\mathrm{d}y}{Q} = \frac{\mathrm{d}u}{R}. \tag{A11.8}$$

This pair of equations, valid on each characteristic curve, enables a general solution of the partial differential equation (A11.1) to be found, as follows. Firstly, the equation defining the family of characteristics is found by integration of (A11.2). As mentioned above, this contains one arbitrary constant, C, say. Next, a second equation involving u is chosen from (A11.8) and integrated, generating a second arbitrary constant, K, say. Note that in order to perform this integration it may be necessary to eliminate x or y using the equation for the characteristics found in the first integral. If $u(x, y)$ were known, the second integration would result in the *same* family of characteristic curves, and hence K and C are not independent. The general solution of the partial differential equation is therefore obtained by setting $K = f(C)$, where f is an unknown function which must be found from additional information, usually the behaviour of $u(x, y)$ on a given curve $y = g(x)$. Note that if $R = 0$ in (A11.1), then (A11.7) implies that $\mathrm{d}u = 0$ on each characteristic curve. Hence $u = K$, a constant, on a characteristic in this case.

More generally, if $u(x_1, x_2, \ldots, x_N)$ is a function of N independent variables $x_1, x_2, \ldots, x_N$ and

$$P_1 \frac{\partial u}{\partial x_1} + P_2 \frac{\partial u}{\partial x_2} + \cdots + P_N \frac{\partial u}{\partial x_N} = R(x_1, x_2, \ldots, x_N, u), \tag{A11.9}$$

where $P_1, P_2, \ldots, P_N$ are functions of $x_1, x_2, \ldots, x_N$, then we have N ordinary differential equations

$$\frac{\mathrm{d}x_1}{P_1} = \frac{\mathrm{d}x_2}{P_2} = \cdots = \frac{\mathrm{d}x_N}{P_N} = \frac{\mathrm{d}u}{R} \tag{A11.10}$$

defining the characteristics. These can, in principle, be integrated and the general solution of (A11.9) is found by making one of the N constants of integration an arbitrary function of the others.

In Chapter 5, Section 5.7, we require the solution $M(\mu, t)$ of the equation

$$\frac{\partial M}{\partial t} = -2\Gamma\mu \frac{\partial}{\partial \mu}[(1 + \mu\bar{n})M]. \tag{A11.11}$$

Writing this in the standard form of (A11.1), we have

$$\frac{\partial M}{\partial t} + 2\Gamma\mu(1 + \mu\bar{n})\frac{\partial M}{\partial \mu} = -2\Gamma\mu\bar{n}M. \tag{A11.12}$$

On a characteristic curve, we therefore have, following (A11.8),

$$\frac{\mathrm{d}t}{1} = \frac{\mathrm{d}\mu}{2\Gamma\mu(1 + \mu\bar{n})} = \frac{\mathrm{d}M}{-2\Gamma\mu\bar{n}M}. \tag{A11.13}$$

From the first equality we have

$$2\Gamma \int \mathrm{d}t = \int \frac{\mathrm{d}\mu}{\mu(1+\mu\bar{n})}, \tag{A11.14}$$

and carrying out the integrations we obtain

$$2\Gamma t + A = \int \left(\frac{1}{\mu} - \frac{\bar{n}}{1+\mu\bar{n}}\right)\mathrm{d}\mu = \ln\left(\frac{\mu}{1+\mu\bar{n}}\right), \tag{A11.15}$$

where A is an arbitrary constant. Taking the exponential of each side of (A11.15) and rearranging, we find that

$$\frac{\mu}{1+\mu\bar{n}}\exp(-2\Gamma t) = C, \tag{A11.16}$$

where $C = \exp(A)$ is an arbitrary constant. Similarly from the second equality in (A11.13) we have

$$\int \frac{\mathrm{d}M}{M} = -\int \frac{\bar{n}}{1+\mu\bar{n}}\,\mathrm{d}\mu. \tag{A11.17}$$

Carrying out the integrations and taking the exponential of each side gives

$$(1+\mu\bar{n})M = K, \tag{A11.18}$$

where K is an arbitrary constant. We can now apply the method described above in which the general solution of the partial differential equation is found by writing $K = f(C)$. We then find from (A11.16) and (A11.18) that the general solution of (A11.11) is

$$M(\mu,t) = \frac{1}{1+\mu\bar{n}} f\left(\frac{\mu}{1+\mu\bar{n}}\exp(-2\Gamma t)\right). \tag{A11.19}$$

Given that $M(\mu,0)$ is known, we find on inserting $t=0$ into (A11.19) that

$$M(\mu,0) = \frac{1}{1+\mu\bar{n}} f\left(\frac{\mu}{1+\mu\bar{n}}\right). \tag{A11.20}$$

To find f in terms of $M(\mu,0)$, we write $x = \mu/(1+\mu\bar{n})$ so that $\mu = x/(1-x\bar{n})$. From (A12.20) we then have

$$f(x) = (1+\mu\bar{n})M(\mu,0) = \frac{1}{1-x\bar{n}} M\left(\frac{x}{1-x\bar{n}},0\right). \tag{A11.21}$$

Substituting this functional form into (A11.19) with x replaced by $\mu\exp(-2\Gamma t)/(1+\mu\bar{n})$, we find that the solution of (A11.11) is

$$M(\mu,t) = \frac{1}{1+\mu\bar{n}[1-\exp(-2\Gamma t)]} M\left(\frac{\mu\exp(-2\Gamma t)}{1+\mu\bar{n}[1-\exp(-2\Gamma t)]},0\right). \tag{A11.22}$$

The second equation in Section 5.7 that we wish to solve is

$$\frac{\partial \chi}{\partial t} = \mathrm{i}W(\xi^* + \xi)\chi - \Gamma\left(\xi\frac{\partial}{\partial \xi} + \xi^*\frac{\partial}{\partial \xi^*}\right)\chi, \tag{A11.23}$$

where $\chi = \chi(\xi, t)$, ξ and ξ^* are treated as independent, and Γ and W are constants. Writing $\xi = x + \mathrm{i}y$, we have that $x = (\xi + \xi^*)/2$ and $y = (\xi - \xi^*)/2\mathrm{i}$. The derivatives with respect to ξ and ξ^* can be written

$$\frac{\partial}{\partial \xi} = \tfrac{1}{2}\left(\frac{\partial}{\partial x} - \mathrm{i}\frac{\partial}{\partial y}\right), \quad \frac{\partial}{\partial \xi^*} = \tfrac{1}{2}\left(\frac{\partial}{\partial x} + \mathrm{i}\frac{\partial}{\partial y}\right), \tag{A11.24}$$

so that (A11.23) becomes

$$\frac{\partial \chi}{\partial t} + \Gamma x\frac{\partial \chi}{\partial x} + \Gamma y\frac{\partial \chi}{\partial y} = 2\mathrm{i}Wx\chi. \tag{A11.25}$$

On the characteristic curves, we therefore have

$$\frac{\mathrm{d}t}{1} = \frac{\mathrm{d}x}{\Gamma x} = \frac{\mathrm{d}y}{\Gamma y} = \frac{\mathrm{d}\chi}{2\mathrm{i}Wx\chi}. \tag{A11.26}$$

From $\Gamma\,\mathrm{d}t = \mathrm{d}x/x$ and $\Gamma\,\mathrm{d}t = \mathrm{d}y/y$, we find on integration that

$$x\exp(-\Gamma t) = K_1, \quad y\exp(-\Gamma t) = K_2 \tag{A11.27}$$

on a characteristic, where K_1 and K_2 are constants. The third relationship, also obtained from (A11.26), is $\mathrm{d}x = \Gamma\,\mathrm{d}\chi/2\mathrm{i}W\chi$ and hence

$$\chi = K_3\exp(2\mathrm{i}Wx/\Gamma). \tag{A11.28}$$

The general solution of (A11.25) is obtained by making K_3 a function of K_1 and K_2 so that

$$\chi(x + \mathrm{i}y, t) = \exp(2\mathrm{i}Wx/\Gamma)f(x\exp(-\Gamma t), y\exp(-\Gamma t)). \tag{A11.29}$$

Since $\chi(x + \mathrm{i}y, 0) = \exp(2\mathrm{i}Wx/\Gamma)f(x, y)$, we have, on substituting in (A11.29) for the functional form of $f(x, y)$,

$$\chi(x + \mathrm{i}y, t) = \exp\left(\frac{2\mathrm{i}Wx}{\Gamma}\right)\exp\left(-\frac{2\mathrm{i}Wx\exp(-\Gamma t)}{\Gamma}\right)\chi((x + \mathrm{i}y)\exp(-\Gamma t), 0). \tag{A11.30}$$

Substituting for x and y in terms of ξ and ξ^*, we obtain the solution

$$\chi(\xi, t) = \exp\left(\frac{\mathrm{i}W}{\Gamma}(\xi + \xi^*)[1 - \exp(-\Gamma t)]\right)\chi(\xi\exp(-\Gamma t), 0). \tag{A11.31}$$

Appendix 12

TRANSFORMATION OF MASTER EQUATIONS INTO PARTIAL DIFFERENTIAL EQUATIONS

A master equation for a field mode can be transformed into a partial differential equation for either the characteristic function or a quasi-probability distribution. To achieve this, we require rules for transforming the actions of $\hat{a}$ and $\hat{a}^\dagger$ acting on either side of ρ.

The characteristic function $\chi(\xi, p)$ in (4.4.1) has the alternative forms

$$\begin{aligned}\chi(\xi, p) &= \mathrm{Tr}[\rho \exp(\xi \hat{a}^\dagger) \exp(-\xi^* \hat{a})] \exp\left[(p-1)|\xi|^2/2\right] \\ &= \mathrm{Tr}[\rho \exp(-\xi^* \hat{a}) \exp(\xi \hat{a}^\dagger)] \exp\left[(p+1)|\xi|^2/2\right]. \end{aligned} \tag{A12.1}$$

The derivatives with respect to ξ and ξ^* are

$$\begin{aligned}\frac{\partial}{\partial \xi} \chi(\xi, p) &= \mathrm{Tr}[\rho \hat{a}^\dagger \exp(\xi \hat{a}^\dagger) \exp(-\xi^* \hat{a})] \exp\left[(p-1)|\xi|^2/2\right] \\ &\quad + \frac{(p-1)}{2} \xi^* \chi(\xi, p) \end{aligned} \tag{A12.2}$$

or

$$\begin{aligned}\frac{\partial}{\partial \xi} \chi(\xi, p) &= \mathrm{Tr}[\hat{a}^\dagger \rho \exp(-\xi^* \hat{a}) \exp(\xi \hat{a}^\dagger)] \exp\left[(p+1)|\xi|^2/2\right] \\ &\quad + \frac{(p+1)}{2} \xi^* \chi(\xi, p), \end{aligned} \tag{A12.3}$$

and

$$\begin{aligned}\frac{\partial}{\partial \xi^*} \chi(\xi, p) &= -\mathrm{Tr}[\hat{a} \rho \exp(\xi \hat{a}^\dagger) \exp(-\xi^* \hat{a})] \exp\left[(p-1)|\xi|^2/2\right] \\ &\quad + \frac{(p-1)}{2} \xi \chi(\xi, p) \end{aligned} \tag{A12.4}$$

or

$$\begin{aligned}\frac{\partial}{\partial \xi^*} \chi(\xi, p) &= -\mathrm{Tr}[\rho \hat{a} \exp(-\xi^* \hat{a}) \exp(\xi \hat{a}^\dagger)] \exp\left[(p+1)|\xi|^2/2\right] \\ &\quad + \frac{(p+1)}{2} \xi \chi(\xi, p), \end{aligned} \tag{A12.5}$$

where we have used the cyclic property of the trace operation. It follows that the replacements

$$\rho\hat{a}^\dagger \to \left(\frac{\partial}{\partial\xi} - \frac{(p-1)}{2}\xi^*\right)\chi(\xi, p), \tag{A12.6}$$

$$\hat{a}^\dagger\rho \to \left(\frac{\partial}{\partial\xi} - \frac{(p+1)}{2}\xi^*\right)\chi(\xi, p), \tag{A12.7}$$

$$\hat{a}\rho \to \left(-\frac{\partial}{\partial\xi^*} + \frac{(p-1)}{2}\xi\right)\chi(\xi, p), \tag{A12.8}$$

and

$$\rho\hat{a} \to \left(-\frac{\partial}{\partial\xi^*} + \frac{(p+1)}{2}\xi\right)\chi(\xi, p) \tag{A12.9}$$

can be used to transform the master equation into an equivalent partial differential equation for $\chi(\xi, p)$. Further, we can regard the actions of $\hat{a}$ and $\hat{a}^\dagger$ on ρ as being equivalent to the actions of the corresponding differential operators on $\chi(\xi, p)$, so that, for example,

$$\hat{a}^\dagger\hat{a}\rho \to \left(\frac{\partial}{\partial\xi} - \frac{(p+1)}{2}\xi^*\right)\left(-\frac{\partial}{\partial\xi^*} + \frac{(p-1)}{2}\xi\right)\chi(\xi, p) \tag{A12.10}$$

and

$$\begin{aligned}\hat{a}\hat{a}^\dagger\rho &\to \left(-\frac{\partial}{\partial\xi^*} + \frac{(p-1)}{2}\xi\right)\left(\frac{\partial}{\partial\xi} - \frac{(p+1)}{2}\xi^*\right)\chi(\xi, p)\\ &= \left[\left(\frac{\partial}{\partial\xi} - \frac{(p+1)}{2}\xi^*\right)\left(-\frac{\partial}{\partial\xi^*} + \frac{(p-1)}{2}\xi\right) + 1\right]\chi(\xi, p),\end{aligned} \tag{A12.11}$$

in accordance with the fact that $[\hat{a}, \hat{a}^\dagger] = 1$. If two operators acting on ρ commute, then so will their corresponding differential operators acting on $\chi(\xi, p)$. For example,

$$\hat{a}(\rho\hat{a}^\dagger) \to \left(-\frac{\partial}{\partial\xi^*} + \frac{(p-1)}{2}\xi\right)\left(\frac{\partial}{\partial\xi} - \frac{(p-1)}{2}\xi^*\right)\chi(\xi, p) \tag{A12.12}$$

is identical to

$$(\hat{a}\rho)\hat{a}^\dagger \to \left(\frac{\partial}{\partial\xi} - \frac{(p-1)}{2}\xi^*\right)\left(-\frac{\partial}{\partial\xi^*} + \frac{(p-1)}{2}\xi\right)\chi(\xi, p). \tag{A12.13}$$

The quasi-probability distribution $W(\alpha, p)$ is the Fourier transform of $\chi(\xi, p)$ as given in (4.5.1). It follows that we can write

$$\alpha W(\alpha, p) = \frac{1}{\pi^2}\int_{-\infty}^{\infty} d^2\xi\chi(\xi, p)\frac{\partial}{\partial\xi^*}\exp(\alpha\xi^* - \alpha^*\xi)$$

$$= -\frac{1}{\pi^2}\int_{-\infty}^{\infty} d^2\xi \exp(\alpha\xi^* - \alpha^*\xi)\frac{\partial}{\partial\xi^*}\chi(\xi, p) \quad \text{(A12.14)}$$

and

$$\alpha^* W(\alpha, p) = -\frac{1}{\pi^2}\int_{-\infty}^{\infty} d^2\xi\chi(\xi, p)\frac{\partial}{\partial\xi}\exp(\alpha\xi^* - \alpha^*\xi)$$

$$= \frac{1}{\pi^2}\int_{-\infty}^{\infty} d^2\xi \exp(\alpha\xi^* - \alpha^*\xi)\frac{\partial}{\partial\xi}\chi(\xi, p), \quad \text{(A12.15)}$$

using integration by parts. Further

$$\frac{\partial}{\partial\alpha}W(\alpha, p) = \frac{1}{\pi^2}\int_{-\infty}^{\infty} d^2\xi\chi(\xi, p)\xi^*\exp(\alpha\xi^* - \alpha^*\xi) \quad \text{(A12.16)}$$

and

$$\frac{\partial}{\partial\alpha^*}W(\alpha, p) = -\frac{1}{\pi^2}\int_{-\infty}^{\infty} d^2\xi\chi(\xi, p)\xi\exp(\alpha\xi^* - \alpha^*\xi). \quad \text{(A12.17)}$$

The differential operators acting on $W(\alpha, p)$ which correspond to the actions on ρ of $\hat{a}$ and $\hat{a}^\dagger$ are then, from (A12.6) to (A12.9),

$$\rho\hat{a}^\dagger \to \left(\alpha^* - \frac{(p-1)}{2}\frac{\partial}{\partial\alpha}\right)W(\alpha, p), \quad \text{(A12.18)}$$

$$\hat{a}^\dagger\rho \to \left(\alpha^* - \frac{(p+1)}{2}\frac{\partial}{\partial\alpha}\right)W(\alpha, p), \quad \text{(A12.19)}$$

$$\hat{a}\rho \to \left(\alpha - \frac{(p-1)}{2}\frac{\partial}{\partial\alpha^*}\right)W(\alpha, p), \quad \text{(A12.20)}$$

and

$$\rho\hat{a} \to \left(\alpha - \frac{(p+1)}{2}\frac{\partial}{\partial\alpha^*}\right)W(\alpha, p). \quad \text{(A12.21)}$$

As with the characteristic function, the action of a number of operators on ρ is transformed into that of the corresponding ordering of differential operators on

$W(\alpha, p)$. For the quasi-probabilities $P(\alpha)$, $W(\alpha)$, and $Q(\alpha)$ corresponding to $p = 1$, 0, and -1, respectively, (A12.18) to (A12.21) become

$$\rho\hat{a}^\dagger \to \alpha^* P(\alpha), \tag{A12.22}$$

$$\hat{a}^\dagger\rho \to \left(\alpha^* - \frac{\partial}{\partial\alpha}\right) P(\alpha), \tag{A12.23}$$

$$\hat{a}\rho \to \alpha P(\alpha), \tag{A12.24}$$

$$\rho\hat{a} \to \left(\alpha - \frac{\partial}{\partial\alpha^*}\right) P(\alpha), \tag{A12.25}$$

$$\rho\hat{a}^\dagger \to \left(\alpha^* + \frac{1}{2}\frac{\partial}{\partial\alpha}\right) W(\alpha), \tag{A12.26}$$

$$\hat{a}^\dagger\rho \to \left(\alpha^* - \frac{1}{2}\frac{\partial}{\partial\alpha}\right) W(\alpha), \tag{A12.27}$$

$$\hat{a}\rho \to \left(\alpha + \frac{1}{2}\frac{\partial}{\partial\alpha^*}\right) W(\alpha), \tag{A12.28}$$

$$\rho\hat{a} \to \left(\alpha - \frac{1}{2}\frac{\partial}{\partial\alpha^*}\right) W(\alpha), \tag{A12.29}$$

$$\rho\hat{a}^\dagger \to \left(\alpha^* + \frac{\partial}{\partial\alpha}\right) Q(\alpha), \tag{A12.30}$$

$$\hat{a}^\dagger\rho \to \alpha^* Q(\alpha), \tag{A12.31}$$

$$\hat{a}\rho \to \left(\alpha + \frac{\partial}{\partial\alpha^*}\right) Q(\alpha), \tag{A12.32}$$

and

$$\rho\hat{a} \to \alpha Q(\alpha). \tag{A12.33}$$

Appendix 13
FOKKER–PLANCK EQUATIONS

Consider the equation

$$\frac{\partial}{\partial t} P(x,t) = \gamma \frac{\partial}{\partial x}[xP(x,t)] + \frac{D}{2}\frac{\partial^2}{\partial x^2} P(x,t) \tag{A13.1}$$

for a time-dependent probability distribution $P(x,t)$ for some real variable x. This is of the Fokker–Planck form: the first-order derivative is the drift term while the second-order derivative is the diffusion term. If we want to find $P(x,t)$ then we need to solve the partial differential equation. The formal solution is

$$\begin{aligned} P(x,t) &= \exp\left(\gamma t \frac{\partial}{\partial x} x + \frac{Dt}{2}\frac{\partial^2}{\partial x^2}\right) P(x,0) \\ &= \exp(\gamma t)\exp\left[-2\gamma t\left(-\frac{x}{2}\frac{\partial}{\partial x} - \frac{D}{4\gamma}\frac{\partial^2}{\partial x^2}\right)\right] P(x,0). \end{aligned} \tag{A13.2}$$

The differential operators $-(D/4\gamma)\partial^2/\partial x^2$ and $-(x/2)\partial/\partial x$ satisfy the commutation relation

$$\begin{aligned} \left[-\frac{D}{4\gamma}\frac{\partial^2}{\partial x^2}, -\frac{x}{2}\frac{\partial}{\partial x}\right] &= \frac{D}{8\gamma}\left(\frac{\partial^2}{\partial x^2} x \frac{\partial}{\partial x} - x\frac{\partial^3}{\partial x^3}\right) \\ &= -\left(-\frac{D}{4\gamma}\frac{\partial^2}{\partial x^2}\right). \end{aligned} \tag{A13.3}$$

Hence, writing

$$\hat{A} = -\frac{D}{4\gamma}\frac{\partial^2}{\partial x^2}, \quad \hat{B} = -\frac{x}{2}\frac{\partial}{\partial x}, \quad \theta = -2\gamma t, \tag{A13.4}$$

we can use theorem (A5.9) to write the formal solution (A13.2) as

$$P(x,t) = \exp(\gamma t)\exp\left(\gamma t x \frac{\partial}{\partial x}\right)\exp\left(\frac{D}{4\gamma}[\exp(2\gamma t) - 1]\frac{\partial^2}{\partial x^2}\right) P(x,0). \tag{A13.5}$$

If we write $P(x,0)=\int_{-\infty}^{\infty}P(x_0,0)\delta(x-x_0)\,\mathrm{d}x_0$ then we find the Green function

$$G(x,t\,|\,x_0,0)=\exp(\gamma t)\exp\left(\gamma tx\frac{\partial}{\partial x}\right)\exp\left(\frac{D}{4\gamma}[\exp(2\gamma t)-1]\frac{\partial^2}{\partial x^2}\right)\delta(x-x_0)$$

$$=\exp(\gamma t)\exp\left(\gamma tx\frac{\partial}{\partial x}\right)\exp\left(\frac{D}{4\gamma}[\exp(2\gamma t)-1]\frac{\partial^2}{\partial x^2}\right)\frac{1}{2\pi}$$

$$\times\int_{-\infty}^{\infty}\exp[\mathrm{i}k(x-x_0)]\,\mathrm{d}k$$

$$=\exp(\gamma t)\exp\left(\gamma tx\frac{\partial}{\partial x}\right)\frac{1}{2\pi}\int_{-\infty}^{\infty}\exp[\mathrm{i}k(x-x_0)]$$

$$\times\exp\left(-k^2\frac{D}{4\gamma}[\exp(2\gamma t)-1]\right)\mathrm{d}k. \tag{A13.6}$$

The integral is only well behaved for $D>0$, that is for positive diffusion, and performing the integration gives

$$G(x,t\,|\,x_0,0)=\left(\frac{\gamma}{\pi D[1-\exp(-2\gamma t)]}\right)^{1/2}\exp\left(\gamma tx\frac{\partial}{\partial x}\right)\exp\left(\frac{-\gamma(x-x_0)^2}{D[\exp(2\gamma t)-1]}\right). \tag{A13.7}$$

The remaining operator can be expanded as

$$\exp\left(\gamma tx\frac{\partial}{\partial x}\right)=\sum_{l=0}^{\infty}\frac{(\gamma t)^l}{l!}\left(x\frac{\partial}{\partial x}\right)^l. \tag{A13.8}$$

The identity

$$\left(x\frac{\partial}{\partial x}\right)^l x^m=\left(x\frac{\partial}{\partial x}\right)^{l-1}mx^m=m^l x^m \tag{A13.9}$$

implies that the action of the operator (A13.8) on a function $F(x)$ represented by its Maclaurin expansion is

$$\exp\left(\gamma tx\frac{\partial}{\partial x}\right)F(x)=\sum_{l=0}^{\infty}\sum_{m=0}^{\infty}\frac{(\gamma t)^l}{l!}F^{(m)}(0)m^l\frac{x^m}{m!}$$

$$=\sum_{m=0}^{\infty}\exp(m\gamma t)F^{(m)}(0)\frac{x^m}{m!}$$

$$=F(x\exp(\gamma t)). \tag{A13.10}$$

It follows that the Green function is

$$G(x,t\,|\,x_0,0)=\left(\frac{\gamma}{\pi D[1-\exp(-2\gamma t)]}\right)^{1/2}\exp\left(\frac{-\gamma[x\exp(\gamma t)-x_0]^2}{D[\exp(2\gamma t)-1]}\right)$$
$$=\left(\frac{\gamma}{\pi D[1-\exp(-2\gamma t)]}\right)^{1/2}\exp\left(\frac{-\gamma[x-x_0\exp(-\gamma t)]^2}{D[1-\exp(-2\gamma t)]}\right),\tag{A13.11}$$

and hence that

$$P(x,t)=\left(\frac{\gamma}{\pi D[1-\exp(-2\gamma t)]}\right)^{1/2}$$
$$\times\int_{-\infty}^{\infty}\exp\left(\frac{-\gamma[x-x_0\exp(-\gamma t)]^2}{D[1-\exp(-2\gamma t)]}\right)P(x_0,0)\,\mathrm{d}x_0.\tag{A13.12}$$

This solution is readily generalized to two dimensions. In Section 5.7 we require the solution of a Fokker–Planck equation for the quasi-probability distribution $W(x,y,p,t)$ for the two real variables x and y having the form

$$\frac{\partial W}{\partial t}=\gamma_x\frac{\partial}{\partial x}(xW)+\gamma_y\frac{\partial}{\partial y}(yW)+\frac{D_{xx}}{2}\frac{\partial^2 W}{\partial x^2}+\frac{D_{yy}}{2}\frac{\partial^2 W}{\partial y^2}.\tag{A13.13}$$

This has positive diffusion if both D_{xx} and D_{yy} are positive. Equation (5.7.16) can be cast into this form by writing $\alpha=x+\mathrm{i}y$ and setting $\gamma_x=\Gamma+g$, $\gamma_y=\Gamma-g$, $D_{xx}=[\Gamma(1-p)-gp]/2$, and $D_{yy}=[\Gamma(1-p)+gp]/2$. For $g>0$, the diffusion is positive for $p<\Gamma/(\Gamma+g)$ and the general solution is then

$$W(x,y,p,t)=\left(\frac{\gamma_x\gamma_y}{\pi^2D_{xx}D_{yy}[1-\exp(-2\gamma_x t)][1-\exp(-2\gamma_y t)]}\right)^{1/2}$$
$$\times\int_{-\infty}^{\infty}\mathrm{d}x_0\int_{-\infty}^{\infty}\mathrm{d}y_0\exp\left(\frac{-\gamma_x[x-x_0\exp(-\gamma_x t)]^2}{D_{xx}[1-\exp(-2\gamma_x t)]}\right)$$
$$\times\exp\left(\frac{-\gamma_y[y-y_0\exp(-\gamma_y t)]^2}{D_{yy}[1-\exp(-2\gamma_y t)]}\right)W(x_0,y_0,p,0).\tag{A13.14}$$

It is often the case that we are interested in the moments of x rather than in the probability (or quasi-probability) distribution itself. Consider the simple Fokker–Planck equation for $P(x,t)$ given in (A13.1). Denoting the nth moment of x at time t as

$$\langle x^n(t)\rangle=\int_{-\infty}^{\infty}x^nP(x,t)\,\mathrm{d}x,\tag{A13.15}$$

we can obtain an ordinary differential equation for this moment since

$$\frac{\mathrm{d}}{\mathrm{d}t}\langle x^n(t)\rangle = \int_{-\infty}^{\infty} x^n \frac{\partial}{\partial t} P(x,t)\,\mathrm{d}x, \tag{A13.16}$$

and using the Fokker–Planck equation (A13.1) we find

$$\frac{\mathrm{d}}{\mathrm{d}t}\langle x^n(t)\rangle = \gamma \int_{-\infty}^{\infty} x^n \frac{\partial}{\partial x}[xP(x,t)]\,\mathrm{d}x + \frac{D}{2}\int_{-\infty}^{\infty} x^n \frac{\partial^2}{\partial x^2} P(x,t)\,\mathrm{d}x. \tag{A13.17}$$

Integration by parts gives

$$\begin{aligned}\frac{\mathrm{d}}{\mathrm{d}t}\langle x^n(t)\rangle &= \gamma[x^{n+1}P(x,t)]_{-\infty}^{\infty} - \gamma\int_{-\infty}^{\infty} nx^nP(x,t)\,\mathrm{d}x \\ &\quad + \frac{D}{2}\left[x^n \frac{\partial}{\partial x}P(x,t)\right]_{-\infty}^{\infty} - \frac{D}{2}\int_{-\infty}^{\infty} nx^{n-1}\frac{\partial}{\partial x}P(x,t)\,\mathrm{d}x \\ &= \gamma[x^{n+1}P(x,t)]_{-\infty}^{\infty} - \gamma\int_{-\infty}^{\infty} nx^nP(x,t)\,\mathrm{d}x \\ &\quad + \frac{D}{2}\left[x^n \frac{\partial}{\partial x}P(x,t)\right]_{-\infty}^{\infty} - \frac{D}{2}[nx^{n-1}P(x,t)]_{-\infty}^{\infty} \\ &\quad + \frac{D}{2}\int_{-\infty}^{\infty} n(n-1)x^{n-2}P(x,t)\,\mathrm{d}x.\end{aligned} \tag{A13.18}$$

If $P(x,t)$ is suitably well behaved, that is tending to zero sufficiently fast as $|x| \to \infty$, we obtain

$$\frac{\mathrm{d}}{\mathrm{d}t}\langle x^n(t)\rangle = -n\gamma\langle x^n(t)\rangle + \frac{D}{2}n(n-1)\langle x^{n-2}(t)\rangle. \tag{A13.19}$$

For $D \geqslant 0$, this is equivalent to a stochastic differential equation or c-number Langevin equation

$$\frac{\mathrm{d}}{\mathrm{d}t}x(t) = -\gamma x(t) + f(t) \tag{A13.20}$$

for a time-dependent variable $x(t)$. Here $f(t)$ is a stochastic Langevin term with ensemble averages

$$\langle f(t)\rangle = 0, \tag{A13.21}$$

$$\langle f(t)f(t')\rangle = D\delta(t-t'). \tag{A13.22}$$

All moments involving an odd number of Langevin terms are zero. Moments involving an even number of Langevin terms are simply a sum over all possible combinations of products of moments of two Langevin terms. For example,

$$\begin{aligned}\langle f(t_1)f(t_2)f(t_3)f(t_4)\rangle &= \langle f(t_1)f(t_2)\rangle\langle f(t_3)f(t_4)\rangle + \langle f(t_1)f(t_3)\rangle\langle f(t_2)f(t_4)\rangle \\ &\quad + \langle f(t_1)f(t_4)\rangle\langle f(t_2)f(t_3)\rangle \\ &= D^2[\delta(t_1-t_2)\delta(t_3-t_4) + \delta(t_1-t_3)\delta(t_2-t_4) \\ &\quad + \delta(t_1-t_4)\delta(t_2-t_3)]. \end{aligned} \tag{A13.23}$$

These, together with the formal solution

$$x(t) = x(0)\exp(-\gamma t) + \int_0^t \exp[-\gamma(t-t')]f(t')\,\mathrm{d}t' \tag{A13.24}$$

of the stochastic differential equation and the fact that $\langle x(0)f(t)\rangle = \langle x(0)\rangle\langle f(t)\rangle = 0$, lead directly to the evolution of the moments $\langle x^n(t)\rangle$. To establish the equivalence between the Fokker–Planck equation and our stochastic differential equation, we generate the equations for the moments $\langle x^n(t)\rangle$ using the stochastic differential equation. The first two of these are

$$\frac{\mathrm{d}}{\mathrm{d}t}\langle x(t)\rangle = -\gamma\langle x(t)\rangle + \langle f(t)\rangle = -\gamma\langle x(t)\rangle \tag{A13.25}$$

and

$$\frac{\mathrm{d}}{\mathrm{d}t}\langle x^2(t)\rangle = -2\gamma\langle x^2(t)\rangle + 2\langle x(t)f(t)\rangle. \tag{A13.26}$$

To find $\langle x(t)f(t)\rangle$, we use the formal solution (A13.24) to obtain

$$\begin{aligned}\langle x(t)f(t)\rangle &= \langle x(0)f(t)\rangle\exp(-\gamma t) + \int_0^t \exp[-\gamma(t-t')]\langle f(t')f(t)\rangle\,\mathrm{d}t' \\ &= D\int_0^t \exp[-\gamma(t-t')]\delta(t-t')\,\mathrm{d}t' \\ &= \frac{D}{2},\end{aligned} \tag{A13.27}$$

where the factor $1/2$ has arisen because the delta function is situated at the end of the range of integration. Hence

$$\frac{\mathrm{d}}{\mathrm{d}t}\langle x^2(t)\rangle = -2\gamma\langle x^2(t)\rangle + D. \tag{A13.28}$$

Equations for higher-order moments of x will involve terms of the form $\langle x^{n-1}(t)f(t)\rangle$. Formally integrating and using the decorrelation of the moments of $f(t)$ leads to $n-1$ identical terms, one for each occurrence of $f(t)$. It follows that

$$\langle x^{n-1}(t)f(t)\rangle = (n-1)\langle x^{n-2}(t)\rangle\langle x(t)f(t)\rangle$$
$$= (n-1)\frac{D}{2}\langle x^{n-2}(t)\rangle, \qquad \text{(A13.29)}$$

and that

$$\frac{\mathrm{d}}{\mathrm{d}t}\langle x^n(t)\rangle = -n\gamma\langle x^n(t)\rangle + \frac{D}{2}n(n-1)\langle x^{n-2}(t)\rangle, \qquad \text{(A13.30)}$$

which is the same as (A13.19) derived from the Fokker–Planck equation.

We can extend the above analysis to two-dimensional probability distributions. Consider the equation

$$\frac{\partial}{\partial t}P(x,y,t) = \frac{\partial}{\partial x}\left[(\gamma_{xx}x + \gamma_{xy}y)P(x,y,t)\right] + \frac{\partial}{\partial y}\left[(\gamma_{yy}y + \gamma_{yx}x)P(x,y,t)\right]$$
$$+\tfrac{1}{2}D_{xx}\frac{\partial^2}{\partial x^2}P(x,y,t) + \tfrac{1}{2}D_{yy}\frac{\partial^2}{\partial y^2}P(x,y,t) \qquad \text{(A13.31)}$$

for a probability (or quasi-probability) distribution $P(x,y,t)$. Again we are interested in moments of the form

$$\langle x^n y^m\rangle = \int_{-\infty}^{\infty}\mathrm{d}x\int_{-\infty}^{\infty}\mathrm{d}y\, x^n y^m P(x,y,t). \qquad \text{(A13.32)}$$

Differentiating with respect to t and substituting for $\dot{P}$ from the Fokker–Planck equation, followed by integration by parts, leads to

$$\frac{\mathrm{d}}{\mathrm{d}t}\langle x^n y^m\rangle = -n\gamma_{xx}\langle x^n y^m\rangle - n\gamma_{xy}\langle x^{n-1}y^{m+1}\rangle$$
$$-m\gamma_{yy}\langle x^n y^m\rangle - m\gamma_{yx}\langle x^{n+1}y^{m-1}\rangle$$
$$+\frac{D_{xx}}{2}n(n-1)\langle x^{n-2}y^m\rangle + \frac{D_{yy}}{2}m(m-1)\langle x^n y^{m-2}\rangle. \qquad \text{(A13.33)}$$

Again, if both $D_{xx} \geqslant 0$ and $D_{yy} \geqslant 0$, then this is equivalent to a stochastic process represented by two c-number Langevin equations

$$\dot{x} = -\gamma_{xx}x - \gamma_{xy}y + f_x(t), \qquad \text{(A13.34)}$$

$$\dot{y} = -\gamma_{yy}y - \gamma_{yx}x + f_y(t). \qquad \text{(A13.35)}$$

We can solve these by formal integration, leading to the solution

$$\mathbf{x}(t) = \exp(-\boldsymbol{\Gamma} t)\mathbf{x}(0) + \int_0^t \exp[-\boldsymbol{\Gamma}(t-t')]\mathbf{f}(t')\,\mathrm{d}t', \tag{A13.36}$$

where

$$\mathbf{x} = \begin{pmatrix} x \\ y \end{pmatrix}, \quad \mathbf{f}(t) = \begin{pmatrix} f_x(t) \\ f_y(t) \end{pmatrix}, \quad \boldsymbol{\Gamma} = \begin{pmatrix} \gamma_{xx} & \gamma_{xy} \\ \gamma_{yx} & \gamma_{yy} \end{pmatrix}. \tag{A13.37}$$

The Langevin terms have ensemble averages

$$\langle f_x(t)\rangle = 0, \tag{A13.38}$$

$$\langle f_y(t)\rangle = 0, \tag{A13.39}$$

$$\langle f_x(t)f_x(t')\rangle = D_{xx}\,\delta(t-t'), \tag{A13.40}$$

$$\langle f_y(t)f_y(t')\rangle = D_{yy}\,\delta(t-t'), \tag{A13.41}$$

$$\langle f_x(t)f_y(t')\rangle = 0. \tag{A13.42}$$

Higher-order moments factorize into products of correlation functions of all possible pairs of Langevin terms as before. For example,

$$\begin{aligned}\langle f_x(t_1)f_x(t_2)f_y(t_3)f_y(t_4)\rangle &= \langle f_x(t_1)f_x(t_2)\rangle\langle f_y(t_3)f_y(t_4)\rangle \\ &\quad + \langle f_x(t_1)f_y(t_3)\rangle\langle f_x(t_2)f_y(t_4)\rangle \\ &\quad + \langle f_x(t_1)f_y(t_4)\rangle\langle f_x(t_2)f_y(t_3)\rangle \\ &= D_{xx}D_{yy}\,\delta(t_1-t_2)\delta(t_3-t_4).\end{aligned} \tag{A13.43}$$

These, together with the formal solution of the stochastic differential equations, allow us to prove that

$$\begin{aligned}\langle x^{n-1}(t)y^m(t)f_x(t)\rangle &= (n-1)\langle x^{n-2}(t)y^m(t)\rangle\langle x(t)f_x(t)\rangle \\ &\quad + m\langle x^{n-1}(t)y^{m-1}(t)\rangle\langle y(t)f_x(t)\rangle \\ &= \frac{D_{xx}}{2}(n-1)\langle x^{n-2}(t)y^m(t)\rangle\end{aligned} \tag{A13.44}$$

and similarly that

$$\langle x^n(t)y^{m-1}(t)f_y(t)\rangle = \frac{D_{yy}}{2}(m-1)\langle x^n(t)y^{m-2}(t)\rangle. \tag{A13.45}$$

Therefore the stochastic differential equations give the equation

$$\frac{\mathrm{d}}{\mathrm{d}t}\langle x^n y^m\rangle = n\langle \dot{x}x^{n-1}y^m\rangle + m\langle \dot{y}x^n y^{m-1}\rangle$$
$$= -n\gamma_{xx}\langle x^n y^m\rangle - n\gamma_{xy}\langle x^{n-1}y^{m+1}\rangle - m\gamma_{yy}\langle x^n y^m\rangle$$
$$- m\gamma_{yx}\langle x^{n+1}y^{m-1}\rangle + \frac{D_{xx}}{2}n(n-1)\langle x^{n-2}y^m\rangle$$
$$+ \frac{D_{yy}}{2}m(m-1)\langle x^n y^{m-2}\rangle \quad \text{(A13.46)}$$

for the moment $\langle x^n y^m\rangle$, in agreement with (A13.33).

More generally, two-dimensional Fokker–Planck equations will involve mixed second-order derivatives so that

$$\frac{\partial}{\partial t}P(x,y,t) = \frac{\partial}{\partial x}\left[(\gamma_{xx}x + \gamma_{xy}y)P(x,y,t)\right] + \frac{\partial}{\partial y}\left[(\gamma_{yy}y + \gamma_{yx}x)P(x,y,t)\right]$$
$$+\left(\frac{D_{xx}}{2}\frac{\partial^2}{\partial x^2} + \frac{D_{yy}}{2}\frac{\partial^2}{\partial y^2} + D_{xy}\frac{\partial^2}{\partial x\,\partial y}\right)P(x,y,t). \quad \text{(A13.47)}$$

Our ability to write this in terms of an equivalent stochastic process again relies on the diffusion being positive. It is possible to find the normal coordinates x' and y' such that the diffusion terms become

$$\frac{D_{xx}}{2}\frac{\partial^2}{\partial x^2} + \frac{D_{yy}}{2}\frac{\partial^2}{\partial y^2} + D_{xy}\frac{\partial^2}{\partial x\,\partial y} = \frac{D_+}{2}\frac{\partial^2}{\partial x'^2} + \frac{D_-}{2}\frac{\partial^2}{\partial y'^2}, \quad \text{(A13.48)}$$

where D_+ and D_- are the eigenvalues of the diffusion matrix

$$\mathbf{D} = \begin{pmatrix} D_{xx} & D_{xy} \\ D_{xy} & D_{yy} \end{pmatrix}. \quad \text{(A13.49)}$$

If these eigenvalues are both positive then we have positive diffusion. The condition for positive diffusion may be simply expressed as the requirement that both D_{xx} and D_{yy} are positive and that $D_{xx}D_{yy} > D_{xy}^2$. The last of these is a statement that the determinant of $\mathbf{D}$ is positive. A pair of stochastic differential equations equivalent to (A13.47) are

$$\dot{x} = -\gamma_{xx}x - \gamma_{xy}y + f_x(t) \quad \text{(A13.50)}$$

and

$$\dot{y} = -\gamma_{yy}y - \gamma_{yx}x + f_y(t), \quad \text{(A13.51)}$$

where the Langevin terms f_x and f_y have the correlation functions

$$\langle f_x(t) f_x(t') \rangle = D_{xx}\delta(t-t'), \tag{A13.52}$$

$$\langle f_y(t) f_y(t') \rangle = D_{yy}\delta(t-t'), \tag{A13.53}$$

and

$$\langle f_x(t) f_y(t') \rangle = D_{xy}\delta(t-t'). \tag{A13.54}$$

These can be derived by employing the normal coordinates x' and y' and using the results (A13.38) to (A13.42) for the normal mode Langevin terms $f_{x'}$ and $f_{y'}$. Higher-order moments can be expressed as sums of products of the correlation functions (A13.52) to (A13.54).

Equations of the Fokker–Planck form most commonly arise in quantum optics as partial differential equations for a quasi-probability distribution $W(\alpha, p, t)$ in the complex variables α and α^*. Consider such an equation of the form

$$\frac{\partial W}{\partial t} = \frac{\partial}{\partial \alpha}[(\gamma_{\alpha\alpha}\alpha + \gamma_{\alpha\alpha^*}\alpha^*)W] + \frac{\partial}{\partial \alpha^*}[(\gamma_{\alpha^*\alpha^*}\alpha^* + \gamma_{\alpha^*\alpha}\alpha)W]$$

$$+ \frac{D_{\alpha\alpha}}{2}\frac{\partial^2 W}{\partial \alpha^2} + \frac{D_{\alpha^*\alpha^*}}{2}\frac{\partial^2 W}{\partial \alpha^{*2}} + D_{\alpha\alpha^*}\frac{\partial^2 W}{\partial \alpha\, \partial \alpha^*}. \tag{A13.55}$$

The requirement that W is a real-valued function imposes the conditions $\gamma_{\alpha^*\alpha^*} = \gamma^*_{\alpha\alpha}$, $\gamma_{\alpha^*\alpha} = \gamma^*_{\alpha\alpha^*}$, $D_{\alpha^*\alpha^*} = D^*_{\alpha\alpha}$, and $D_{\alpha\alpha^*}$ is real. We can again obtain the stochastic differential equations equivalent to this Fokker–Planck equation by re-expressing (A13.55) in terms of the real variables $x = (\alpha + \alpha^*)/2$ and $y = (\alpha - \alpha^*)/2\mathrm{i}$. In terms of these, the Fokker–Planck equation is

$$\frac{\partial W}{\partial t} = \frac{\partial}{\partial x}\left[\left(\frac{x}{2}(\gamma_{\alpha\alpha} + \gamma_{\alpha\alpha^*} + \gamma^*_{\alpha\alpha} + \gamma^*_{\alpha\alpha^*}) + \frac{\mathrm{i}y}{2}(\gamma_{\alpha\alpha} - \gamma_{\alpha\alpha^*} - \gamma^*_{\alpha\alpha} + \gamma^*_{\alpha\alpha^*})\right)W\right]$$

$$+ \frac{\partial}{\partial y}\left[\left(-\frac{\mathrm{i}x}{2}(\gamma_{\alpha\alpha} + \gamma_{\alpha\alpha^*} - \gamma^*_{\alpha\alpha} - \gamma^*_{\alpha\alpha^*}) + \frac{y}{2}(\gamma_{\alpha\alpha} - \gamma_{\alpha\alpha^*} + \gamma^*_{\alpha\alpha} - \gamma^*_{\alpha\alpha^*})\right)W\right]$$

$$+ \tfrac{1}{8}(D_{\alpha\alpha} + D^*_{\alpha\alpha} + 2D_{\alpha\alpha^*})\frac{\partial^2 W}{\partial x^2} + \tfrac{1}{8}(2D_{\alpha\alpha^*} - D_{\alpha\alpha} - D^*_{\alpha\alpha})\frac{\partial^2 W}{\partial y^2}$$

$$- \frac{\mathrm{i}}{4}(D_{\alpha\alpha} - D^*_{\alpha\alpha})\frac{\partial^2 W}{\partial x\, \partial y}. \tag{A13.56}$$

Comparing this with the two-dimensional Fokker–Planck equation (A13.47), we find that the conditions for positive diffusion reduce to $D_{\alpha\alpha^*} > |D_{\alpha\alpha}|$. If we have positive diffusion then we can write the equivalent stochastic differential equations

$$\dot{\alpha} = -\gamma_{\alpha\alpha}\,\alpha - \gamma_{\alpha\alpha^*}\,\alpha^* + f_\alpha(t) \tag{A13.57}$$

and

$$\dot{\alpha}^* = -\gamma^*_{\alpha\alpha}\,\alpha^* - \gamma^*_{\alpha\alpha^*}\,\alpha + f_{\alpha^*}(t) \tag{A13.58}$$

for the complex variables α and α^*. The Langevin terms have the correlation functions

$$\langle f_\alpha(t) f_\alpha(t')\rangle = D_{\alpha\alpha}\,\delta(t-t'), \tag{A13.59}$$

$$\langle f_{\alpha^*}(t) f_{\alpha^*}(t')\rangle = D^*_{\alpha\alpha}\,\delta(t-t'), \tag{A13.60}$$

and

$$\langle f_\alpha(t) f_{\alpha^*}(t')\rangle = D_{\alpha\alpha^*}\,\delta(t-t'). \tag{A13.61}$$

Appendix 14
CUBIC EQUATIONS

In Chapter 6, we require the solutions of the cubic equation

$$x^3 + Ax^2 + Bx + C = 0. \tag{A14.1}$$

We first remove the quadratic term using the substitution

$$x = y - \tfrac{1}{3}A, \tag{A14.2}$$

so that (A14.1) becomes

$$y^3 + \left(B - \tfrac{1}{3}A^2\right)y + \left(C + \tfrac{2}{27}A^3 - \tfrac{1}{3}AB\right) = 0. \tag{A14.3}$$

The solutions of this equation can be found by writing

$$y = \lambda \cos\theta. \tag{A14.4}$$

Substituting this form into (A14.3) and dividing by λ^3 gives

$$\cos^3\theta + \frac{1}{\lambda^2}\left(B - \tfrac{1}{3}A^2\right)\cos\theta + \frac{1}{\lambda^3}\left(C + \tfrac{2}{27}A^3 - \tfrac{1}{3}AB\right) = 0. \tag{A14.5}$$

Comparing (A14.5) with the trigonometric identity

$$\cos^3\theta - \tfrac{3}{4}\cos\theta - \tfrac{1}{4}\cos 3\theta = 0, \tag{A14.6}$$

we see that

$$\frac{1}{\lambda^2}\left(B - \tfrac{1}{3}A^2\right) = -\tfrac{3}{4} \tag{A14.7}$$

and

$$\frac{1}{\lambda^3}\left(C + \tfrac{2}{27}A^3 - \tfrac{1}{3}AB\right) = -\tfrac{1}{4}\cos 3\theta. \tag{A14.8}$$

Solving (A14.7) for λ, we find

$$\lambda = \pm\tfrac{2}{3}(A^2 - 3B)^{1/2} \tag{A14.9}$$

and hence from (A14.8) that

$$\cos 3\theta = \mp\frac{(27C + 2A^3 - 9AB)}{2(A^2 - 3B)^{3/2}}. \tag{A14.10}$$

Solving (A14.10) for θ and substituting this solution together with (A14.9) into (A14.4), we then find from (A14.2) that

$$x = -\tfrac{1}{3}A \pm \tfrac{2}{3}(A^2 - 3B)^{1/2}\cos\left[\tfrac{1}{3}\cos^{-1}\left(\mp\frac{(27C + 2A^3 - 9AB)}{2(A^2 - 3B)^{3/2}}\right)\right]. \tag{A14.11}$$

The three roots of (A14.1) are the three distinct values of (A14.11).

SELECTED BIBLIOGRAPHY

Allen, L. and Eberly, J. H. (1975). *Optical resonance and two-level atoms*, Wiley, New York. Reprinted by Dover.

Barnett, S. M. and Pegg, D. T. (1989). On the Hermitian optical phase operator. *Journal of Modern Optics*, **36**, 7–19.

Blow, K. J., Loudon, R., Phoenix, S. J. D., and Shepherd, T. J. (1990). Continuum fields in quantum optics. *Physical Review A*, **42**, 4102–14.

Cahill, K. E. and Glauber, R. J. (1969). Ordered expansions in boson amplitude operators. *Physical Review*, **177**, 1857–81.

Cahill, K. E. and Glauber, R. J. (1969). Density operators and quasiprobability distributions. *Physical Review*, **177**, 1882–1902.

Carmichael, H. (1993). *An open systems approach to quantum optics*. Springer-Verlag, Berlin.

Cohen-Tannoudji, C. (1994). *Atoms in electromagnetic fields*. World Scientific, Singapore.

Cohen-Tannoudji, C., Dupont-Roc, J., and Grynberg, G. (1989). *Photons and atoms: introduction to quantum electrodynamics*. Wiley, New York.

Cohen-Tannoudji, C., Dupont-Roc, J., and Grynberg, G. (1992). *Atom-photon interactions: basic processes and applications*. Wiley, New York.

Compagno, C., Passante, R., and Persico, F. (1995). *Atom-field interactions and dressed atoms*. Cambridge University Press, Cambridge.

Gardiner, C. W. (1991). *Quantum noise*. Springer-Verlag, Berlin.

Glauber, R. J. (1963). Coherent and incoherent states of the radiation field. *Physical Review*, **131**, 2766–88.

Klauder, J. R. and Sudarshan, E. C. G. (1968). *Fundamentals of quantum optics*. Benjamin, New York.

Knight, P. L. and Allen, L. (1983). *Concepts of quantum optics*. Pergamon, Oxford.

Knight, P. L. and Milonni, P. W. (1980). The Rabi frequency in optical spectra. *Physics Reports*, **66**, 21–107.

Loudon, R. (2000). *The quantum theory of light* (3rd edn). Oxford University Press, Oxford.

Loudon, R. and Knight, P. L. (1987). Squeezed light. *Journal of Modern Optics*, **34**, 709–59.

Louisell, W. H. (1973). *Quantum statistical properties of radiation*. Wiley, New York. Republished 1990.

Mandel, L. and Wolf, E. (1995). *Optical coherence and quantum optics*. Cambridge University Press, Cambridge.

Meystre, P. and Sargent, M. (1990). *Elements of quantum optics*. Springer-Verlag, Berlin.

Milonni, P. W. (1994). *The quantum vacuum: an introduction to quantum electrodynamics*. Academic Press, San Diego.

Pegg, D. T. and Barnett, S. M. (1989). Phase properties of the quantized single-mode electromagnetic field. *Physical Review A*, **39**, 1665–75.

Pegg, D. T. and Barnett, S. M. (1997). Quantum optical phase. *Journal of Modern Optics*, **44**, 225–64.

Perelomov, A. (1985). *Generalized coherent states and their applications*. Springer-Verlag, Berlin.

Peřina, J. (1991). *Quantum statistics of linear and nonlinear optical phenomena*. Kluwer, Dordrecht.

Sargent, M., Scully, M. O., and Lamb, W. E. (1974). *Laser physics*. Addison-Wesley, New York.
Schubert, M. and Wihelmi, B. (1986). *Nonlinear optics and quantum electrodynamics*. Wiley, New York.
Shore, B. W. (1990). *The theory of coherent atomic excitation*. Wiley, New York.
Shore, B. W. and Knight, P. L. (1993). The Jaynes–Cummings model. *Journal of Modern Optics*, **40**, 1195–1238.
Stenholm, S. (1984). *Foundations of laser spectroscopy*. Wiley, New York.
Walls, D. F. and Milburn, G. J. (1995). *Quantum optics*. Springer-Verlag, Berlin.

INDEX

9 780198 563617

Printed in the United States
By Bookmasters